Two week loan

turn on or before the la
ped below.
are made for late retu

The Molecular Biology
of Down Syndrome

Edited by
G. Lubec

Journal of Neural Transmission
Supplement 57

SpringerWienNewYork

Prof. Dr. G. Lubec
Universitäts-Kinderklinik, Wien
Österreich

© 1999 Springer-Verlag/Wien

Printed in Austria

Printing: A. Holzhausens Nfg., A-1070 Wien
Graphic design: Ecke Bonk
Printed on acid-free and chlorine-free bleached paper
SPIN: 10743082

CIP data applied for

With 88 (partly coloured) Figures

ISSN 0303-6995
ISBN 3-211-83377-3 (soft cover) Springer-Verlag Wien New York
ISBN 3-211-83378-1 (hard cover) Springer-Verlag Wien New York

Preface

There is abundant literature on Down Syndrome but work on Down Syndrome brain is limited.

New technologies of biochemistry and molecular biology including gene hunting methods are powerful tools for brain research but have been hardly used. Studies on Down Syndrome brain are hampered by a multitude of factors as e.g. availability of brain samples, absence of a valid animal model, heterogeneity of Down Syndrome neuropathology, etc. and it is therefore not surprising that only few groups are involved in this area of brain research.

It is the aim of this book to challenge research on molecular biology and biochemistry of Down Syndrome brain by presenting differing current concepts and contributing by the demonstration of most recent research data, which are in part not proven or generally accepted or are even fragmentary but may represent innovative approaches or may provide first evidence for new concepts or are challenging current opinions on Down Syndrome brain biology and pathology. The book is concepted as a reflection of the frontiers in Down Syndrome brain research and food for thought rather than simply as an updating or state of the art.

Moreover, the book as it is, shows that we are still at the beginning of work on Down Syndrome brain and is approaching neuroscientists to work and cooperate on the subjects stressed in this publication. The concerted action of many scientists is needed for having a change to solve the problem and therefore the book was also a first appeal to the scientifique community to focus on Down Syndrome brain research and a first approach to draw the attention to the intriguing problems in the field.

Although Down Syndrome is known for ages we still do not know whether it's phenotype is due to aneuploidy (and involving chromosomes other than chromosome 21) or to trisomy 21 per se and we still do not even know whether trisomy 21 is of infectious origin!

Technology i. e. genomics and proteomics, is far advanced and we should (have been?) be able to clarify the many open basic questions using the already available biochemical, genetic, molecular biological or virological methods.

This book is to catch the spirit of Down Syndrome brain research, a prerequisite to initiate and stimulate dedicated work on Down Syndrome brain.

Vienna, October 1999   **G. Lubec**

Note added in proof

   While editing this book we have put together and generated by gene hunting some preliminary evidence for a viral etiology of trisomy 21: subtractive hybridization revealed a sequence with 42 % homology to ICP 4, an infectious cell protein induced by herpes virus infection.

This book is dedicated to Dietrich Mateschitz, Salzburg, Austria.

Contents

Listed in Current Contents / Life Sciences

J Neural Transm (1999) [Suppl] 57: 1–19

Molecular abnormalities of the brain in Down Syndrome: relevance to Alzheimer's neurodegeneration

S. M. de la Monte

Pathology and Neuropathology, and Massachusetts General Hospital,
Harvard Medical School, Boston, MA, U.S.A.

Summary. Down syndrome is caused by over-expression of genes located within a segment of chromosome 21, termed the Down locus. Down syndrome is associated with developmental abnormalities of the central nervous system that result in mental retardation and age-dependent Alzheimer-type neurodegeneration. Some of the neurodegenerative lesions, including Aβ amyloid deposition, apoptotic cell death, and aberrant dendritic arborization, are in part due to constitutively increased expression of genes that encode the amyloid precursor protein, superoxide dismutase I, and S100-beta, and located within the Down locus. However, neurodegeneration in Down syndrome is also associated with aberrant expression of genes that are not linked to the Down locus, including the growth associated protein, GAP-43, nitric oxide synthase 3, neuronal thread protein, and pro-apoptosis genes such as p53, Bax, and interleukin-1β-converting enzyme. Increased expression of these non-Down locus genes correlates with proliferation of dystrophic neurites and apoptotic cell death, two important correlates of cognitive impairment in Alzheimer's disease. This article reviews the functional importance of abnormal gene expression in relation to Alzheimer-type neurodegeneration in brains of individuals with Down syndrome.

Abbreviations

DS Down Syndrome, *AD* Alzheimer's disease, *SODI* superoxide dismutase I, *APP* amyloid precursor protein, *GAP-43* growth-associated protein-43, *NTP* neuronal thread protein, *NOS* nitric oxide synthase.

Overview

Down syndrome (DS) is the most frequent genetic cause of mental retardation, affecting up to 1 in 700 pregnancies. The mental retardation is characterized by deficits in short-term and long-term memory, language skills, cognition, and learning. The brains are characteristically small, rounded, foreshortened, and exhibit a steep rise of the occipital lobes, extreme narrowing of the superior temporal gyri, incomplete opercularization with exposure of the insular cortex, and reduced secondary sulcation. These abnormalities are largely due to diminished and malformed growth of the frontal and temporal lobes. Histologically, DS brains show widespread abnormalities in cortical lamination, irregular clustering of neuronal cell bodies, and muted dendritic arborization. Abnormalities in neuronal cytoarchitecture are accompanied by functional disturbances in glial cells manifested by increased levels of S100 protein expression in astrocytes and delayed myelination. In addition, the microglia are prominent and activated, and vascular dysplasias with focal calcifications can be foun§ in the basal ganglia. Although the gyral and dendritic spine abnormalities are believed to represent the neuroanatomic substrate of mental retardation in DS [1], the potential contributions of impaired glial cell function, vascuolopathy, and cytokine activation to maldevelopment and impaired cognitive function have not yet been determined. Since the neuronal cytoarchitecture and dendritic spines appear normal during fetal development and infancy, and only subsequently become abnormal, the defect in dendritic process ramification reflects disturbances in gene expression or intracellular signaling that arise postnatal [2].

Down Syndrome locus

The Down Syndrome phenotype of dysmorphic features, hypotonia, and psychomotor delay is primarily (>95%) caused by Trisomy 21, but a small percentage (5%) is due to translocation of the 21q22 band, also known as the Down locus. Therefore, over-expression of genes residing at the 21q22 segment is considered necessary and sufficient to cause Down syndrome. Several genes important for central nervous system (CNS) function have been mapped to the Down locus, including the amyloid precursor protein (APP, nexin II), protein S-100 beta, and superoxide dismutase I. In addition, a number of genes with unknown functions have recently been mapped to the DS locus. Functional analysis of newly identified developmentally regulated genes that map to the DS locus may help provide information about the cellular and molecular basis of mental retardation in Down syndrome. Examples of potential interest include: the DS cell adhesion molecule which is a matrix gene that shares homology with the immunoglobulin superfamily [3]; tetratricopeptide repeat (TPRD) which is ubiquitously expressed in the telencephalon, mesencephalon, and metencephalon during development [4]; KCNJ15, a gene that encodes a potassium channel belonging to the family of inward rectifier K^+ channels [5]; and hmc 18h10, a homolog of rat PEP-19, a neuron-specific polypeptide that is expressed in the cerebellar Purkinje cells and developmentally regulated [6].

Alzheimer neurodegeneration in Down Syndrome

In considering the molecular pathology of DS, a matter of critical interest is that virtually all individuals with Down syndrome who survive beyond 35 years of age develop neuropathologic changes that closely resemble

Alzheimer's disease (AD) [7–10], although only a subset develop progressive dementia with clinical symptoms of AD [11, 12]. Brains from middle-aged individuals with DS manifest progressive accumulations of senile plaques, Aβ amyloid deposits, and neurofibrillary tangles. These neurodegenerative changes are accompanied by loss of neurons and proliferation of dystrophic neurites, as occur in AD. Therefore, DS probably represents the best natural model of pre-programmed AD-type neuropathology, suggesting that detailed understanding of the molecular abnormalities of the CNS in Down syndrome will ultimately provide important clues about the pathogenesis of AD neurodegeneration. This review will focus on abnormalities in gene expression and the accumulation of gene products in brains of individuals with DS as they relate to the premature development of AD neuropathology. Although the major emphasis will be on genes that map to the DS locus, several genes that are located in other regions of the human genome, that have demonstrated roles in AD neurodegeneration, and that are aberrantly expressed in DS brains prior to, or in close association with the progressive accumulation of AD lesions will also be discussed.

Superoxide dismutase I

CuZn-superoxide dismutase (Cu/Zn SOD; SOD1) is a key enzyme in the metabolism of oxygen free radicals. The SOD1 gene maps to the DS locus on Chromosome 21. Individuals with DS have three copies of the SOD1 gene, and correspondingly, the levels of SOD1 activity in all tissues are increased by approximately 50% relative to control. Increased SOD1 activity can generate free-radical stress through over-production of hydrogen peroxide. Under normal circumstances, increased SOD activity with free radical production in the CNS activates adaptive responses such as increased glutathione peroxidase activity, but in DS, such adaptive responses are often inadequate [13]. In the absence of sufficient free-radical scavenger agents, cells are rendered more vulnerable to oxidative stress. This phenomenon could account for the early onset of apoptotic cell loss and activation of pro-apoptosis genes in the brains of individuals with DS [14].

The transgenic mouse model of SOD1 over-expression has helped to elucidate the role of SOD1 in the pathogenesis of neurodevelopmental and neurodegenerative abnormalities in DS. In the CNS, one of the major consequences of SOD1 over-expression is impaired structure of synaptic terminals and attendant proliferation of dystrophic neurites. For example, transgenic mice that over-express the human SOD1 gene at levels comparable to those observed in DS exhibit withdrawal and destruction of terminal axons and associated development of multiple small terminals [25, 26]. Ultrastructural studies demonstrated reduced ratios of terminal axon areas to postsynaptic membranes with dysmorphic secondary folds, similar to the findings in DS and normal aging [25, 26].

One of the more exciting discoveries in the last several years was that some forms of familial amyotrophic lateral sclerosis (ALS) are associated with mutations in the SOD1 gene [15, 16]. Subsequent studies in transgenic mice demonstrated that over-expression of mutant SOD1 results in motor neuron disease [17–19]. Finally, in sporadic cases of AD, increased superoxide anion production and free radical generation are detectable in peripheral blood monocytes, although the mechanism does not involve mutation or over-expression of the SOD1 gene [20]. Therefore, over-expression of either wildtype or mutant SOD1, or increased SOD1 activity can contribute to the pathogenesis of neurodegeneration.

The mechanisms by which over-expression or mutation of the SOD1 gene results in neurodegeneration with neurite retraction, degeneration, and sprouting are not known. However, since the neurodegeneration in DS and in the transgenic mouse model of ALS does not occur until adulthood, aging-associated factors must govern the onset of altered molecular and cellular function and activate the cascade of neurodegeneration. One prevailing theory is that aging and age-associated diseases are precipitated, propagated, or caused by mitochondrial DNA damage due to superoxide and free-radical injury. With increasing age, mitochondrial DNA mutations increase [21–24]. Progressive mitochondrial DNA damage could render CNS cells more vulnerable to oxidant free-radical injury. Age-associated reduced tolerance to oxidative stress could expose underlying previously compensated defects such as mutations in the SOD1 gene, over-expression of SOD1, or the lack of adequate free-radical scavenger mechanisms. Moreover, over-expression of the SOD1 gene or increased SOD1 enzyme activity could result in constitutive oxidative stress leading to further mitochondrial DNA damage and impaired mitochondrial function, thereby accelerating the aging process. Impaired mitochondrial function reduces the ability of neurons to maintain synaptic connections and SOD1 over-expression or increased enzyme activity can cause neurite retraction due to impaired uptake of biogenic amines leading to a state akin to trophic factor withdrawal [27]. Therefore, over-expression of SOD1 in DS brains probably contributes to the reduced arborization of neuritic terminals noted in young children, and also may have a role in synaptic disconnection and aberrant neuritic sprouting associated with AD-type neurodegeneration.

SOD1 over-expression is likely to perturb function of non-neuronal CNS cells as well. Glial cells and cerebral microvasculature elements are important potential targets of increased SOD1 activity. For example, SOD1 over-expression in oligodendrocytes could contribute to the delayed myelination during development [28] and white matter atrophy associated with neurodegeneration in DS [29]. In addition, over-expression of SOD1 in protoplasmic astrocytes and oligodendrocytes could compromise membrane integrity, exposing neuronal cell processes to a toxic extracellular environment. With regard to the microvasculature, increased SOD1 expression and attendant increased production of hydrogen peroxide and oxidant free radicals could impair the integrity of the blood brain barrier. However, further systematic investigation is required to understand the consequences of SOD1 over-expression in glial cells and the cerebral microvasculature in relation to neurodegeneration.

Amyloid precursor protein

The amyloid precursor protein (APP, nexin II) gene encodes a transmembrane protein that when proteolytically cleaved, yields 3–5 KD Aβ amyloid peptides. Aβ amyloid accumulates mainly as extracellular insoluble fibrillar deposits in brains with AD and in middle-aged individuals with DS. The APP gene maps to the DS locus and consequently, DS brains have significantly elevated levels of APP mRNA [30]. In both DS and AD, Aβ amyloid deposits are localized in cells, vessels, and senile plaques. A number of studies have characterized the distribution of Aβ amyloid in developing DS brains and demonstrated that over-expression of the APP gene results in Aβ amyloid deposition in plaques and cerebral vessels in a predictable age-dependent manner.

Despite over-expression of the APP gene from birth, no significant abnormalities in Aβ amyloid deposition are detectable in young children with DS. Using antibodies generated to distinct epitopes of the APP molecule, DS and normal control fetal, neonatal, and infant brains were demonstrated to have similar levels and distributions of APP immunoreactivity corresponding to the amino-terminal (secretory) and C-terminal fragments and to Aβ 1–28 [31]. During childhood, this Aβ amyloid immunoreactivity disappears in both DS and control brains. These findings suggest that APP expression may have an important function during normal growth and development of the CNS. For example, APP expression in glial cells may have a role in developmental CNS myelination.

However, DS brains are distinguished from normal control brains by the progressive age-dependent accumulations of heterogeneous Aβ amyloid deposits in plaques and cerebrovascular Aβ amyloid deposition. From the neonatal period to adulthood, the Kunitz protease inhibitor (KPI) fragment of APP accumulates in the media of cerebral arteries. Beginning in the second decade, cortical plaques, initially consisting of Aβ1–42 (ending at amino acid 42) deposits, become more heterogeneous with respect to the immunoreactive profiles and associated neurodegenerative changes. Beginning in the second decade, the percentage of Aβ1–42-positive plaques with apolipoprotein E immunoreactivity increases progressively with age. Beginning in the fourth decade, Aβ1–40 deposits are detectable and progressively increase with age, as do the densities of senile plaques with amino-terminal APP immunoreactivity distributed at the periphery and Aβ1–28 immunoreactivity localized within or around the cores [31]. Therefore, Aβ amyloid plaque formation begins with deposition of Aβ1–42 peptides, and with advancing age, there is progressive deposition of more heterogeneous populations of Aβ species as well as Apolipoprotein E, although the number and density of Aβ1–42 positive plaques does not substantially increase with age [32]. The progressive diversification of Aβ1–42-positive plaques due to accumulation of Aβ1–40, amino-terminal APP, and Aβ1–28 coincides with the proliferation of degenerating neurites around senile plaques. Given the abundant and early deposition of Aβ amyloid and associated progressive AD-type neurodegeneration, it would be of interest to determine the extent to which

APP over-expression and Aβ amyloid depositition contribute to cognitive impairment in Down syndrome.

In AD, with or without pre-existing DS, neuronal dysfunction and cognitive impairment are correlated with cell loss, neurite degeneration, and reduced synaptic density. Filamentous Aβ amyloid deposits are abundantly distributed in the neuropil as plaques and around blood vessels as congophilic amyloid angiopathy. The filamentous, insoluble Aβ amyloid deposits are derived from the carboxyl terminus of APP molecules cleaved within the transmembrane domain. Intermolecular interactions of the hydrophobic regions of the Aβ amyloid peptides promote the formation of large aggregates under physiologic conditions [33]. A pathogenic role for fibrillar Aβ amyloid deposits in AD neurodegeneration is suggested by their conspicuous localization in terminal zones of degenerating neurons that develop neurofibrillary tangles, and in vitro experiments demonstrating that Aβ amyloid can be neurotoxic and promote apoptosis [34, 35]. However, the potential roles of Aβ amyloid angiopathy and APP over-expression in glial cells in relation to neurodevelopmental or neurodegenerative changes in DS have not been determined.

In DS, over-expression of the APP gene is responsible for the earlier and more abundant deposition of Aβ amyloid compared with normal aged individuals and patients with sporadic AD. It is not known why Aβ deposition increases with aging or in association with sporadic AD since the mechanism does not involve gene over-expression or gene mutation. One possible mechanism under intense investigation involves acquired abnormalities in cellular enzymatic processing of APP. Another potential factor is disregulation of gene expression due to abnormal activation of the promoter. The APP promoter was mapped to the proximal region (-39 to -49) of the APP gene [36], and recently identified as an upstream stimulatory factor unrelated to the fos/jun complex or AP-4 factor [37]. The APP promoter shares similarity with housekeeping gene promoters due to the presence of a GC-rich region around the transcription start site and the lack of a TATA box. Gene transfer experiments showed the GC-rich region contains overlapping binding sites for different transcription factors whose binding is mutually excluded. Therefore, an imbalance between different transcription factors could cause APP over-expression in otherwise normal individuals [38]. The localization of APP at synaptic sites in the brain provides a mechanism by which disregulation of APP expression could impair synaptic plasticity [38] and contribute to neurodegeneration. One potential therapeutic strategy might be to attack genes that specifically up-regulate the expression of APP.

S100 Beta protein

S100-β is a calcium-binding protein that is synthesized by CNS astrocytes. S100-β has trophic effects on neuronal and glial cells, resulting in neurite extension and glial cell proliferation. S100-β is abundantly expressed in hypertrophic reactive astrocytes, and consequently it immunoreactivity is increased up to 20-fold in brain regions with severe AD neurodegeneration, including

DS + AD. The gene that encodes S100-β protein is located within the DS brains with locus, and the expression levels of the corresponding mRNA and protein are ten-fold higher in brains of individuals with Down syndrome compared with normal controls [39]. Increased levels of S100-β expression can induce neurite extension and proliferation. Therefore, over-expression of the S100-β gene in DS brains could contribute to the excessive proliferation of cortical neuropil neurites that are detectable beginning in second decade of life. Experimental evidence for this pathologic response was demonstrated in transgenic mice bearing multiple copies of the S100-β gene since the brains exhibit proliferation of neurites similar to that which occurs in DS [2]. Moreover, over-expression of the S100-β gene in transgenic mice results in neuronal cytoskeletal and behavioral signs of altered aging [40]. With AD neurodegeneration, S100-β protein expression with aberrant neurite proliferation may increase in reaction to Aβ 1–40 amyloid deposits since Aβ1–40 amyloid is mitogenic and induces S100-β expression in glial cells [41]. Therefore, Aβ 1–40 amyloid deposits may indirectly mediate aberrant proliferation of neurites in DS and AD brains.

Genes of unknown significance in relation to the Down syndrome phenotype

The DS locus on sub-band q22.2 of chromosome 21 contains multiple genes that have not been characterized, but may contribute to the trisomy 21 phenotype. Within the past few years, a number of genes localized to the DS locus and with potential interest in relation to abnormal CNS development and function in DS have been partly characterized including: the DS cell adhesion molecule [3], tetratricopeptide repeat TPRD gene [4], AVP vasopressin-homology gene [42], Bach1 transcription factor [43], mitochondrial NADH:ubiquinone oxidoreductase [44], KCNJ15/Kir4.2 gene that encodes a potassium channel inward rectifier K^+ (Kir) protein [5], and the PEP-19 gene that encodes a neuron-specific polypeptide [6]. The DS cell adhesion molecule (DSCAM) is a novel member of the immunoglobulin (Ig) superfamily and represents a new class of neural cell adhesion molecules. Since the brain is the major organ of DSCAM gene expression throughout development and in the mature organism, DSCAM over-expression may contribute to the central and peripheral nervous system defects in DS [3]. The TPRD gene has significant homology to matrix proteins, and therefore its predominant over-expression in the CNS during development may have relevance to the CNS malformations in DS [4]. The AVP vasopressin-homology gene is over-expressed in the temporal lobes of fetuses with DS and in sporadic AD [42]. Vasopressin peptide may be functionally important for memory and learning. Therefore, over-expression of AVP in DS and AD may have relevance to the pathogenesis of neurodegeneration [42]. The Bach1 transcription factor contains a basic leucine zipper and BTB-zinc finger domains that are directly involved in DNA binding for transcription regulation [43]. Mitochondrial NADH:ubiquinone oxidoreductase (Complex I) gene is ubiquitously expressed but the functional relevance to the DS phenotype is unknown [44]. The KCNJ15/Kir4.2 gene encodes a potassium channel inward rectifier protein that is expressed in the adult brain [5]. The PEP-19 gene encodes a neuron-specific polypeptide that is expressed in several regions of the CNS, and PEP19 probes specifically mark Purkinje cells in the cerebellum [6]. Despite their localization within the DS locus, the contributions of NADH ubiquinone oxidoreductase, KCNJ15/Kir4.2, and PEP19 to the DS phenotype have not been determined, but remain promising targets of future research due to the abundant expression of these genes in the CNS.

Non-Down syndrome locus abnormalities in gene expression

Although the over-expression of genes that map within the DS locus undoubtedly contribute to the DS phenotype, it is important to consider the potential role that non-DS locus genes may have on development or neurodegeneration in DS. Conceivably, the expression and regulation of other genes may be adversely and secondarily affected by over-expression of DS locus genes. For example, genes modulated with neuritic growth, including the growth associated protein, GAP-43 (neuromodulin) [45], nitric oxide synthase 3 (NOS3) [46], and neuronal thread protein [NTP] [47] and which do not map to the DS locus are also aberrantly expressed in brains of individuals with DS beginning years prior to the onset of AD neurodegeneration.

Neuritic sprouting and neuritic dystrophy

The prominent neuropathologic changes observed in DS during development and AD-type neurodegeneration include reduced arborization of neuritic terminals [28] and proliferation of dystrophic neurites [46, 47]. The resulting impaired synaptic connections contribute to mental retardation and cognitive decline. Neurodegenerative cortical neuritic pathology in DS and AD is readily detected by immunostaining with antibodies to phospho-tau or phospho-neurofilament (Fig. 1A), and consistently associated with aberrant expression of GAP-43 [45] (Fig. 1E, 1F), NOS3 [46, 48] (Fig. 1C, 1D), and NTP [44, 47] (Fig. 1B). The genes that encode GAP-43, NOS3, and NTP are normally modulated with neuritic growth during development and regeneration, and are aberrantly expressed in DS brains years prior to the onset of AD.

In AD, increased GAP-43 immunoreactivity is localized in proliferated neuropil neurites in the cerebral cortex [45, 49], the perikarya of cortical neurons, and white matter glial cells [45]. In AD, increased GAP-43 immunoreactivity is associated with constitutively elevated levels of its mRNA in some neurons and many white matter glial cell processes (glial sprouts) (Fig. 1F), suggesting either up-regulated transcription or enhanced mRNA stability. Identical abnormalities have been detected in brains of individuals with DS beginning in the second decade [45]. Since GAP-43 is highly expressed during neurite outgrowth, its aberrant expression in DS prior to the

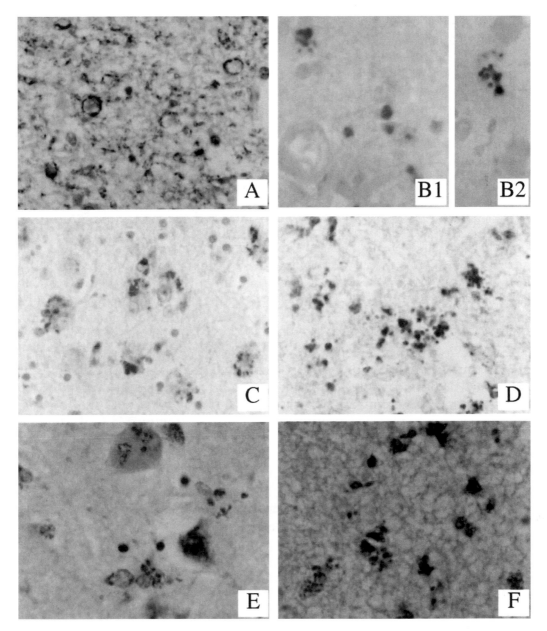

Fig. 1. Neuritic and glial sprouting in brains with DS + AD, detected with monoclonal antibodies to phospho-neurofilament (**A**), neuronal thread protein (**B1,B2**), nitric oxide synthase 3 (**C,D**), and the growth associated protein, GAP-43 (**E,F**). Immunoreactivity was revealed with avidin-biotin horseradish peroxidase labeling system using diamino-benzidine as the chromogen (brown precipitate). The sections were counterstained with hematoxylin to provide a light blue contrast for unlabeled structures. Neuritic sprouts are vesicular (**A**) or coarse granular to punctate in morphology (**B,C,E**). Glial sprouts are mainly punctate and often juxtaposed to glial cell nuclei (**D,E**)

onset of AD suggests an important role for this molecule in the pathogenesis of neurite proliferation with AD-type neurodegeneration [45].

Nitric oxide synthase enzyme activity generates nitric oxide through oxidation of a guanidino nitrogen of L-arginine [50]. Three isoforms of NOS exist: NOS1, also known as neuronal NOS, and NOS3 (endothelial) are constitutively activated, and both are expressed in CNS neurons [51, 52]. NOS2 (inducible) is activated by calcium flux induced by lipopolysaccharide and cytokines [53]. In the nervous system, NO has dual effects in neuronal cells depending upon the source of NO and the levels of enzyme activity. NO can either promote neurite outgrowth and sustain viability of neurons [54], or function as a neurotoxin and promote apoptosis [55, 56]. Moreover, in vitro studies suggest that over-expression of NOS3 in neurons can cause p53-mediated apoptosis [14].

AD is associated with aberrant NOS3 expression with high levels of immunoreactivity localized in terminals of degenerating/proliferating neurites [46]. The same abnormality occurs in DS beginning in the second decade of life [48]. Of interest is that the regional and cellular distributions of abnormal NOS3 expression in AD and DS brains are virtually identical to those of GAP-43, and the immunoreactivities co-localize within the same cell processes (de la Monte and Bloch, unpublished observation). This suggests that in AD and DS brains, aberrant expression of GAP-43 and NOS3 may be linked to the same alterations in intracellular signaling. Increased NOS3 immunoreactivity and mRNA expression occur in cortical neurons and glial cells that exhibit increased apoptosis and degeneration in AD [46]. In addition, recent evidence suggests that NOS3 expression may also be altered in the cerebral vasculature in AD and DS (de la Monte and Bloch, unpublished observation). Further studies along these lines demonstrated co-localization of NOS3 with p53, a pre-apoptosis gene, in neurons, glial cells, cerebrovascular smooth muscle cells, and dystrophic neurites (Fig. 2).

Neuronal thread proteins (NTPs) are a family of molecules expressed in the brain and modulated with neuritic sprouting during development, regeneration, and neurodegeneration. Experimentally induced neurite outgrowth by stimulation with growth factors, retinoic acid, or phorbol ester myristate increases expression of NTP molecules [47, 57–61]. Developmental and regenerative neuritic sprouting are mainly associated with increased levels of the ~15kD and ~18kD NTP molecules, whereas AD neurodegeneration is associated with increased levels of the ~21kD and ~42kD NTP molecules. Experimental over-expression of the AD7c-NTP cDNA results in both increased neuritic sprouting and apoptosis [62]. Elevated levels of the 42kD AD7c-NTP species are detectable in cerebrospinal fluid (CSF) and correlate with severity of dementia in AD. Analysis of DS brains demonstrated increased levels of AD7c-NTP beginning in the second decade and prior to the development of significant AD-type neurodegeneration [47]. With increasing age, DS brains exhibit increased AD7c-NTP immunoreactivity with levels approaching those in sporadic AD, whereas in normal controls, AD7c-NTP expression does not increase with age [47]. Recent preliminary studies demonstrated elevated levels of AD7c-NTP protein in CSF and urine in

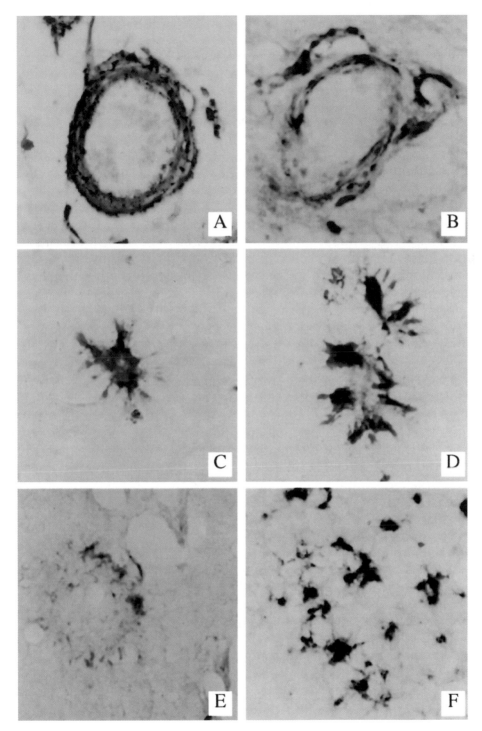

Fig. 2. Increased NOS3 (brown) immunoreactivity associated with p53 (blue) apoptosis gene expression in medium sized cerebral meningeal vessels (**A,B**) and neuropil neurites (**F**). In Panels **A** and **B**, the p53 immunoreactivity is localized in nuclei of smooth muscle cells, and the NOS3 immunoreactivity is present in the cytoplasm of endothelial and smooth muscle cells. **C,D** Small cortical NOS3-immunoreactive vessels surrounded by p53-immunoreactive cell processes, suggesting associated degeneration of perivascular glial cell processes. **E** Lack of NOS3 immunoreactivity in neuritic senile plaques with abundant p53-immunoreactive (degenerated) dystrophic neurites. **F** Co-localization of NOS3 with p53 in dystrophic cortical neurites (brown + blue appears black)

individuals with DS and clinical evidence of AD-type dementia [Ghanbari et al., unpublished].

Activation of pro-apoptosis pathways in Down Syndrome

Apoptosis is now a recognized mechanism of cell loss in DS, AD, and other neurodegenerative diseases [14, 63–66]. Increased apoptosis in DS and AD neurodegeneration has been detected mainly with in situ end-labeling assays [14], and by demonstrating higher expression of pro-apoptosis genes including p53, CD95/Apol-Fas receptor, and Bax [14, 63–66]. The p53 tumor suppressor gene encodes a phosphoprotein that binds to DNA and functions as a transcription factor resulting in cell cycle arrest at the G1-S boundary in response to DNA damage. Wild-type p53, when accumulated in the nucleus, has a demonstrated role in apoptosis in neurons and glial cells [14, 66, 67]. The mechanism involves down-regulation of Bcl-2, up-regulation of Bax, and increased transcription of c-jun. The Bcl family of proto-oncogenes encodes membrane-associated proteins that either promote or inhibit apoptosis. Bcl-2 suppresses apoptosis by forming heterodimers with the death-promoting molecules, Bax and Bad. Therefore, reduced levels of Bcl-2 or increased levels of Bax and Bad can lead to increased apoptosis. Low levels of Bcl-2 can also increase apoptosis by promoting p53 translocation to the nucleus. Finally, increased cell loss and susceptibility to apoptosis in DS and AD are reflected by above normal-levels of DNA repair enzymes. For example, the increased levels of the ERCC2- p80 and ERCC3 p89 DNA repair proteins detected in the temporal and frontal lobes of individuals with DS suggest gene activiation in response to ongoing DNA damage [68].

Since apoptosis is undoubtedly an important mechanism of cell loss, it is critical to understand the mechanisms of pro-apoptosis gene activation or survival gene inhibition. In this regard, aberrant expression of SOD1, NOS3, Aβ-amyloid, and AD7c-NTP are likely to play a role. Over-expression of the SOD1 gene and excess production of superoxide and oxygen free-radicals could increase susceptibility of CNS cells to oxidative stress-induced apoptosis. High levels of NO can lead to peroxynitrite and free radical formation. In this regard, experimental over-expression of NOS3 in CNS-derived neuronal cells causes p53- and CD95-mediated apoptosis [69]. The recent findings of high levels of NOS3 immunoreactivity co-localized with p53 in dystrophic neurites in the cerebral cortex (Fig. 2F) and glial cells distributed in both cortex and white matter suggest that aberrant NOS3 expression may lead to apoptosis of both cell types. In contrast, NOS3 immunoreactivity is not associated with p53-immunoreactive dystrophic neurites in plaques (Fig. 2E). High levels of NOS3 immunoreactivity also co-localized with p53 in cerebrovascular smooth muscle cells DS and AD [46, 48] (Fig. 2A and B), suggesting a role for NOS3 in the pathogenesis of CNS vasculopathy in AD-type neurodegeneration. Finally, NOS3-immunoreactive small intracortical vessels are frequently surrounded by p53-immunoreactive dystrophic cell processes (Fig. 2C and D) that are probably glial in origin. This relationship suggests impairment of the blood-brain barrier in relation to NOS3-associated CNS vasculopathy. CNS vasculopathy and micro-ischemic lesions have demonstrated importance to the clinical progression of AD neurodegeneration [70]. Therefore, aberrantly increased NOS3 expression may mediate apoptosis of neurons, glial cells, and vascular cells in the CNS in both DS and AD.

A causal role for Aβ-amyloid deposition in neurodegeneration is suggested by the associated (peri-plaque) proliferation of p53- and Bax-immunoreactive neurites in AD and DS brains [65, 66]. In addition, we have recently demonstrated co-localization of Aβ-amyloid with p53 in cerebrovascular smooth muscle cells (Fig. 3). Intense levels of Aβ-amyloid immunoreactivity in vascular smooth muscle were associated with p53 expression in endothelial (Fig. 3A) and smooth muscle (Fig. 3C and D) cells of large and

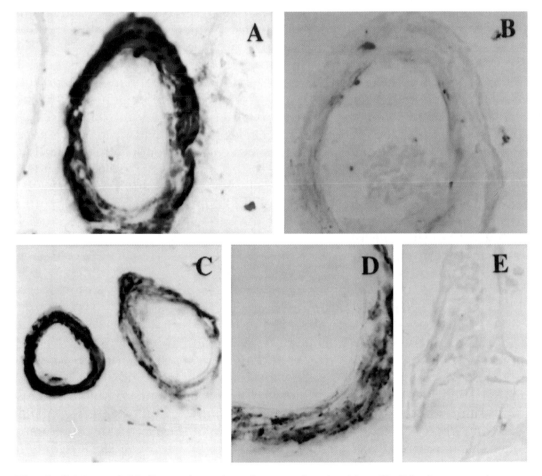

Fig. 3. βA4 amyloid (brown) angiopathy associated with p53 (blue) apoptosis gene expression in smooth muscle and endothelial cells in the cerebral cortex. **A** Intense levels of βA4 immunoreactivity in vascular smooth muscle associated with low levels of p53 in endothelial cells. **B** Increased p53 immunoreactivity in a cerebral vessel devoid of βA4 immunoreactivity. **C,D** Co-localization of βA4 and p53 immunoreactivity in cerebrovascular smooth muscle cells. **E** Control cortical vessel lacking both βA4 and p53 immunoreactivity

medium size cortical and leptomeningeal vessels. However, increased p53 immunoreactivity was frequently detected in endothelial cells in the absence of Aβ-amyloid immunoreactivity (Fig. 3B). Although Aβ amyloid peptide can experimentally induce apoptosis or excitotoxic death of neurons [34, 35], the in vivo observations suggest that apoptosis gene expression may precede Aβ-amyloid deposition in cerebral vessels.

Increased AD7c-NTP immunoreactivity is distributed in neurons with neurodegeneration and in dystrophic neuropil neurites in AD [57, 62]. Experimental over-expression of the AD7c-NTP cDNA causes both apoptosis and neuritic sprouting in neuronal cells [62]. Recent studies have demonstrated that apoptosis associated with over-expression of AD7c-NTP is mediated by increased levels of p53, whereas AD7c-NTP-induced neuritic sprouting is correlated with phospho-tau accumulation and up-regulation of NOS-3 [57, 62].

p53- and CD95-mediated apoptosis is effected by caspase activation. Caspases are a family of cysteine proteases which when activated, cause nuclease release with cleavage of DNA and proteolytic cleavage of cytoskeletal and membrane proteins. Increased levels of caspase-1 (interleukin-1β converting enzyme; ICE) immunoreactivity are detectable in young DS brains years prior to the onset of AD neurodegeneration [71]. The increased caspase-1 immunoreactivity is observed mainly in glial cells [71], corresponding with the more abundant glial cell apoptosis and apoptosis gene expression in both DS and AD [14]. One recurring theme is that apoptosis, degeneration, and aberrant gene expression in CNS glial cells are under-appreciated aspects of the molecular pathology of DS and AD. Given the reduced white matter volume in developing DS brains, and the white matter atrophy with fiber loss observed in early, pre-clinical AD, future studies should define the role of glial cell pathology in the pathogenesis of AD neurodegeneration.

Trisomy 16 Down Syndrome mouse model

The discovery of genetic homology between human chromosome 21 and mouse chromosome 16 led to the development of the murine trisomy 16 model of Down syndrome [72]. Trisomy 16 mice exhibit numerous developmental abnormalities including neurochemical and neuroanatomical alterations similar to those that occur in DS. The mouse chromosome 16 contains a region (16C3.3–4) that is homologous to the DS locus on human chromosome 21. Therefore, it was possible to generate a DS mouse model of segmental trisomy 16 (Ts65Dn mice) in which genes located in the region corresponding to the Down locus are over-expressed. Although Ts65DN mice do not show early-onset of AD neuropathology, they do demonstrate impaired performance in complex learning tasks requiring the integration of visual and spatial information [73].

In murine (m) trisomy 16, four distinct developmental gene over-expressions relevant to CNS function have been demonstrated: amyloid precursor protein (APP), preprosomatostatin, SOD1, and growth-associated protein 43 (GAP-43) [74]. The mAPP gene encodes cerebrovascular amyloid peptide and is located in close proximity to both SOD1 and the protooncogene, Ets-2 on chromosome 16 [72]. As observed in DS, APP over-expression in murine trisomy 16 increases steadily throughout fetal and postnatal development [74]. CNS cultures taken from trisomy 16 fetal mice express increased levels of the 210kDa (high-molecular weight) neurofilament subunit [75]. Over-expression of the gene that encode the 210kD neurofilament subunit in transgenic mice is sufficient to cause degeneration of spinal motor neurons [76].

Conclusions

Although the CNS pathology of DS and the propensity to develop AD neurodegeneration have been recognized for decades, only in recent years have the molecular mechanisms of mental retardation and AD-type neurodegeneration begun to be elucidated. Much of the focus has been justifiably placed on the analysis of genes that map to the DS locus since that region of Chromosome 21 is necessary and sufficient to produce the DS phenotype. The identification of the murine DS homology domain and generation of the DS mouse model has helped determine the role of SOD1 and APP over-expression in relation to neurodegeneration. However, it is important to recognize that other gene abnormalities not directly related to the DS locus contribute to the pathogenesis of AD-type neurodegeneration in DS. Future studies will likely define the cascade of neurodegeneration in relation to aberrant expression of genes located either within or outside of the Down syndrome locus.

References

1. Sustrova M, Sarikova V (1997) Down's syndrome — effect of increased gene expression in chromosome 21 on the function of the immune and nervous system. Bratisl Lek Listy 98(4): 221–228
2. Becker LE et al (1993) Association of phenotypic abnormalities of Down syndrome with an imbalance of genes on chromosome 21. APMIS [Suppl] 40: 57–70
3. Moffett P et al (1996) Characterization of msim, a murine homologue of the Drosophila sim transcription factor. Genomics 35(1): 144–155
4. Tsukahara F et al (1998) Molecular characterization of the mouse mtprd gene, a homologue of human TPRD: unique gene expression suggesting its critical role in the pathophysiology of Down syndrome. J Biochem (Tokyo) 123(6): 1055–1063
5. Gosset P et al (1997) A new inward rectifier potassium channel gene (KCNJ15) localized on chromosome 21 in the Down syndrome chromosome region 1 (DCR1). Genomics 44(2): 237–241
6. Chen H, Bouras C, Antonarakis SE (1996) Cloning of the cDNA for a human homolog of the rat PEP-19 gene and mapping to chromosome 21q22.2–q22.3. Hum Genet 98(6): 672–677
7. Schapiro MB et al (1988) Dementia in Down's syndrome: cerebral glucose utilization, neuropsychological assessment, and neuropathology. Neurology 38(6): 938–942

8. Mann DM (1988) The pathological association between Down syndrome and Alzheimer disease. Mech Ageing Dev 43(2): 99–136

9. Lai F, Williams RS (1989) A prospective study of Alzheimer disease in Down syndrome. Arch Neurol 46(8): 849–853.

10. Mann DM (1988) Alzheimer's disease and Down's syndrome. Histopathology 13(2): 125–137

11. Devenny DA et al (1996) Normal ageing in adults with Down's syndrome: a longitudinal study. J Intellect Disabil Res 40(Pt 3): 208–221

12. Brugge KL et al (1994) Cognitive impairment in adults with Down's syndrome: similarities to early cognitive changes in Alzheimer's disease. Neurology 44(2): 232–238

13. Antila E, Westermarck T (1989) On the etiopathogenesis and therapy of Down syndrome. Int J Dev Biol 33(1): 183–188

14. de la Monte SM et al (1998) P53- and CD95-associated apoptosis in neurodegenerative diseases. Lab Invest 78(4): 401–411

15. Ince PG et al (1998) Amyotrophic lateral sclerosis associated with genetic abnormalities in the gene encoding Cu/Zn superoxide dismutase: molecular pathology of five new cases, and comparison with previous reports and 73 sporadic cases of ALS. J Neuropathol Exp Neurol 57(10): 895–904

16. Chou SM (1997) Neuropathology of amyotrophic lateral sclerosis: new perspectives on an old disease. J Formos Med Assoc 96(7): 488–498

17. Dal Canto MC, Gurney ME (1995) Neuropathological changes in two lines of mice carrying a transgene for mutant human Cu,Zn SOD, and in mice overexpressing wild type human SOD: a model of familial amyotrophic lateral sclerosis (FALS). Brain Res 676(1): 25–40

18. Dal Canto MC, Gurney ME (1994) Development of central nervous system pathology in a murine transgenic model of human amyotrophic lateral sclerosis. Am J Pathol 145(6): 1271–1279

19. Morrison BM et al (1998) Time course of neuropathology in the spinal cord of G86R superoxide dismutase transgenic mice. J Comp Neurol 391(1): 64–77

20. Margaglione M et al (1995) Cu/Zn superoxide dismutase in patients with non-familial Alzheimer's disease. Aging (Milano) 7(1): 49–54

21. Ozawa T (1998) Mitochondrial DNA mutations and age. Ann NY Acad Sci 854: 128–154

22. Ozawa T (1997) Oxidative damage and fragmentation of mitochondrial DNA in cellular apoptosis. Biosci Rep 17(3): 237–250

23. Richter C, Suter M, Walter PB (1998) Mitochondrial free radical damage and DNA repair. Biofactors 7(3): 207–208

24. Richter C (1995) Oxidative damage to mitochondrial DNA and its relationship to ageing. Int J Biochem Cell Biol 27(7): 647–653

25. Groner Y et al (1990) Down syndrome clinical symptoms are manifested in transfected cells and transgenic mice overexpressing the human Cu/Zn-superoxide dismutase gene. J Physiol 84(1): 53–77

26. Groner Y et al (1994) Cell damage by excess CuZnSOD and Down's syndrome. Biomed Pharmacother 48(5–6): 231–240

27. Elroy-Stein O, Groner Y (1988) Impaired neurotransmitter uptake in PC12 cells overexpressing human Cu/Zn-superoxide dismutase–implication for gene dosage effects in Down syndrome. Cell 52(2): 259–267

28. Becker L et al (1991) Growth and development of the brain in Down syndrome. Prog Clin Biol Res 373: 133–152

29. de la Monte SM, Hedley-Whyte ET (1990) Small cerebral hemispheres in adults with Down's syndrome: contributions of developmental arrest and lesions of Alzheimer's disease. J Neuropathol Exp Neurol 49(5): 509–520

30. Oyama F et al (1994) Down's syndrome: up-regulation of beta-amyloid protein precursor and tau mRNAs and their defective coordination. J Neurochem 62(3): 1062–1066

31. Arai Y et al (1997) Developmental and aging changes in the expression of amyloid precursor protein in Down syndrome brains. Brain Dev 19(4): 290–294

32. Lemere CA et al (1996) Sequence of deposition of heterogeneous amyloid beta-peptides and APO E in Down syndrome: implications for initial events in amyloid plaque formation. Neurobiol Dis 3(1): 16–32

33. Beyreuther K et al (1991) Mechanisms of amyloid deposition in Alzheimer's disease. Ann NY Acad Sci 640: 129–139

34. Loo DT et al (1993) Apoptosis is induced by beta-amyloid in cultured central nervous system neurons. Proc Natl Acad Sci USA 90(17): 7951–7955

35. Forloni G et al (1996) Apoptosis-mediated neurotoxicity induced by beta-amyloid and PrP fragments. Mol Chem Neuropathol 28(1–3): 163–171

36. Lahiri DK, Robakis NK (1991) The promoter activity of the gene encoding Alzheimer beta-amyloid precursor protein (APP) is regulated by two blocks of upstream sequences. Brain Res Mol Brain Res 9(3): 253–257

37. Kovacs DM et al (1995) The upstream stimulatory factor functionally interacts with the Alzheimer amyloid beta-protein precursor gene. Hum Mol Genet 4(9): 1527–1533

38. Beyreuther K et al (1993) Regulation and expression of the Alzheimer's beta/A4 amyloid protein precursor in health, disease, and Down's syndrome. Ann NY Acad Sci 695: 91–102

39. Marks A et al (1996) Accumulation of S100 beta mRNA and protein in cerebellum during infancy in Down syndrome and control subjects. Brain Res Mol Brain Res 36(2): 343–348

40. Whitaker-Azmitia PM et al (1997) Transgenic mice overexpressing the neurotrophic factor S-100 beta show neuronal cytoskeletal and behavioral signs of altered aging processes: implications for Alzheimer's disease and Down's syndrome. Brain Res 776(1–2): 51–60

41. Pena LA, Brecher CW, Marshak DR (1995) beta-Amyloid regulates gene expression of glial trophic substance S100 beta in C6 glioma and primary astrocyte cultures. Brain Res Mol Brain Res 34(1): 118–126

42. Labudova O et al (1998) Brain vasopressin levels in Down syndrome and Alzheimer's disease. Brain Res 806(1): 55–59

43. Blouin JL et al (1998) Isolation of the human BACH1 transcription regulator gene, which maps to chromosome 21q22.1. Hum Genet 102(3): 282–288

44. de Coo RF et al (1997) Molecular cloning and characterization of the human mitochondrial NADH: oxidoreductase 10-kDa gene (NDUFV3). Genomics 45(2): 434–437

45. de la Monte SM, Ng SC, Hsu DW (1995) Aberrant GAP-43 gene expression in Alzheimer's disease. Am J Pathol 147(4): 934–946

46. de la Monte SM, Bloch KD (1997) Aberrant expression of the constitutive endothelial nitric oxide synthase gene in Alzheimer disease. Mol Chem Neuropathol 30(1–2): 139–159

47. de la Monte SM et al (1996) Developmental patterns of neuronal thread protein gene expression in Down syndrome. J Neurol Sci 135(2): 118–125

48. Sohn YK et al (1999) Neuritic sprouting with aberrant expression of the nitric oxide synthase 3 gene in neurodegenerative diseases. J Neurol Sci (in press)

49. Masliah E et al (1991) Patterns of aberrant sprouting in Alzheimer's disease. Neuron 6: 729–739

50. Schmidt H, Walter U (1994) NO at work. Cell 78: 919–925

51. Bredt D, Snyder S (1994) Transient nitric oxide synthase expression in neurons of embryonic cerebral cortical plate, sensory ganglia, and olfactory epithelium. Neuron 13: 301–313

52. Dinerman J et al (1994) Endothelial nitric oxide synthase localized to hippocampal pyramidal cells: implications for synaptic plasticity. Proc Natl Acad Sci USA 91: 4214–4218

53. Merrill J et al (1993) Microglial cell cytotoxicity of oligodendrocytes is mediated through nitric oxide. J Immunol 151: 2132–2141
54. Peunova N, Enikolopov G (1995) Nitric oxide triggers a switch to growth arrest during differentiation of neuronal cells. Nature 375: 68–73
55. Stamler (1994) Redox signaling: nitration and related target interactions of nitric oxide (Review). Cell 78: 931–936
56. Xia Y et al (1996) Nitric oxide synthase generates superoxide and nitric oxide in arginine-depleted cells leading to peroxynitrite-mediated cellular injury. Proc Natl Acad Sci USA 93: 6770–6774
57. de la Monte SM et al (1996) Profiles of neuronal thread protein expression in Alzheimer's disease. J Neuropathol Exp Neurol 55: 1038–1050
58. de la Monte SM, Xu YY, Wands JR (1996) Neuronal thread protein gene modulation with sprouting: relevance to Alzheimer's disease. J Neurol Sci 138: 26–35
59. de la Monte SM, Garner W, Wands JR (1997) Neuronal thread protein gene modulation with cerebral infarction. J Cereb Blood Flow Metab 17: 623–635
60. Xu YY, Wands JR, de la Monte SM (1993) Characterization of thread proteins expresed in neuroectodermal tumors. Cancer Res 53: 3832–3829
61. de la Monte SM, Wands JR (1992) Neuronal thread protein over-expression in brains with Alzheimer's disease. J Neurol Sci 32: 733–742
62. de la Monte SM et al (1997) Characterization of the AD7c-NTP cDNA and its expression in Alzheimer's disease. J Clin Invest 160: 2093–2104
63. Nishimura T et al (1995) Fas antigen expression in brains of patients with Alzheimer-type dementia. Brain Res 695(2): 137–145
64. Su JH et al (1994) Immunohistochemical evidence for apoptosis in Alzheimer's disease. Neuroreport 5(18): 2529–2533
65. Su JH, Deng G, Cotman CW (1997) Bax protein expression is increased in Alzheimer's brain: correlations with DNA damage, Bcl-2 expression, and brain pathology. J Neuropathol Exp Neurol 56(1): 86–93
66. de la Monte SM, Sohn YK, Wands JR (1997) Correlates of p53- and Fas (CD95)-mediated apoptosis in Alzheimer's disease. J Neurol Sci 152(1): 73–83
67. Kitamura Y et al (1997) Changes of p53 in the brains of patients with Alzheimer's disease. Biochem Biophys Res Commun 232(2): 418–421
68. Hermon M et al (1998) Expression of DNA excision-repair-cross-complementing proteins p80 and p89 in brain of patients with Down Syndrome and Alzheimer's disease. Neurosci Lett 251(1): 45–48
69. de la Monte SM et al (1999) Nitric oxide synthase 3 over-expression is sufficient to cause apoptosis and molecular abnormalities observed in Alzheimer's disease. J Neuropathol Exp Neurol (in press)
70. Etienne D et al (1998) Cerebrovascular pathology contributes to the clinical progression of Alzheimer's disease. J Alz Dis 1: 119–134
71. Griffin WS et al (1989) Brain interleukin 1 and S-100 immunoreactivity are elevated in Down syndrome and Alzheimer disease. Proc Natl Acad Sci USA 86(19): 7611–7615
72. Reeves RH et al (1987) Genetic linkage in the mouse of genes involved in Down syndrome and Alzheimer's disease in man. Brain Res 388(3): 215–221
73. Reeves RH et al (1995) A mouse model for Down syndrome exhibits learning and behaviour deficits [see comments]. Nature Genet 11(2): 177–184
74. O'Hara BF et al (1989) Developmental expression of the amyloid precursor protein, growth-associated protein 42, and somatostatin in normal and trisomy 16 mice. Brain Res Dev Brain Res 49(2): 300–304
75. Plioplys AV (1988) Expression of the 210kDa neurofilament subunit in cultured central nervous system from normal and trisomy 16 mice: regulation by interferon. J Neurol Sci 85(2): 209–222

76. Xu Z et al (1993) Increased expression of neurofilament subunit NF-L produces morphological alterations that resemble the pathology of human motor neuron disease. Cell 73(1): 23–33

Author's address: Dr. S. M. de la Monte, Departments of Medicine and Pathology, Rhode Island Hospital, Brown University School of Medicine, 55 Claverick Street, Room 419, Providence, RI 02403, USA, e-mail: delamonte@hotmail.com

J Neural Transm (1999) [Suppl] 57: 21–40
© Springer-Verlag 1999

Reduced aldehyde dehydrogenase levels in the brain of patients with Down Syndrome

G. Lubec[1], **O. Labudova**[1], **N. Cairns**[2], **P. Berndt**[3], **H. Langen**[3], and **M. Fountoulakis**[3]

[1] Department of Pediatrics, University of Vienna, Austria
[2] Institute of Psychiatry, Brain Bank, London, United Kingdom
[3] F. Hoffmann-La Roche, Ltd., Basel, Switzerland

Summary.   Aldehyde dehydrogenase (ALDH) is a key enzyme in fructose, acetaldehyde and oxalate metabolism and represents a major detoxification system for reactive carbonyls and aldehydes. In the brain, ALDH exerts a major function in the metabolism of biogenic aldehydes, norepinephrine, dopamine and diamines and γ-aminobutyric acid. Subtractive hybridization studies in Down Syndrome (DS) fetal brain showed that mRNA for ALDH are downregulated. Here we studied the protein levels in the brain of adult patients. The proteins from five brain regions of 9 aged patients with DS and 9 controls were analyzed by two-dimensional (2-D) gel electrophoresis and identified by matrix-assisted laser desorption ionization mass spectrometry. ALDH levels were reduced in the brain regions of at least half of the patients with Down Syndrome, as compared to controls. The decreased ALDH levels in the DS brain may result in accumulation of aldehydes which can lead to the formation of plaques and tangles reflecting abnormally cross-linked, insoluble and modified proteins, found in aged DS brain. Furthermore, we constructed a 2-Dmap including approximately 120 identified human brain proteins.

Introduction

Aldehyde dehydrogenase (ALDH) is a key enzyme in fructose, acetaldehyde and oxalate metabolism and a major detoxification system for reactive carbonyls and aldehydes (Epstein, 1992; Quintanilla and Tampier, 1995; Shiohara et al., 1984). Several isoenzymes of ALDH are expressed in mammals, including human brain (Stewart et al., 1996; Zimatkin and Karpuk, 1996; Hague et al., 1997; Pietruszko et al., 1984; Maring et al., 1985; Ryzlak et al., 1988). The regional distribution of the ALDH forms has been described (Weiner and Ardelt, 1984; Zimatkin et al., 1992). In brain, ALDH fulfills a major function in the metabolism of biogenic aldehydes (Nilsson and Tottman, 1987; Tottmar, 1985; Nilsson and Tottmar, 1985; Hafer et al., 1987).

ALDH has been shown to be the most important enzyme for the handling of the dopamine metabolite 3,4-dihydroxyphenylacetaldehyde (Colzi et al., 1996). ALDH is predominantly localized in dopamine rich areas (Hafer et al., 1987; McCaffery and Drager, 1994; Petersen, 1985; Zimatkin, 1991). Kawamura and coworkers have shown that ALDH is also related to neuronal metabolism of norepinephrine, metabolizing the aldehyde formed after deamination by monoamine oxidase to 3,4-dihydroxymandelic acid (1997). Furthermore, inhibition of ALDH resulted in significantly increased diamine oxidase activity in brain (Ruggeri, 1986). In addition to its role in mono- and diamine metabolism, ALDH was found to be involved in another brain neurotransmitter system, the dehydrogenattion of γ-amino-butyraldehyde a metabolite of γ-aminobutyric acid (GABA) (Kikonyogo and Pietruszko, 1996).

Applying subtractive hybridization we have found mRNA for ALDH downregulated in Down Syndrome (DS) fetal brain. Here we studied the protein expression levels in adult patients with DS in comparison with the normal brain.

Methods

Subtractive hybridization

Subtractive hybridization was performed as described (Labudova and Lubec, 1998). Fetal brain samples of two fetuses with DS and two age and sex matched controls, 23rd week of gestation, were obtained from the Brain Bank of the Institute of Psychiatry (Denmark Hill, London, UK). Hippocampus (gyrus parahippocampalis) was taken into liquid nitrogen and ground for the isolation of mRNA. Isolation of mRNA was performed using the Quick Prep Micro mRNA purification kit (Pharmacia Biotech Inc., Uppsala, Sweden). One microgram of mRNA from each preparation was quality-checked by cDNA cloning kit (Gibco, Life Technologies, Eggenstein, Germany, cat. 18248-013) using the incorporation of [α-^{32}P]dATP (Amersham, Buckinghamshire, UK) with subsequent electrophoresis on 1% agarose, followed by autoradiography. The Reflection film (Dupont, Germany) was exposed to the gel for 2h at room temperature.

Construction of the subtractive library

Ten μg each of mRNA from brain of DS and control were biotinilated using UV irradiation at 360nm, according to the instructions supplied in the subtractor kit (Invitrogen, Leek, Netherlands). One μg of mRNA-pools each from the DS brain sample was subjected to reverse transcriptase reaction (subtractor kit, Invitrogen) and the cDNA-pools were hybridized with the corresponding biotinilated mRNAs from controls. The subtractive hybridization mixture was incubated with streptavidin according to the subtractor kit given above and thus the biotinilated molecules (non-induced biotinilated mRNAs and the hybrid [biotinilated mRNAs/cDNAs]) complexed. The streptavidine complexes were removed by repeated phenol-chloroform extraction and subtracted cDNAs were separated from the aqueous phase by alcohol precipitation (subtractor kit).

In order to amplify and clone subtracted cDNAs, they were ligated with Not I linkers followed by Not I-digestion. The Not I-linked cDNAs were ligated to Not I site of sPORT

1 cloning vector (cDNA cloning kit, Gibco). To enable visualization of subtracted cDNAs, the cloned cDNAs were amplified using universal primers

I 5'-GTAAAACGACGGCCAGT-3'
II 5'-ACAGCTATGACCATG-3'

from multiple cloning site of the sPORT-1 vector (cDNA cloning kit, Gibco). Amplified cDNAs were analyzed by electrophoresis on 1% agarose.

Cloning of subtracted Not I-linked cDNAs

Not I-linked cDNAs ligated with sPORT 1 vector were used for the transformation of highly competent INFalpha F *E. coli* cells (Invitrogen, Leek, Netherlands). Plated clones were analysed by plasmid isolation kit (Quiagen, Hilden, Germany) and digestion with Eco RI/Hind III. Recombinant clones were sequenced by K. Granderath, MW-Biotech (Ebersberg, Germany). Homologies were determined by computer assisted comparison of data from the genebank sequence library: fastA@ebi.ac.uk (GBALL, Gen Bank Heidelberg, Germany). Subtractive hybridization was performed cross-wise, i.e. DS sample mRNA subtraction from control and vice versa at the 1:3 level (DS mRNAs:control mRNAs).

Brain samples for determination of protein levels

The brain regions temporal, frontal, occipital, parietal cortex and cerebellum of karyotyped patients with DS (n = 9; 3 females, 6 males; 56.1 ± 7.1 years old) and controls (n = 9; 5 females, 4 males; 72.6 ± 9.6 years old) were used for the studies at the protein level. The brain samples have been characterized (Seidl et al., 1997; Mirra et al., 1991; Tierney et al., 1988). Post mortem brain samples were obtained from the MRC London Brain Bank for Neurodegenerative Diseases, Institute of Psychiatry). In all DS brains there was evidence of abundant beta A plaques and neurofibrillary tangles. The controls were brains from individuals with no history of neurological or psychiatric illness. The major cause of death was bronchopneumonia in DS and heart disease in controls. Post mortem interval of brain dissection in DS and controls was 30.6 ± 17.5 and 34.8 ± 15.0 h, respectively. Tissue samples were stored at −70°C and the freezing chain was never interrupted.

Two-dimensional gel electrophoresis

Brain tissue was suspended in 0.5 ml of sample buffer consisting of 40 mM Tris, 5 M urea (Merck, Darmstadt, Germany), 2 M thiourea (Sigma, St. Louis, MO, USA), 4% CHAPS (Sigma), 10 mM 1,4-dithioerythritol (Merck), 1 mM EDTA (Merck) and a mixture of protease inhibitors, 1 mM PMSF and 1 µg of each pepstatin A, chymostatin, leupeptin and antipain. The suspension was sonicated for approximately 30 s and centrifuged at 10,000 × g for 10 min and the supernatant was centrifuged further at 150,000 × g for 45 min. The protein content in the supernatant was determined by the Coomassie blue method (Bradford, 1976).

The 2-D gel electrophoresis was performed essentially as reported (Langen et al., 1997). Samples of approximately 1.5 mg were applied on immobilized pH 3–10 nonlinear gradient strips (IPG, Pharmacia Biotechnology, Uppsala, Sweden), at both, the basic and acidic ends of the strips. The proteins were focused at 300 V for 1 h, after which the voltage was gradually increased to 3,500 V within 6 h. Focusing was continued at at 5,000 V for 48 h. The second-dimensional separation was performed on 9–16% linear gradient polyacrylamide gels (chemicals from Serva, Heidelberg, Germany and Bio-Rad, Hercules, CA, USA). The gels were stained with colloidal Coomassie blue (Novex, San Diego, CA, USA) for 48 h, destained with water and scanned in a Molecular Dynamics Personal densitometer. The images were processed using Photoshop (Adobe) and PowerPoint (Microsoft) software. Protein spots were quantified using the ImageMaster 2D Elite software (Amersham Pharmacia Biotechnology).

Matrix-assisted laser desorption ionization mass spectroscopy (MALDI-MS)

MALDI-MS analysis was performed as described (Fountoulakis and Langen, 1997) with minor modifications. Briefly, spots were excised, destained with 50% acetonitrile in 0.1 M ammonium bicarbonate and dried in a speedvac evaporator. The dried gel pieces were reswollen with 3 µl of 3 mM Tris-HCl, pH 8.8, containing 50 ng trypsin (Promega, Madison, WI, USA) and after 15 min, 3 µl of water were added. One µl was applied onto the dried matrix spot. The matrix consisted of 15 mg nitrocellulose (Bio-Rad) and 20 mg α-cyano 4 hydroxycinnamic acid (Sigma) dissolved in 1 ml acetone:isopropanol (1:1, v/v). 0.5 µl of the matrix solution was applied on the sample target. Specimen were analyzed in a time-of-flight PerSeptive Biosystems mass spectrometer (Voyager Elite, Cambridge, MA, USA) equipped with a reflectron. An accelerating voltage of 20 kV was used. Calibration was internal to the samples. The peptide masses were matched with the theoretical peptide masses of all proteins from all species of the SWISS-PROT database. For protein search, monoisotopic masses were used and a mass tolerance of 0.0075% was allowed.

Results

Subtractive hybridization

The corresponding downregulated sequence found by subtractive hybridization in clone 189 is given in Fig. 1 and reveals a homology of 100% with the human aldehyde dehydrogenase (hsALDH1, accession number K03000) (30). A series of homologies between 80 and 97% was found with aldehyde dehydrogenases from other species as *Rhizobium meliloti* (U39940), *Pichia angusta* (U40996), *Cladosporium herbarum* (X78228), *E. coli* (X52905, S67313, S67318, M77739, M35859), *E. buchholzi* (X95396), *Leishmania tarentolae* (Z316698), rat liver, mitochondrial (X14977), mouse embryonic cells (S71509), and mus musculus ALDH (U07235).

Two dimensional protein map of brain proteins

Human brain protein extracts from five regions of the brain were separated by 2-D gels and the protein spots were visualized following stain with colloidal Coomassie blue. Figure 2 shows an example of brain proteins from the parietal lobe of one control, separated on a 2-D gel. The protein spots were analyzed by MALDI-MS, following in-gel digestion. The peptide masses were matched with the theoretical peptide masses of all proteins of the SWISS-PROT database. We used internal standards to correct the measured peptide

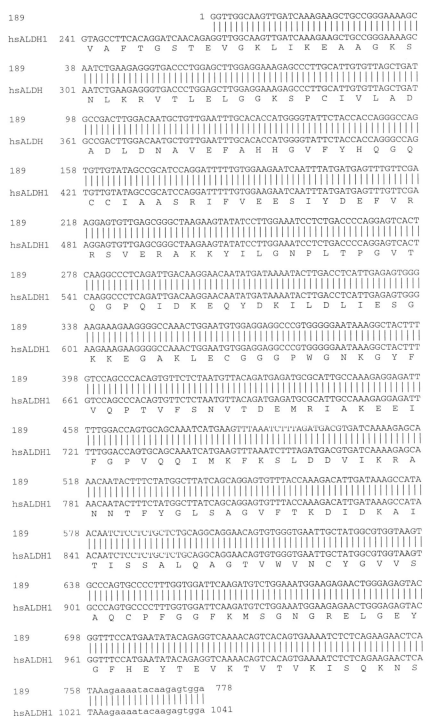

Fig. 1. Alignment of the nucleic acid and protein sequence found by subtractive hybridization from our clone 189 with the sequence obtained from Gen Bank Heidelberg. An 100% homology was found in the open reading frame (capitals) and in the nontranslated region (small letters) clearly showing the assignment of the found sequence to hsALDH1

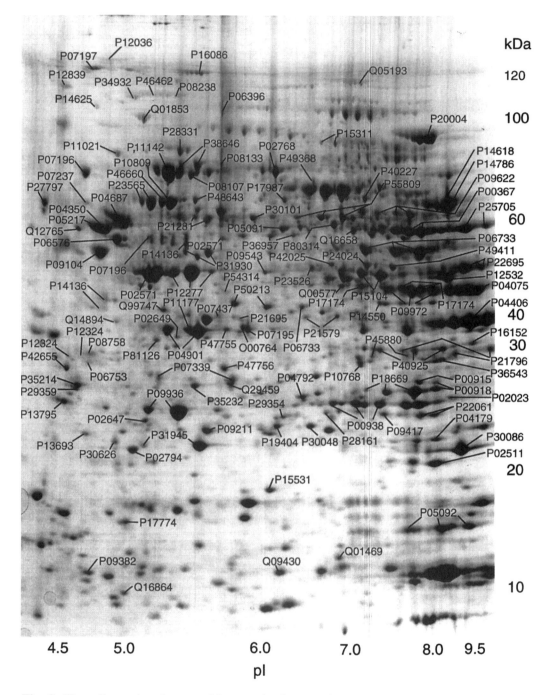

Fig. 2. Two-dimensional map of human brain proteins. The proteins from the parietal lobe of cortex of a control were extracted and separated on a 3–10 nonlinear IPG strip, followed by a 9–16% SDS-polyacrylamide gel, as stated under Methods. The gel was stained with Coomassie blue. The spots were analyzed by MALDI-MS. The proteins identified are designated with their SWISS-PROT accession numbers. The names of the proteins are listed in Table 1

masses, reducing, thus, the windows of mass tolerance and increasing the confidence of identification. Approximately 120 proteins were identified (Table 1). In Fig. 2, the proteins identified are labeled with their SWISS-PROT accession numbers. On the two-dimensional map, many proteins are represented by multiple spots, as for example the proteins P00367, P00938, P04406, P18669 and others. Presently, we do not know the reason(s) or biological significance of the observed heterogeneities.

Quantification of ALDH

The 2-D gels with proteins from the corresponding brain regions of 9 patients with DS and 9 controls were compared with each other, in order to detect differences at the protein level. Several differences were observed. Fig. 3 shows a 2-D gel image of a control donor, in which one protein is represented by a relatively strong spot (Fig. 3A) and of a DS patient, in which the corresponding spot is not visible (Fig. 3B).

For identification, the protein spot was excised from the 2-Dgel (Fig. 3A) and after tryptic digestion, the masses of the peptides generated were

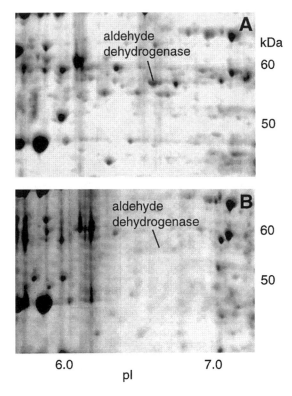

Fig. 3. Partial two-dimensional gel images of human brain proteins from the parietal cortex lobe, from a control (**A**), and a patient with Down syndrome (**B**). The gels were prepared and the protein was analyzed as described under legend to Fig. 2. The spot corresponding to aldehyde dehydrogenase is indicated (in B practically invisible)

Table 1. Proteins from the parietal region of human brain were extracted and separated by 2-D gels as stated under Methods. The proteins were identified by MALDI-MS, following trypsin digestion. The identified protein spots are characterized by their SWISS-PROT accession numbers and are indicated in Fig. 1. The theoretical and the approximate observed Mr and pI values, as well as the matching and total peptides and the protein sequence coverage by the matching peptides are given

Number	Abbr. Name	Protein Name	Mr		pI		MALDI-MS		
			Theor.	Observ.	Theor.	Observ.	Peptides Matching	Total	Sequence Coverage
O00764	PDXK_HUMAN	Pyridoxine kinase (EC 2.7.1.35)	35,307	31,000	6.1	6.1	8	20	33
P00367	DHE3_HUMAN	Glutamate dehydrogenase 1	61,701	50,000	7.8	9	11	20	24
P00915	CAH1_HUMAN	Carbonic anhydrase I (EC 4.2.1.1)	28,778	26,000	7.1	8.2	11	20	54
P00918	CAH2_HUMAN	Carbonic anhydrase II (EC 4.2.1.1)	29,153	25,000	7.5	8.4	10	18	48
P00938	TPIS_HUMAN	Triose-phosphate isomerase (EC 5.3.1.1) (tim)	26,806	26,000	6.9	7.5	9	20	47
P02023	HBB_HUMAN	Hemoglobin beta chain	15,971	27,000	7.3	8	10	20	71
P02511	CRAB_HUMAN	Alpha crystallin b chain	20,146	21,000	7.4	8.3	7	20	44
P02571	ACTG_HUMAN	Actin cytoplasmic 2 (gamma actin)	42,107	40,000	5.3	5.3	7	20	19
P02647	APA1_HUMAN	Apolipoprotein A-I (APO-AI)	30,758	24,000	5.6	5.2	7	18	22
P02768	ALBU_HUMAN	Serum albumin	71,317	66,000	6.2	6.4	11	20	20
P02794	FRIH_HUMAN	Ferritin heavy chain	21,252	22,000	5.4	5.1	6	20	46
P04075	ALFA_HUMAN	Fructose-biphosphate aldolase (EC 4.1.2.13)	39,720	40,000	8.1	9	10	17	36
P04179	SODM_HUMAN	Superoxide dismutase	24,877	22,000	8.4	8.4	5	20	34
P04264	K2C1_HUMAN	Keratin, type II cytoskeletal 1 (cytokeratin 1)	66,018	50,000	8.3	6.8	6	18	15
P04350	TBB5_HUMAN	Tubulin beta-5 chain	50,055	48,000	4.7	4.8	6	20	14
P04406	G3P2_HUMAN	Glyceraldehyde 3-phosphate dehydrogenase (EC 1.2.1.12)	36,070	32,000	8.7	8.7	7	21	29
P04687	TBA1_HUMAN	Tubulin alpha-1 chain	50,809	50,000	4.9	5	7	20	21
P04901	GBB1_HUMAN	Guanine nucleotide-binding protein beta subunit 1	38,151	31,000	5.9	5.7	6	20	19

P05091	DHAM_HUMAN	Aldehyde dehydrogenase (ALDHI) (EC 1.2.1.3)	56,858	50,000	7	6.5	7	18	14
P05092	CYPH_HUMAN	Peptidyl-prolyl cis-trans isomerase A (EC 5.2.1.8) (PPIase)	18,097	15,000	7.8	8.4	5	20	31
P05217	TBB2_HUMAN	Tubulin beta-2 chain	50,255	50,000	4.6	4.8	10	20	23
P06396	GELS_HUMAN	Gelsolin	86,043	95,000	6.2	5.9	9	20	16
P06576	ATPB_HUMAN	ATP synthase beta chain (EC 3.6.1.34)	56,524	48,000	5.2	5	17	19	34
P06733	ENOA_HUMAN	Alpha enolase (EC 4.2.1.11)	47,350	43,000	7.4	7.8	11	20	29
P06753	TPM3_HUMAN	Tropomyosin α chain	32,855	28,000	4.5	4.8	5	18	18
P07195	LDHH_HUMAN	L-lactate dehydrogenase h chain (EC 1.1.1.27) (ldh-b)	36,769	31,000	6	6.1	8	19	25
P07196	NFL_HUMAN	Neurofilament triplet L protein (nf-l)	61,665	66,000	4.5	4.7	12	20	22
P07197	NFM_HUMAN	Neurofilament triplet M protein (NF-M)	102,000	130,000	4.7	4.7	15	20	19
P07237	PDI_HUMAN	Protein disulfide isomerase	57,467	57,000	4.7	4.7	8	20	17
P07339	CATD_HUMAN	Cathepsin D (EC 3.4.23.5)	45,036	27,000	6.5	5.3	7	20	14
P07437	TBB1_HUMAN	Tubulin beta-1 chain	50,240	33,000	4.6	5.8	8	20	19
P08107	HS71_HUMAN	Heat shock 70kDa protein 1 (HSP70.1)	70,294	65,000	5.4	5.6	9	19	17
P08133	ANX6_HUMAN	Annexin VI (lipocortin VI)	76,036	70,000	5.4	5.8	13	20	19
P08238	HS9B_HUMAN	Heat shock protein HSP90-beta (HSP 90)	83,453	105,000	4.8	5.5	8	20	13
P08758	ANX5_HUMAN	Annexin V (lipocortin V)	35,840	29,000	4.8	4.7	9	20	28
P09104	ENOG_HUMAN	Gamma enolase (EC 4.2.1.11)	47,467	43,000	4.8	4.8	10	21	27
P09211	GTP_HUMAN	Glutathione S-transferase P (gstp1-1) (EC 2.5.1.18)	23,438	23,000	5.4	5.8	6	17	36
P09382	LEG1_HUMAN	Galectin-1 (lactose-binding lectin 1)	14,917	12,000	5.2	4.9	5	20	45
P09417	DHPR_HUMAN	Dihydropteridine reductase (EC 1.6.99)	26,015	22,000	7.4	7.7	6	22	33
P09543	CN37_HUMAN	2',3'-Cyclic nucleotide 3'-phosphodiesterase (EC 3.1.4.37)	47,947	41,000	9.8	6.3	6	18	16
P09622	DLDH_HUMAN	Dihydrolipoamide dehydrogenase (EC 1.8.1.4)	54,686	55,000	7.7	8	7	17	13

(continued)

Table 1. *Continued*

Number	Abbr. Name	Protein Name	Mr Theor.	Mr Observ.	pI Theor.	pI Observ.	MALDI-MS Peptides Matching	MALDI-MS Peptides Total	Sequence Coverage
P09936	UBL1_HUMAN	Ubiquitin carboxyl-terminal hydrolase isozyme L1 (EC 3.1.2.15)	25,150	24,000	5.3	5.5	7	19	57
P09972	ALFC_HUMAN	Fructose-bisphosphate aldolase C (EC 4.1.2.13)	39,699	36,000	6.8	8.1	12	19	40
P10768	ESTD_HUMAN	Esterase D (EC 3.1.1.1)	31,955	28,000	7	7.4	9	18	51
P10809	P60_HUMAN	Mitochondrial matrix protein P1 (hsp-60)	61,187	58,000	5.6	5.5	10	20	26
P11021	GR78_HUMAN	78 kDa glucose regulated protein (grp 78)	72,185	77,000	4.9	5	10	19	18
P11142	HS7C_HUMAN	Heat shock cognate 71 kDa protein	71,082	70,000	5.3	5.4	6	19	14
P11177	ODPB_HUMAN	Pyruvate dehydrogenase E1 component (EC 1.2.4.1) (pdhe1-b)	39,536	31,000	6.9	5.6	11	20	29
P12036	NFH_HUMAN	Neurofilament triplet H protein (NF-H)	111,940	140,000	6.1	5	10	20	11
P12277	KCRB_HUMAN	Creatine kinase beta chain (EC 2.7.3.2)	42,902	40,000	5.4	5.6	13	20	43
P12324	TPMN_HUMAN	Tropomyosin (TM30-NM)	29,242	28,000	4.6	4.7	6	18	25
P12532	KCRU_HUMAN	Creatine kinase, ubiquitous mitochondrial (EC 2.7.3.2)	47,406	40,000	8.3	8.4	10	20	29
P12839	NFM_RAT	Neurofilament triplet m protein (nf-m)	95,716	110,000	4.6	4.6	14	20	18
P13795	SN25_HUMAN	Synaptosomal associated protein 25 (snap-25)	23,528	23,000	4.5	4.6	7	20	33
P14136	GFAP_HUMAN	Glial fibrillary acidic protein (gfap)	49,906	47,000	5.3	5.5	13	20	29
P14550	ALDX_HUMAN	Alcohol dehydrogenase (NADP+) (EC 1.1.1.2)	36,760	35,000	6.8	7.6	5	18	18

P14618	KPY1_HUMAN	Pyruvate kinase M1 (EC 2.7.1.40)	58,280	55,000	7.6	8.7	6	18	14
P14625	ENPL_HUMAN	Endoplasmin (94kDa glucose regulated protein) (GRP94)	92,696	100,000	4.6	4.8	9	20	13
P14786	KPY1_HUMAN	Pyruvate kinase. M2 isozyme (EC 2.7.1.40)	58,316	55,000	7.8	8.5	9	20	20
P15104	GLNA_HUMAN	Glutamine synthetase (EC 6.3.1.2)	42,664	39,000	6.9	7.5	6	19	20
P15311	EZRI_HUMAN	Ezrin (p81) (cytovillin)	69,338	91,000	6.2	7	8	18	12
P15531	NDKA_HUMAN	Nucleoside diphosphate kinase A (EC 2.7.4.6)	17,308	18,000	6.1	6.2	7	21	45
P16086	SPCN_RAT	Spectrin alpha chain, non-erythroid	118,671	125,000	5.7	5.7	7	20	6
P16152	DHCA_HUMAN	Carbonyl reductase (NADPH) (EC 1.1.1.184)	30,509	30,000	8.3	9.4	7	18	28
P17174	AATC_HUMAN	Aspartate aminotransferase (transaminase A) (EC 2.6.1.1)	46,334	37,000	7.2	7.4	11	20	34
P17774	GLMB_HUMAN	GLIA maturation factor β (GMF-β)	16,742	15,000	5	5.1	5	18	37
P17987	TCPA_HUMAN	T-complex protein 1, alpha subunit (TCP-1 alpha)	60,818	58,000	6	6.4	11	20	22
P18669	PMGB_HUMAN	Phosphoglycerate mutase, brain form (EC 5.4.2.1) (pgam-b)	28,768	25,000	7.2	8	7	20	35
P19404	NUHM_HUMAN	NADH-ubiquinone dehydrogenase 24kDa subunit (EC 1.6.99.3)	27,631	24,000	8.1	6.5	6	20	26
P20004	ACON_BOVIN	Aconitate hydratase (aconitase) (EC 4.2.1.3)	86,045	85,000	7.9	8.3	8	20	11
P21281	VAT2_HUMAN	Vacuolar ATP synthase subunit B, brain isoform (EC 3.6.1.34)	56,823	53,000	5.6	5.8	8	20	16
P21579	SYT1_HUMAN	Synaptotagmin I (p65)	47,884	32,000	8.2	7	10	18	23
P21695	GPDA_HUMAN	Glycerol-3-phosphate dehydrogenase (EC 1.1.1.8)	38,065	32,000	6.1	6	5	20	15
P21796	POR1_HUMAN	Voltage-dependent anion-selective channel protein 1 (VDAC1)	30,736	29,000	9	0	8	15	45
P22061	PIMT_HUMAN	Protein-beta-aspartate methyltransferase (EC 2.1.1.77)	24,642	24,000	7.3	8.2	7	20	42
P22695	UCR2_HUMAN	Ubiquinol-cytochrome C reductase complex protein 2 (EC 1.10.2.2.)	48,610	40,000	9.1	9.4	7	17	18

(continued)

　G. Lubec et al.

Table 1. *Continued*

Number	Abbr. Name	Protein Name	Mr Theor.	Mr Observ.	pI Theor.	pI Observ.	Matching	Total	Sequence Coverage
P23526	SAHH_HUMAN	Adenosylhomocysteinase (EC 3.3.1.1)	48,254	40,000	6.4	6.8	8	18	20
P23565	AINX_RAT	Alpha-internexin (alpha-inx) (rat)	56,252	56,000	5	5.3	11	21	20
P25705	ATPA_HUMAN	ATP synthase alpha chain (EC 3.6.1.34)	59,827	50,000	9.9	9.6	14	20	31
P28161	GTM2_HUMAN	Glutathione S-transferase mu-2 (EC 2.5.1.18)	25,768	22,000	6.3	6.5	8	18	36
P28331	NUAM_HUMAN	NADH-ubiquinone oxidoreductase 75 kDa subunit (EC 1.6.99.3)	80,548	80,000	6	5.6	11	19	22
P29359	143G_BOVIN	14-3-3 protein gamma (kcip-1) (bovine)	28,160	28,000	4.6	4.7	8	20	28
P30048	TDXM_HUMAN	Thioredoxin-dependent peroxide reductase	28,017	23,000	7.8	6.8	5	18	21
P30086	PBP_HUMAN	Phosphatidylethanolamine-binding protein	21,026	22,000	7.6	8.8	10	20	75
P30101	ER60_HUMAN	Protein disulfide isomerase er-60 (EC 5.3.4.1) (erp60)	57,145	49,000	6.3	6.2	10	20	18
P30626	SORC_HUMAN	Sorcin (CP-22)	21,947	20,000	5.3	5	5	20	23
P31930	UCR1_HUMAN	Ubiquinol-cytochrome-C reductase complex core protein I	53,269	42,000	6.3	5.8	11	20	30
P31945	NKFB_HUMAN	Natural killer cell enhancing factor B (NKEF-B)	22,048	22,500	5.8	5.7	5	20	40
P34932	HS74_HUMAN	Heat shock 70 kDa protein 4 (HSP70RY)	79,857	105,000	5	5.2	8	20	13
P35214	143G_RAT	14-3-3-Protein gamma (protein kinase C inhibitor protein-1)	28,324	28,000	4.6	4.7	6	20	23

Aldehyde dehydrogenase in Down Syndrome

P35232	PHB_HUMAN	Prohibitin	29,842	27,000	5.5	5.6	8	18	33
P36543	VATE_HUMAN	Vacuolar ATP synthase subunit E, brain isoform (EC 3.6.1.34)	26,185	28,000	8.4	8.4	7	18	19
P36957	ODO2_HUMAN	Dihydrolipoamide succinyltransferase component E2	48,951	48,000	9.2	6.5	6	18	15
P38646	GR75_HUMAN	75 kDa glucose regulated protein (grp 75)	74,018	69,000	6.2	6.5	8	20	15
P40227	TCPZ_HUMAN	T-complex protein 1, zeta subunit (TCP-1 zeta)	58,443	58,000	6.7	7.4	7	20	13
P40925	MDHC_HUMAN	Malate dehydrogenase (EC 1.1.1.37)	36,500	29,000	7.4	8.1	6	20	21
P42024	ACTZ_HUMAN	α-Centractin (centrosome-associated actin homolog)	42,700	40,000	6.6	7.5	8	18	26
P42025	ACTY_HUMAN	β-Centractin	42,380	40,000	6.4	7	6	18	15
P42655	143E_HUMAN	14-3-3 protein epsilon (protein kinase C inhibitor protein-1)	29,326	31,000	4.5	4.5	13	20	49
P45880	POR2_HUMAN	Voltage-dependent anion-selective channel protein 2 (VDAC2)	38,638	28,000	6.7	7.5	8	20	30
P46462	TERA_RAT	Transitional endoplasmic reticulum ATPase	89,976	110,000	5	5.2	10	20	14
P46660	AINX_MOUSE	Alpha-internexin (alpha-inx) (mouse)	56,007	58,000	5	5.5	12	20	21
P47755	CAZ2_HUMAN	F-actin capping protein alpha-2 subunit (CAPZ)	33,156	31,000	5.7	5.9	6	18	39
P47756	CAPB_HUMAN	F-actin capping protein β subunit (CAPZ)	30,951	28,000	5.8	5.9	5	18	22
P48643	TCPE_HUMAN	T-complex protein 1, ε subunit (TCP-1-ε) (KIAA0098)	60,088	58,000	5.4	5.6	7	18	15
P49368	TCPG_HUMAN	T-complex protein 1, γ subunit (cct-gammma)	60,862	63,000	6.6	6.8	12	20	20
P49411	EFTU_HUMAN	Elongation factor Tu (p43)	49,852	40,000	7.6	7.8	12	18	34
P50213	IDHA_HUMAN	Isocitrate dehydrogenase (NAD) (EC 1.1.1.41)	40,022	35,000	6.9	6.1	10	19	25
P54314	GBB5_MOUSE	Guanine nucleotide-binding protein beta subunit 5 (mouse)	39,504	36,000	6	5.9	6	20	24

(continued)

Table 1. *Continued*

Number	Abbr. Name	Protein Name	Mr Theor.	Mr Observ.	pI Theor.	pI Observ.	Matching	Total	Sequence Coverage
P55809	SCOT_HUMAN	Succinyl-CoA:3-ketoacid-coenzyme A transferase (EC 2.8.3.5)	56,578	52,000	7.4	7.1	9	20	26
P80314	TCPB_MOUSE	T-complex protein 1, beta subunit (TCP-1 beta)	57,753	50,000	6.4	7.1	8	20	22
P81126	SNAB_BOVIN	Beta-soluble NSF attachment protein (SNAP-β)	33,875	30,000	5.3	5.3	7	18	28
Q00577	PUR_HUMAN	Transcriptional activator protein PUR-alpha	35,003	36,000	6.4	7.2	6	18	27
Q01469	FABE_HUMAN	Fatty acid-binding protein (E-FABP)	15,496	13,000	7	7.1	5	18	31
Q01853	TERA_MOUSE	Transitional endoplasmic reticulum ATPase (TER ATPase)	89,935	95,000	5	5.3	11	20	11
Q05193	DYN1_HUMAN	Dynamin-1	97,745	115,000	7.3	7.3	17	20	15
Q09430	PRO2_BOVIN	Profilin II (bovine)	12,011	12,000	9.3	6.5	5	14	41
Q12765	Y193_HUMAN	Hypothetical protein KIAA0193	39,352	50,000	4.4	4.7	5	18	16
Q16658	FASC_HUMAN	Fascin (actin binding protein)	55,123	50,000	7.2	7.4	8	18	16
Q16864	VATF_HUMAN	Vacuolar ATP synthase subunit F (EC 3.6.1.34)	13,349	11,000	5.3	5.1	5	17	31
Q99747	SNAG_HUMAN	Gamma-soluble NSF attachment protein (SNAP-γ)	35,066	35,000	5.2	5.3	9	20	31

determined by MALDI-MS. One protein entry was selected with 9 matching out of 18 total peptides. The protein was unambiguously identified as ALDH, mitochondrial form (SWISS-PROT accession number P05091). Figure 4 shows the mass spectrum, the determined monoisotopic matching masses and the deduced amino acid sequences of the peptides.

The spots representing ALDH were quantified in the 2-D gels of samples from the same brain region in DS and controls using the ImageMaster 2D Elite software. ALDH levels were determined as a percentage of total proteins present in the gel part considered. Figure 5 shows the ALDH levels in the five brain regions normalized to the average levels of controls (C). In general, the ALDH levels were reduced in all brain parts of patients with DS in comparison with the controls, and in many instances no clear spot corresponding to the protein could be detected (as in the example shown in

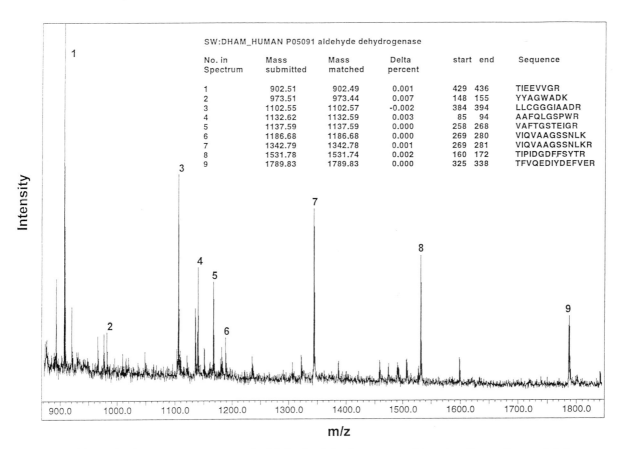

SW:DHAM_HUMAN P05091 aldehyde dehydrogenase

No. in Spectrum	Mass submitted	Mass matched	Delta percent	start	end	Sequence
1	902.51	902.49	0.001	429	436	TIEEVVGR
2	973.51	973.44	0.007	148	155	YYAGWADK
3	1102.55	1102.57	-0.002	384	394	LLCGGGIAADR
4	1132.62	1132.59	0.003	85	94	AAFQLGSPWR
5	1137.59	1137.59	0.000	258	268	VAFTGSTEIGR
6	1186.68	1186.68	0.000	269	280	VIQVAAGSSNLK
7	1342.79	1342.78	0.001	269	281	VIQVAAGSSNLKR
8	1531.78	1531.74	0.002	160	172	TIPIDGDFFSYTR
9	1789.83	1789.83	0.000	325	338	TFVQEDIYDEFVER

m/z

Fig. 4. MALDI mass spectrum of aldehyde dehydrogenase. Proteins from the parietal lobe of cortex were separated by 2-D gels. Following separation, the spot was excised from the gel, the protein in-gel digested with trypsin and analyzed as described under Methods. Protein search in the SWISS-PROT database yielded one entry with 9 matching peptides out of 18 and a sequence coverage of 20% by the matching peptides. The numbers indicate the matching peptides. *Inset*: The found and theoretical masses of the tryptic fragments of the protein, the percent mass difference, the start and end residues and the amino acid sequence of the matching peptides are indicated

G. Lubec et al.

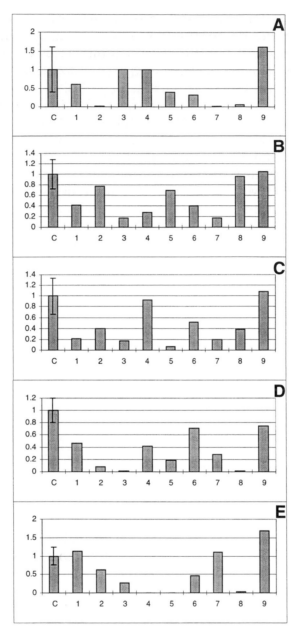

Fig. 5. Quantification of ALDH levels in the brain regions of patients with DS. The proteins from the frontal (**A**), parietal (**B**), temporal (**C**), occipital (**D**) cortex lobes and the cerebellum (**E**) of the patients with DS and the control group were separated on 2-D gels and visualized following stain with colloidal Coomassie blue. In partial gel images, including ALDH and the neighboring proteins, the intensities of the spots representing ALDH were quantified, compared to the total proteins present. The quantification was performed using the ImageMaster 2D software. In each donor, the percentage of the volume of the spot representing ALDH was determined. The ALDH percentage of each DS patient was normalized to the mean value of the controls and the relative values are indicated. C, control samples; the bars indicate the standard deviation of ALDH levels in the control group. The numbers represent patients in the DS group

Fig. 3B). However, the reduction was not the same in all regions. In particular, in the cerebellum, the decrease was less significant compared to the other regions. The intensity of the spot representing ALDH was variable in the control group as well. In certain cases, this resulted in an increased value of the standard deviation (such as in Fig. 5A). We observed differences in the ALDH expression level in the five regions of the brain of the same patient (Fig. 5). Similar types of variations have been observed for many other brain proteins as well (data not shown).

Two brain samples from the cerebellum of two patients with DS were missing (Fig. 5E, patients 4 and 5). In the other cases, where ALDH levels appear to be very low (for example, Fig. 3A, patients 2, 7, 8), the spots representing ALDH were either weak or invisible. Because the spots representing ALDH were usually weak, a certain imprecision in the determination of the expression levels may have interfered. Except of the ALDH levels, a comparison between the two 2-D images (Fig. 3) shows the presence of additional differences in the protein expression level which, however, were inconsistent. In certain of the 2-D gels with DS or control brain samples, some of the differences seen in Fig. 3 were not present (data not shown).

Discussion

We investigated the ALDH protein levels in individual brain regions using two-dimensional gel electrophoresis and constructed a two-dimensional map of human brain proteins which may be useful to researchers involved in the investigation of brain disorders. ALDH protein levels were reduced in the DS brain, however, the cause of ALDH reduction remains unclear. As ALDH(s) is a key enzyme in fructose and oxalate metabolism and is detoxifying reactive carbonyls (Quintanilla and Tampier, 1995), the decrease of aldehyde metabolizing capacities could lead to an increase of alkenals like acetaldehyde, glycolaldehyde, glyceraldehyde etc. The generation of oxidative stress by accumulating aldehydes would fit to one of the current concepts of DS pathogenesis (Busciglio and Yankner, 1995). Decreased ALDH in our cohort of patients with DS showing Alzheimer's disease (AD) pathology would also explain previous observations of increased "advanced glycation end products", a series of poorly defined compounds derived from nonenzymatic glycation by sugar-aldehydes in the Maillard reaction, in tangles and plaques (Harrington and Colaco, 1994; Smith et al., 1995; Vitek et al., 1994; Yan et al., 1995). Increased tissue aldehydes could also be incriminated directly for the induction of cellular toxicity in DS brain. Mechanisms leading to toxic effects are summarized in an excellent review (Esterbauer et al., 1991).

The present study on aged DS brain regions revealed that ALDH levels were decreased in a high percentage of the patients as compared to controls. In addition to ALDH, we observed more differences between control and DS brain (Fig. 3). Differences in the protein expression level should be expected if one takes into consideration that the groups under investigation included members of different age, sex and disease history. Moreover, patients with DS are even clinically heterogeneous in terms of learning, memory, hypotonia and motor functions and so they are also in their behavior. Even psychometric data vary greatly in trisomy 21 (Epstein, 1992). Such variations may result in differential protein expression in brain.

We cannot exclude that some of the differences seen in Fig. 3 are disease-related, which have not been studied up to now. Many proteins located in the region of ALDH have not been identified yet. As our knowledge about the identity and levels of the proteins represented by visible spots in the 2-D map increases, additional cases of disease-related differential protein expression may be elucidated. The large spots seen at the lower left end of Fig. 3B are due to increased levels of glial fibrillary acidic protein, a known marker of neuronal decay, in the brain of patients suffering from Down Syndrome (manuscript submitted).

In summary, we found decreased ALDH levels in the DS brain. ALDH deficiency may result in the accumulation of aldehydes in the brain with the consequences of increased protein-cross-linking and modification, increased oxidative stress and direct cytotoxic effects. We are currently studying brain aldehydes and ALDH enzyme kinetics in patients with DS to further substantiate our finding of deficient ALDH, which was never reported before.

Acknowledgments

The authors at the University of Vienna are highly indebted to the Red Bull Company, Salzburg, Austria, for generous financial support. The skillful technical assistance of J.-F. Juranville is appreciated.

References

Bradford M (1976) A rapid and sensitive method for the quantitation of microgram quantities of protein utilizing the principle of protein-dye binding. Anal Biochem 72: 248–254

Busciglio J, Yankner BA (1995) Apoptosis and increased generation of reactive oxygen species in Down's syndrome neurons in vitro. Nature 378: 776–779

Colzi A, Musolino A, Iuliano A, Fornai F, Bonuccelli U, Corsini GU (1996) Identification and determination of 3,4-dihydroxyphenylacetylacetaldehyde, the dopamine metabolite in in vivo dialysate from rat striatum. J Neurochem 66: 1510–1517

Epstein CJ (1992) Down Syndrome (Trisomy 21). In: Scriver CR, Beaudet AL, Sly WS, Valle D (eds) The metabolic and molecular basis of inherited diseases. McGraw Hill, New York, pp 749–794

Esterbauer H, Schaur RJ, Zollner H (1991) Chemistry and biochemistry of 4-hydroxynonenal, malondialdehyde and related aldehydes. Free Radic Biol Med 11: 81–128

Fountoulakis M, Langen H (1997) Identification of proteins by matrix — assisted laser desorption ionization-mass spectroscopy following in-gel digestion in low-salt, nonvolatile buffer and simplified peptide recovery. Anal Biochem 250: 153–156

Hafer G, Agarwal DP, Goeddee HW (1987) Human brain aldehyde dehydrogenase: activity with DOPAL and isoenzyme distribution. Alcohol 4: 413–418

Hague NS, LeBlanc CJ, Isacson O (1997) Differential dissecion of the rat E16 ventral mesencephalon and survival and reinnervation of the 6-OHDA — lesioned striatum by a subset of aldehyde dehydrogenase positive TH neurons. Cell Transplant 6: 239–248

Harrington CR, Colaco CALS (1994) Alzheimer's disease, a glycation connection. Nature 370: 247–248

Kawamura M, Kopin IJ, Kador PF, Sato S, Tjurmina O, Eisenhofer G (1997) Effects of aldehyde dehydrogenase / aldose reductase inhibition on neuronal metabolism of norepinephrine. J Auton Nerv Syst 66: 145–148

Kikonyogo A, Pietruszko R (1996) Aldehyde dehydrogenase from adult brain that dehydrogenastes gamma-aminobutyraldehyde: purification, characterization, cloning and distribution. Biochem J 316: 317–324

Labudova O, Lubec G (1998) cAMP upregulates the transposable element mys-1: a possible link between signaling and mobile DNA. Life Sci 62: 431–437

Langen H, Roeder D, Juranville J-F, Fountoulakis M (1997) Effect of the protein application mode and the acrylamide concentration on the resolution of protein spots separated by two-dimensional gel electrophoresis. Electrophoresis 18: 2085–2090

McCaffery P, Drager UC (1994) High levels of retinoic acid-generating dehydrogenase in the meso-telencephalic dopamine system. Proc Natl Acad Sci USA 91: 7772–7776

Maring JA, Deitrich RA, Little R (1985) Partial purification and properties of human brain aldehyde dehydrogenase. J Neurochem 45: 1903–1910

Mirra SS, Heyman A, McKeel D, Sumi S, Crain BJ (1991) The consortium to establish a registry for Alzheimer disease (CERAD). II. Standardisation of the neuropathological assessment of Alzheimer's disease. Neurology 41: 479–486

Nilsson GE, Tottmar O (1985) Biogenic aldehydes in brain: characterization of a reaction betwreen rat brain tissue and indole-3-acetaldehyde. J Neurochem 45: 744–75

Nilsson GE, Tottmar O (1987) Biogenic aldehydes in brain: on their preparation and reactions with brain tissue. J Neurochem 48: 1566–1572

Petersen DR (1985) Aldehyde dehydrogenase and aldehyde reductase in isolated bovine brain microvessels. Alcohol 2: 79–83

Pietruszko R, Meier J, Major LF, Saini N, Manz H, Hawley RJ (1984) Human brain: aldehyde dehydrogenase activity and isoenzyme distribution in different areas. Alcohol 1: 363–367

Quintanilla ME, Tampier L (1995) Acetaldehyde metabolism by brain mitochondria from UchA and UchB rats. Alcohol 12: 519–524

Ruggeri P, Saija A, Costa G, Caputi AP (1986) Influence of several aldehyde dehydrogenase and aldehyde reductase inhibitors on diamine oxidase in rat brain. Res Commun Chem Pathol Pharmacol 51: 205–209

Ryzlak MT, Pietruszko R (1988) Human brain "high-Km" aldehyde dehydrogenase: purification, characterization, and identification as NAD^+-dependent succinic semialdehyde dehydrogenase. Arch Biochem Biophys 266: 386–396

Seidl R, Greber S, Schuller E, Bernert G, Cairns N, Lubec G (1997) Evidence against increased oxidative DNA damage in Down Syndrome. Neurosci Lett 235: 137–140

Shiohara E, Tsukuda M, Chiba S, Yamazaki H, Nishiguchi K, Miyamoto R, Nakanishi S (1984) Subcellular aldehyde dehydrogenase activity and acetaldehyde oxidation by isolated intact mitochondria of rat brain and liver after acetaldehyde treatment. Toxicol 30: 25–30

Smith MA, Sayre LM, Monnier VM, Perry G (1995) Radical AGEing in Alzheimer's disease. Trends Neurosci 18: 172–176

Stewart MJ, Malek K, Crabb DW (1996) Distribution of messenger RNAs for aldehyde dehydrogenase 1, aldehyde dehydrogenase 2 and aldehyde dehydrogenase 5 in human tissues. J Invest Med 44: 42–46

Tierney MC, Fisher RH, Lewis AJ, Torzitto ML, Snow WG, Reid DW, Nieuwstraaten P, Van Rooijen LAA, Derks HJGM, Van Wijk R, Bischop A (1988) The NINCDA-

ADRDA work group criteria for the clinical diagnosis of probable Alzheimer's disease. Neurology 38: 359–36

Totmar O (1985) Biogenic aldehydes: metabolism, binding to brain membranes and electrophysiological effects. Prog Clin Biol Res 183: 51–66

Vitek MP, Bhattacharya K, Glendening JM, Stopa E, Vlassara H, Bucala R, Manogue K, Cerami A (1994) Advanced glycation end products contribute to amyloidosis in Alzheimer's disease. Proc Natl Acad Sci USA 91: 4766–4770

Weiner H, Ardelt B (1984) Distribution and properties of aldehyde dehydrogenase in regions of the rat brain. J Neurochem 42: 109–115

Yan SD, Yan SF, Chen X, Fu J, Chen M, Kuppusamy P, Smith D, Perry G, Godman C, Nawroth P, Zweier JL, Stern D (1995) Non-enzymatically glycated tau in Alzheimer's disease induces neuronal oxidant stress resulting in cytokine gene expression and release of amyloid beta-peptide. Nature Med 1: 693–699

Zimatkin SM (1991) Histochemical study of aldehyde dehydrogenase in the rat CNS. J Neurochem 56: 1–11

Zimatkin SM, Karpuk YG (1996) Regional and cellular distribution of mitochondrial high-affinity aldehyde dehydrogenase in the rat brain. Neurosci Behav Physiol 26: 225–230

Zimatkin SM, Rout UK, Koivusalo M, Buehler R, Lindros KO (1992) Regional distribution of low Km mitochondrial aldehyde dehydrogenase in the rat central nervous system. Alcohol Clin Exp Res 16: 1162–1167

Authors' address: Dr. M. Fountoulakis, F. Hoffmann — La Roche, Ltd., PRPG-T, Building 93-444, CH-4070 Basel, Switzerland, e-mail: michael.fountoulakis@roche.com

J Neural Transm (1999) [Suppl] 57: 41–60
© Springer-Verlag 1999

The Down Syndrome critical region

B. L. Shapiro

Departments of Oral Science and Laboratory Medicine and Pathology, and
Institute of Human Genetics, University of Minnesota, Minneapolis, MN, U.S.A.

Summary. Since the early 1970's numerous attempts have been made to learn whether specific segments of chromosome 21, when triplicated, are responsible for the clinical condition Down syndrome (DS). Studies were reported in which positive or negative clinical diagnoses of DS were made in the presence of partial trisomy of one or another segment of the chromosome. The distal half of the long arm of 21 (21q22) possesses most of the gene transcribing sites of the chromosome. It was this region that was thought to contain loci essential to production of the clinical syndrome. Subsequent studies identified subregions of this band as "minimal" or "critical" sites necessary and sufficient to produce the clinical condition. A major problem with these assignments was that different investigators defined different critical/minimal regions. In 1994 evidence was presented in which regions of most of the long arm of chromosome 21 were said to contribute to the DS phenotype. Soon after, a report described a child with DS and partial tetrasomy of the short arm and proximal long arm of 21, segments clearly distinct from the previously identified critical areas. Thus the clinical diagnosis of DS can be made in the presence of partial aneuploidy of nearly all segments of chromosome 21. It must be concluded that no evidence exists that individual loci on 21 are singularly responsible for specific phenotypic abnormalities in DS. Without exception, each of the clinical findings associated with DS is a multifactorial trait. The analysis of each trait in DS should thus be similar to analyses of the same traits in the general population with a focus on the way aneuploidy affects expression of multifactorial characteristics.

Introduction

Down syndrome (DS) is the clinical expression of the most common postnatally viable human chromosomal abnormality. In the years following the realization that it resulted from triplication of all [trisomy (Lejeune et al., 1959)] or portions [translocation (Polani, 1960)] of chromosome 21, attempts have been made to determine whether (and if so, which) specific segments of 21 were responsible for the clinical condition. This quest was based on the notion (for which no evidence exists) that identifiable phenotypic expressions of the condition were specific consequences of specific loci on the chromosome. The earliest reports relating clinical findings to triplicated segments of 21 suffered from occurring in the period prior to (before 1971) the application of histochemical banding techniques to human chromosomes; until recently the unavailability of molecular markers combined with banding left some question as to truly triplicated segments. The citation by some authors of several presumably informative cases was inappropriate because the cases were mosaic (Ilbery et al., 1961; Hongell and Airaksinen, 1972; Verma et al., 1977; Cervenka et al., 1977) or because the chromosomal origin of the extra chromosomal material was assumed, not demonstrated, to be no. 21 based on clinical findings of "mild" or "partial" DS (Dent et al., 1963; Dekaban and Zelson, 1968; Subrt and Prchlikova, 1970).

The association of particular segments of #21 and the clinical phenotype of DS was based on the following assumptions: In subjects with partial trisomy of #21: (i) if the presence of a segment in triplicate is correlated with clinical DS, then loci on this segment are associated with the pathogenesis of DS; and (ii) if clinical DS is not apparent in the presence of a particular triplicated segment, then loci on that segment are not involved with the pathogenesis of clinical DS. These assumptions were based on presumptions that the clinical findings in DS are specific to DS and are a direct expression of triplicated loci on chromosome 21. That is, the single mutant gene — direct phenotypic expression paradigm of classical genetics was invoked, e.g. "the belief that it will ultimately be possible to explain an aneuploid phenotype — indeed to predict it — from a knowledge of the functions of the individual genes present in the unbalanced region" (Epstein, 1988). This "belief" contrasts with the assertion of Lewontin (1992) that "even if I knew the complete molecular specification of every gene in an organism . . . and the complete sequence of the environments, I could not specify the organism" (p. 26). What must be kept in mind (but most often has not) in attempting to relate specific chromosomal segments or loci to particular phenotypic expressions is that i) no single finding in DS (except, probably, mental retardation) occurs in all affected subjects; ii) without exception every abnormality except for the extra chromosome or chromosomal segment [in most cases other than phenocopies in which unequivocal clinical diagnosis of DS is made in the absence of any detectable cytogenetic abnormality (McCormick et al., 1989)] occurs in the general population, albeit much less commonly; and iii) much overlap of clinical findings exists among the human autosomal aneuploid states and teratogen-induced malformation syndromes (Shapiro, 1994).

Chromosome bands

About twenty five years ago techniques were developed for staining human chromosomal preparations (Craig and Bickmore, 1993; Sumner, 1994) which theretofore were monochromatic structures roughly distinguishable only by size, shape and sometimes by particular structures such as satellites. Methods such as G (Giemsa) and R (reverse) banding produced alternating transverse dark and light bands. G and R patterns are essentially the same but are reversed so that G-positive bands coincide with R-negative bands. These techniques permitted for the first time unequivocal identification of each human chromosome pair. In the very early years of banding it was noted that those human trisomies compatible with live birth (trisomy of 13, 18 or 21) had a relatively small portion of negative G bands (Ganner and Evans, 1971) and it was inferred that positive G (negative R) bands are deficient in genes. This notion was supported by the observation that sequences complementary to mRNA appeared localized in negative G bands (Yunis et al., 1977) and more recently that hypersensitivity to DNase — regarded as a marker for potentially active genes — is localized to negative G bands (Kerem et al., 1984; de la Torres et al., 1992). Craig and Bickmore (1993) demonstrated that of 1,023 genes assigned to single chromosome bands 80% coincided with R bands and 20% with G bands. It was reasonable to assume that altered doses of segments of G negative bands would have more profound effects on the developing organism than G dark regions.

Soon after the advent of chromosomal banding techniques it became evident that the major portion of G negative (R positive) regions (Paris Conference, 1971) on #21 were located on the distal half of the long arm (21q22) (Fig. 1). Many reports described the presence of this, the 21q22 region, in partial trisomies and clinical DS while most reports noted the absence of "classical" clinical DS with triplication of the short arm and proximal long arm of 21. Use of sophisticated molecular probes for localization of specific loci and correlation with clinical findings led to proposals that one or another specific subsegment on 21q22 was responsible for specific clinical findings in DS. After attempts of more than twenty years by numerous groups to narrow the region of chromosome segment 21q22 responsible for the clinical condition of DS, Korenberg et al. (1994) implicated nearly all of the long arm in generating different aspects of the DS phenotype. Most recently the DS phenotype has been described in a subject with tetrasomy of all but 21q22 (Daumer-Haas et al., 1994), the generally accepted DS pathogenetic segment. In this paper I shall review the evolution of proposed sites of chromosome 21 which, when in triplicate, have been presumed responsible for the characteristic signs in DS. Band 21q22 possesses most of the gene transcribing chromatin of 21 and it is not surprising that this band (and its subsegments) has most often been implicated as the portion of 21 associated with DS. Although numerous attempts have been made to exclude the role of all but some subband of 21q22 in the pathogenesis of DS, it will be seen that no narrow region of chromosome 21 can be said to be exclusively causative in development of the DS phenotype (Shapiro, 1997).

Triplication of 21q21–21q22

Of the five cases of Aula et al. (1973) "with clinical features suggesting Down's syndrome" and a partial trisomy of chromosome 21, three were

44 B. L. Shapiro

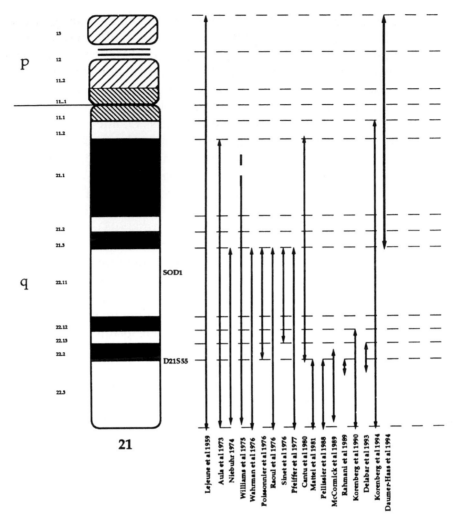

Fig. 1. Giemsa banding (G-bands) of human chromosome 21. Arrows depict, in chronological order from left to right, proposed sites of segments or loci which when in triplicate give rise to the phenotype of Down syndrome. The heavy line on the right is of a case with partial tetrasomy

nonmosaic and therefore potentially informative for determining effects of excess segments of chromosome 21. Cases 1 and 2 were thought to result from interstitial deletions of the long arm of 21 with a loss of the G positive region. These deletion cases showed fewer clinical findings associated with DS then did Case 4 which showed a translocation to #15 of the distal segment of the long arm of 21 including the G positive band (Fig. 1). Aula et al. (1973) concluded that "... trisomy of the (G) band segment and the distal portion of the long arm of chromosome 21 is required for the expression of a typical Down's syndrome. Partial trisomy of the long arm without the (G) band segment is associated with a clinical disease showing many, but not all of the typical stigmata of Down's syndrome, but with less severe mental retardation." [In their review, Park et al. (1987) chose to designate one of these deletion cases of Aula et al. (1973) as negative and the other as partial for the DS phenotype.] Williams et al. (1975) studied three generations of a family in which a reciprocal 10;21 translocation occurred. Using G banding they estimated that the breakpoint occurred in the very proximal part of band q21. Members of the family with partial trisomy of the long arm of 21 distal to the proximal portion of q21 resulted in the DS phenotype; members with trisomy of the short arm, the centromere and proximal long arm were mentally retarded but did not show the physical DS phenotype. Williams et al. (1975) concluded that the "main determinants of the Down phenotype are localized to the distal q21–q22 region of the long arm of chromosome 21, proximal to the subterminal C band." Pueschel et al. (1980), after describing their patient with an 8;21 translocation, concluded that "only additional chromosomal material of a specific segment of the lower arm of chromosome #21 will produce the Down syndrome phenotype." In the patient of Emberger et al. (1980) with a "phenotype of trisomy 21," most of the long arm of 21 was triplicated. Jenkins et al. (1983) described a "case of 'atypical' DS with a partial trisomy 21 in which the proposita does not have all of the clinical features of DS" [my emphasis because no individual with DS has *all* of the potential clinical features of the condition (Shapiro, 1983)]. The chromosomal break was thought to be at or distal to band 21q11 and it was concluded that trisomy for the distal segment of 21q whether it be more or less than 21q22 is responsible for the DS phenotype. Other reports narrowed the critical region thought to generate the DS phenotype to band 21q22.

Triplication of 21q22

Niebuhr (1974), apparently unaware of the paper by Aula et al. (1973), proposed that the distal segment of the long arm (21q22) may be pathogenetic in DS. He based this suggestion on clinical variability in reported cases with tandem translocations in which phenotypic severity appeared to be correlated with the extent of triplication of the long arm of 21. Three reports described translocation of the distal G pale (R dark) band of 21 where the breakpoint was thought to be the junction of the proximal G dark (q21) and distal G pale (q22) bands (Wahrman et al., 1976; Raoul et al., 1976; Pfeiffer et al., 1977) resulting in triplication of 21q22 → qter (Fig. 1). It was concluded that "(t)risomy of 21q22 causes the specific anomalies of Down syndrome" (Pfeiffer et al., 1977) and that "(t)riplication of the pale band or part of it, is responsible for production of the mongoloid phenotype" (Wahrman et al., 1976). The phenotype of the patient of Raoul et al. (1976) was said to be "that of classical trisomy 21." Wahrman et al. (1976) concluded that the distal pale band of the long arm . . . 21q22 . . . is the "critical mongoloid segment."

Habedank and Rodewald (1982), who stated that the "responsibility of the band 21q22 *for all the phenotypical features* of classic trisomy 21 has been well documented . . ." (my emphasis), described 3 sibs with a 4;21 translocation resulting in triplication of 21q22.2 to 21qter. They attributed the "moderate"

DS phenotype in their patients to the exclusion of 21q22.1 from the triplicated segment. "By the exclusion of SOD1 (q.v.) in them, we can confirm the nonessentiality of the sub-band 21q22.1 and of the SOD1 excess for most of the Down"s syndrome stigmata including the mental retardation. However, the sub-band 21q22.1 in triplicate might be required for the completion of the full syndrome..." (Habedank and Rodewald, 1982). Park et al. (1987) similarly stated that "band 21q22 (is) the causal duplicated segment in the Down syndrome." They concluded also that determinants for the syndrome are located in the terminal portion of 21q (q22 to qter). Based on their review of the literature, Shabtai et al. (1990) concluded that: "... trisomy for the proximal portion of the long arm of chromosome 21 does not result in the DS phenotype, the relevant portion being probably only band q22... Duplication of band 21q22 alone appears responsible for most of the typical manifestations of trisomy 21 (DS)." At about this time, reports were published in which attempts were made to further narrow the "critical" or "minimal" subsegments of 21q22 responsible for clinical DS.

Triplication of 21q22 subsegments

Proximal 21q22 (and SOD1)

The patient of Poissonnier et al. (1976) with a phenotype "strongly reminiscent of trisomy 21 syndrome" possessed a translocation resulting in partial trisomy of 21q22.1–q22.2 and disomy for q22.3. They concluded that the characteristic features of DS, particularly mental retardation and the facial features are due to band 21q22.1 and perhaps 21q22.2, exclusive of telomeric q22.3. Cantu et al. (1980) described an infant with clinical DS who, with a 21;21 translocation, was trisomic for much of 21q. They discounted the potential contribution to the phenotype of their patient of other portions of 21 which were triplicated when they concluded that "... subband 21q22.1 is responsible for the minimal features required to diagnose DS." Because of the correlations among monosomic or trisomic segments of #21, erythrocyte superoxide dismutase (SOD1) activity and clinical DS phenotype, Sinet et al. (1976) proposed that trisomy of subband 21q22.1 is "responsible for a great part of the clinical features of trisomy for the whole chromosome 21."

The gene for SOD1 had been assigned to chromosome 21 by somatic cell hybrid studies [Tan et al., 1973; Baltimore Conference, 1975). It was localized to distal 21q21 (Leschot et al., 1981) or proximal 21q22 (Sinet et al., 1976; Wulfsberg et al., 1983; Jenkins et al., 1983; Kirklionis and Sergovich, 1986; Miyazaki et al., 1987; Korenberg et al., 1990; Daumer-Haas et al., 1994) or at the junction of these bands (Huret et al., 1987; Delabar et al., 1987). A large number of studies reported increased erythrocyte SOD1 activity in patients with trisomy 21 (Sichitui et al., 1974; Sinet et al., 1974, 1975, 1976; Frants et al., 1975; Priscu and Sichitui, 1975; Kedziora et al., 1976; Crosti et al., 1976; Gilles et al., 1976; Feaster et al., 1977). The % of control mean SOD1 activity in cells from subjects with trisomy 21 ranged between 138 and 150, consistent with a phenotypic expression of gene dose. Increased SOD1 activity of ± 150% was found also in platelets (Sinet et al., 1975), fibroblasts, lymphocytes and polymorphonuclear leukocytes (Feaster et al., 1977), T lymphocytes, non T lymphocytes, and macrophages (Baeteman et al., 1983). Monosomic (#21) erythrocytes (Sinet et al., 1976) and cultured fibroblasts (Feaster et al., 1977) expressed about 60% of control activity. The dose effects were said to support the localization of the SOD1 locus on 21 and the proposal first made by Sinet et al. (1976) that SOD1 may play a central and crucial role in the pathogenesis of DS. Huret et al. (1987) and Delabar et al. (1987) described two patients with typical DS features but (even with prometaphase preparations) no cytologic evidence of partial trisomy. Using a SOD1 specific DNA probe and in situ hybridization they determined that submicroscopic duplications at the interface between distal 21q21 and q22.1, probably the proximal third of 21q22.1, which contained the SOD1 locus "should contain a large part of the genetic information involved in the pathogenesis of Down syndrome." In contrast, Jeziorowska et al. (1988) found 10 of 69 cases with trisomy 21 and "typical" DS to have red blood cell SOD1 activity no different from 50 control subjects.

Sinet et al. (1976) and Jeziorowska et al. (1982) found increased SOD1 activity in individuals with DS and partial trisomies due to translocations. Other investigators found that subjects with translocation and "typical" (Leonard et al., 1986) or "unequivocally. . . . Down's syndrome" (Pellissier et al., 1988) had SOD1 activity no different from controls (Kedziora et al., 1976, 1979; Mattei et al., 1981; Habedank and Rodewald, 1982; Leonard et al., 1986; Miyazaki et al., 1987; Jeziorowska et al., 1988; Pellissier et al., 1988; Korenberg et al., 1990; Delabar et al., 1993; Daumer-Haas et al., 1994). Mattei et al. (1981) and Pellissier et al. (1988) described two patients with clinical DS but duplication of only band 21q22.3 and normal SOD1 activity. The patient of Mattei et al. (1981) with trisomy of 21q223 presented "many of the classical clinical signs of trisomy 21 with normal SOD1 activity." Pellissier et al. (1988) described a three generation family in which four individuals presented with clinical DS. Using quantitative Southern blotting the duplicated region in these individuals did not include DNA sequences for SOD1. They concluded that "SOD1 is not necessary for the development of the phenotypic features of DS seen in this family." In the family of Korenberg et al. (1990) with the phenotypic features of DS but two copies of the SOD1 locus and normal SOD1 activity it was concluded that overexpression of SOD1 is not necessary for the expression of clinical DS. Delabar and Sinet and coworkers (1993) after phenotypic and molecular analysis of 10 patients with partial trisomy 21 concluded that the chromosomal region including the SOD1 locus "does not significantly contribute to most Down syndrome features." In the case of partial tetrasomy of 21pter → q22.1 with typical DS the SOD1 locus was present in the diploid copy number (Daumer-Haas et al., 1994). Leschot et al. (1981) described a patient with partial trisomy 21 who had increased SOD1 but no DS.

The locus for SOD1 is on band 21q22.1 (OMIM, 1995). Increased SOD1 activity is expressed in most cases in cells in which the chromosomal segment containing the SOD1 locus is triplicated. In a significant proportion of cases in which clinical DS was apparent the gene dosage was not reflected in increased cellular enzyme activity suggesting that triplicated SOD1 loci are not necessary for the development of DS. Despite the initial promise of the pathogenetic contribution of the SOD1 locus and its environs to the DS phenotype and the extensive experimental work attempting to demonstrate its role [most notably by Groner and coworkers (1994)], it must be concluded that SOD1 is not a molecular marker for DS and is not crucial to its development.

Distal 21q22 (and D21S55)

In contrast to assertions implicating proximal 21q22 as crucial to generation of the DS phenotype were reports of patients with partial trisomy of distal segments of 21q22 who were readily diagnosed as having DS. McCormick et al. (1989) used single-copy DNA sequences mapped on chromosome 21 to determine copy number in sixteen patients some of whom had partial trisomy of 21. Phenotypically the patients fell into three groups: partial trisomy with DS features; partial trisomy without a clinical diagnosis of DS; and normal karyotype with features compatible with DS ("phenocopies"). Trisomy for locus D21S13 (distal 21q22.2 or proximal 21q22.3) through locus D21S58 (junction of 21q21 and 21q22) was excluded from significant contribution to many DS features when present in three copies since patients with this partial trisomy had three copies for some or all of this region without having a clinical diagnosis of DS. McCormick et al. (1989) concluded that the "minimal chromosome region necessary in triplicate to result in the Down syndrome phenotypes in the patients characterized includes the area from locus D21S55 to locus COL6A1" (distal 21q22.2 or very proximal 22.3 to subterminal q22.3).

Two patients with partial trisomy due to translocation (21;21 and 11;21) were described by Pellissier et al. (1988). Their patients were diagnosed with DS although they both possessed partial trisomy only for 21q22.3. In situ hybridization with specific probes previously localized to subbands 21q22.1 and 21q22.3 was applied to chromosome preparations and it was concluded "that the clinical picture of Down syndrome is mainly connected to trisomy for band 21q223" (Pellissier et al., 1988) in agreement with Mattei et al. (1981) who examined a child with many features which permitted the clinical diagnosis of DS. In this case high resolution preparations were necessary to conclude that the chromosomal rearrangement could be interpreted as a probable duplication of sub-band 21q223 (Mattei et al., 1981).

Korenberg et al. (1990) examined a three generation family with four individuals with partial trisomy 21 who expressed many of the characteristic features of DS. By combining molecular characterization with in situ hybridization and high resolution cytogenetic analysis using probes for 21q22.1, 21q22.2 and 21q22.3 they were able to exclude the amyloid precursor and SOD1 genes "from the region of band 21q22.1 involved in the generation of the DS features. . . . All sequences located in 21q22.3 were present in three copies in the affected individuals, whereas those located proximal to this region were present in only two copies." They included a caveat that the region of triplication defined in this family may not be responsible for the "complete phenotype in DS" but did assert that, "Except for a possible phenotypic contribution from the deletion of chromosome band 4q35 (to which very distal 21q22.1 to 21qter was translocated), these data provide a molecular definition of the minimal region of chromosome 21 which, when duplicated, generates the facial features, heart defect, a component of mental retardation, and probably several of the dermatoglyphic changes of DS," and "This region may include parts of bands 21q22.2 and 21q22.3, but must exclude . . . most of band 21q22.1" (Korenberg et al., 1990). In 1992 Korenberg et al. extended these observations. They described clinical, cytogenetic and molecular analyses of two patients with DS who had partial trisomy 21 with different but overlapping segments. They defined regions which they proposed "contain the genes" for the congenital heart disease and duodenal stenosis seen in DS and identified a chromosomal overlap region "that may be sufficient for the development of the facial and some of the other physical features of DS." The region from DNA marker D21S55 [very distal 21q22.2 or very proximal 21q22.3 (Cox and Shimizu, 1991)] to the telomere was proposed as containing genes for congenital heart disease in DS; the region from D21S8 [21q21.2 (Cox and Shimizu, 1991)] to D21S15 [21q22.3 (Cox and Shimizu, 1991) was said to be the region that may be involved in the development of duodenal stenosis. The location of the region associated with the facial features of DS was thought to be "either band q22.2 or the adjacent region in band q22.1." I should note here that they acknowledged that their analyses did not "exclude genes located in other regions from contributing to a particular phenotypic feature." (Korenberg et al., 1992).

Since they suspected that the duplication of a specific region of chromosome 21 could be responsible for "the main features" of DS, Rahmani et al. (1989, 1990) used probes specific to different segments of 21q to define and localize this region in two patients reported previously (Poissonnier, 1976; Mattci, 1981) with clinical DS who had trisomy of different segments of 21. They assessed the copy number of five DNA sequences and concluded that locus D21S55 on 21q22.2–21q22.3 was the only subsegment duplicated in both cases. It was concluded that although other genes outside the D21S55 region may also play a role "(t)his region, located on the proximal part of 21q22.3 is postulated to contain genes the overexpression of which plays a major role in the pathogenesis of Down syndrome." Several years later the same group (Delabar et al., 1993) expanded their suspicion of the role of the D21S55 region as the major segment responsible for DS after studying ten patients with partial trisomy of different segments of 21. Thirty three clinical features were assessed. They compared minimal overlap regions with positive findings in at least two subjects. Thirteen of these features mapped to the single region D21S55 located on sub-band 21q22.2 or very proximal 21q22.3. Six other features mapped to a region spanning sub-band 21q22.2 to the proximal quarter of 21q22.3. "Thus, the clustering of most of the Down syndrome features within a relatively small region of chromosome 21 suggests that only one or a few genes belonging to this region are required to produce this complex phenotype" (Delabar et al., 1993). In a cogent comment on this work Korenberg (1994) opined that: "Sinet and coworkers... utilize overlap exclusively to create the DS map. Because 9 of their 10 subjects include the region D21S55, it is difficult to conclude otherwise than this region is important in all of the features... the inclusion of only a single case with a non-D21S55 duplication limits the evaluation of a DS chromosomal region and excludes the clear contribution of other regions to the DS phenotypes...".

In a series of reports from many laboratories examining karyotype-phenotype associations in partial 21 trisomics, segments of 21 potentially critical in the production of DS had been narrowed from the entirety of 21 to the middle G-dark (q21) and G-light (q22) bands to q22. The pathogenetic centrality of 21q22 was then reduced to its subsegments with conflicting conclusions. Different investigators attributed to different subsegments of q22 a key role in abnormal phenogenesis in DS: q22.1, q22.2 and even q22.3. In several papers, including that just cited, Korenberg and coworkers (1994) began broadening potentially pathogenetic regions of 21q thought to be responsible for phenotypic expression of trisomy 21 that had so recently been narrowed.

Triplication of 21q11.1 → qter

The conclusions of Korenberg et al. (1994) concerning stretches of chromosome 21 thought to be significant in production of the DS phenotype were a dramatic departure from the increasingly narrow focus [including their own (Korenberg et al., 1990)] of the previous twenty years. They studied cells from sixteen individuals with partial trisomy 21 for whom they had detailed clinical evaluations. They determined molecular breakpoints using fluorescent in situ hybridization and Southern blot dosage analysis of 32 markers unique to chromosome 21. They analyzed 25 phenotypic features and assigned regions "likely to contain the genes responsible" (Korenberg et al., 1994).

By invoking phenomena such as penetrance and expressivity in individuals with full and partial trisomy and using chromosomal overlap procedures of partial trisomies, they created "a phenotypic map" of chromosome 21. The expectation was that a trait will not be expressed at a frequency above that in the general population in individuals whose triplicate region does not include the "candidate region." For nearly all traits thus "mapped" the potential location of chromosomal region(s) responsible for a given phenotype spanned from 21q11.2 to 21qter. They concluded that: "This study provides evidence for a significant contribution of genes outside the *D21S55* region to the DS phenotypes, including the facies, microcephaly, short stature, hypotonia, abnormal dermatoglyphics, and mental retardation ... and augurs against a single DS chromosomal region responsible for most of the DS phenotypic features" (Korenberg et al., 1994). Much of the clinical leg of this study depended on the use of concepts such as penetrance and expressivity for complex phenotypes such as those analyzed. Dependence on these concepts may be questionable even in the context of explaining single gene traits for which they where originally coined (Timofeeff-Ressovsky, 1931). Even with single gene traits the use of these terms may "become dangerous, once we forget that they do not explain a biologic mechanism, but rather are labels for our ignorance" (Vogel and Motulsky, 1986). Little justification exists for invoking these concepts in the absence of independent evidence of action of a specific locus or loci. Whatever the assumptions, the work of Korenberg et al. (1994) reopened the potential contribution of most of the long arm of 21 to a role in the pathogenesis of the DS phenotype. Rather than attributing the DS phenotype to a small restricted subsegment of 21q22 as had previous groups including their own, Korenberg and her colleagues (1994) proposed to "associate particular regions (of 21q) with specific phenotypes." As I discuss next, this expansion from 21q22 and subsegments of it thought necessary for the development of DS to most of 21q apparently extends to additional segments of 21.

Triplication of 21pter → q21

A number of cases have been described in which subjects possessed partial trisomy of segments of chromosome 21 exclusive of the proposed critical band 21q22. Of these reported cases with partial trisomy of 21p and proximal 21q the two patients of Daniel (1979) expressed no clinical findings and the only clinical problem noted in two others was mental retardation (Raoul et al., 1976; Williams et al., 1975). A fifth case showed minor anomalies that did not suggest even slight features of DS (Shabtai et al., 1990). A diagnosis of DS could not be made for several other cases but the patients expressed multiple abnormalities that are often part of the DS phenotype (Pfeiffer et al., 1977; Hagemeijer and Smit, 1977; Kitsiou-Tzeli et al., 1984; Williams et al., 1990; Petersen et al., 1990). Neu et al. (1971) described a prebanding case in which the extra chromosome appeared to be missing most of the long arm. Clinical findings of DS were diminished in degree and number but four physicians made the diagnosis of DS on clinical grounds. In the case of de la Chapelle et al. (1973) where more than half of the distal portion of 21q was not in triplicate the proposita had "features compatible with 21 trisomy" but the overall phenotype did not present the "clear-cut features resembling those in Down's syndrome." They speculated that inconsistent DS features as well as anomalies not usually associated with DS may have been due to the concomitant trisomy of most of the long arm of chromosome 4. Similarly, in the case of Sanchez et al. (1977) (trisomic for all but 21q22) clinical manifestations were said to be compatible with DS but it was not possible to make a firm clinical diagnosis of DS because she lacked some of the "most discriminating findings" of DS. Sanchez et al. (1977) did conclude that trisomy 21pter → q21.3 was not devoid of phenotypic effects. Park et al. (1987) reviewed some of the previous cases of translocations involving triplication of portions of chromosome 21. According to them, "In all cases where there was no duplication of any of segment 21q22, Down syndrome stigmata were not seen." In their own cases Park et al. (1987) described two subjects with free proximal trisomy 21. Both subjects demonstrated mental retardation, short stature, hypotonia, developmental delay, microcephaly and hypoplastic nails. The first case was also affected by hyperextensible joints, seizures, small hands and feet, brachydactyly, flat feet and strabismus; the second subject was trigonocephalic. In light of the presence of these signs, why would these investigators suggest that these subjects were considered to be "without manifestations of the Down syndrome" (Park et al., 1987)?

Although for most of the cases with triplication of 21pter to q21 diagnosis of DS could not be made clinically, most of them expressed multiple abnormalities each of which is commonly seen in DS. Except for the five subjects for whom a diagnosis of DS could not be made, partial trisomy for segments other than the so-called "critical region" does lead to clinical expression of signs that contribute to the DS phenotype and have been attributed by some investigators to specific segments on 21q22. It must be emphasized that none of the physical findings of DS associated with full trisomy 21 is found in all cases; most are not even found in a majority of cases (Smith and Berg, 1976). [For example, neither epicanthal folds nor transverse palmar creases were among the 10 most discriminating signs for DS listed by Jackson et al. (1976)]. That extra copies of chromosomal segments other than those on 21q22 can result in a fully developed case of DS was most dramatically revealed in the extremely unusual case of Daumer-Haas, Korenberg et al. (1994) in whom they found a partial tetrasomy for 21pter → q21.1 due to an extra isodicentric chromosome 21:

"Down syndrome was suspected immediately after birth because of the distinct craniofacial appearance including brachycephaly, third fontanel, upward slanting, short palpebral fissures, round face, and protruding tongue. In addition, diastasis recti, umbilical hernia, joint laxity, brachydactyly, and a single transverse palmar crease on both hands were noted ... (also) ulnar loops on all fingertips and a distal triradius ... definite hypotonia was present at birth ... at 18 months (she) showed a mental and motor age of 12 months" (Daumer-Haas et al., 1994).

Using a battery of probes unique to chromosome 21 and in situ hybridization the patient was shown to be disomic for the entire distal half of 21q, i.e. 21q22. The authors concluded that their results "suggest that genes lying outside the currently defined DCR (Down syndrome critical region) in 21q22.2–22.3, i.e. 21pter → q21, contribute significantly to the phenotype of Down syndrome" (Daumer-Haas et al., 1994).

Discussion

Clinical signs which permit the diagnosis of DS have been reported by different investigators to be secondary to triplication of the following segments of chromosome 21: 21q22, 21q22.1 and 22.2, 21q22.1, 21q22.3, the SOD1 locus and the single subregion identified by the DNA marker, D21S55, located on subband 21q22.2 or very proximal 21q22.3. Indeed, the so-called "Down Syndrome Critical Region (DCR or DSCR)" has been said to be mapped to 21q22.3, "the chromosomal region that, if triplicated, results in the phenotypic characteristics of Down syndrome" (OMIM, 1998). It is odd that so many investigators have assumed that a very small segment or even locus of 21q22 would be the critical region for signs associated with DS. One wonders, for example, what would lead a group to assert that, "(a)lthough chromosome 21 contains a large number of genes, it is probable that only a few contribute to the major features of D.S." (Fuentes et al., 1995). In reality, no rationale exists for dismissing transcribing loci throughout chromosome 21 from some effect on the phenotype of DS. It comes as no surprise, therefore, that after almost 25 years the search for a minimal region on chromosome 21 responsible for producing DS has come nearly full circle back to most of the entire chromosome. Findings shown in Fig. 1, particularly the analyses of Korenberg et al. (1994) and the partial tetrasomy case of Daumer-Haas et al. (1994), in whom clinical DS was found in the presence of two extra portions of 21 other than the distal half of the long arm, make the label of a DSCR untenable. A "minimal" or "critical" region cannot be different regions identified by different investigators. It must be concluded that no critical region of chromosome 21 exists which can be said to be singularly responsible for clinical findings in DS. Of course the distal of the long arm of chromosome 21 has the major effect in trisomy 21 since this is where [G negative bands (Craig and Bickmore, 1993)] the greatest concentration of transcribing DNA on this chromosome occurs. For this reason, it is understandable why partial trisomy of 21q distal would have effects more profound than that of other portions of 21.

Korenberg et al. (1994) whose laboratories had been among those contributing to the critical region notion, reversed that trend when they proposed that different segments over most of 21q were responsible for different phenotypic expressions of DS. They suggested that DS should be considered a "contiguous gene syndome." Contiguous gene syndrome describes rare conditions caused by microdeletion of two or more consecutive loci in which deletion of genes at closely linked loci results in diverse phenotypic effects, sometimes showing features of several monogenic diseases. The use of this phrase for triplication of nearly an entire long arm of a chromosome is without precedent and clearly not justified in the context of DS. In any case, the subject with DS described subsequently by Daumer-Haas, Korenberg and their coworkers (1994) possessed no segment overlap between her partial tetrasomy and the cases of partial trisomy on which claims of DS critical regions were based. A different set of loci involved in the partial tetrasomy case would of course speak against a contiguous gene explanation for DS.

As already mentioned, it is reasonable that the distal half of the long arm of chromosome 21 has been considered to be of pathogenetic importance in development of the DS phenotype since this segment comprises at least 85%

of the transcribing portions of 21. On the other hand, the finding that tetrasomy for 21p → 21q proximal produces the DS phenotype when trisomy for this segment usually does not, further supports the idea that it may not be specific loci but rather the amount of superfluous transcribing genetic material that contributes significantly to the phenotype of DS (Shapiro, 1983). It is relevant, in this context, that more commonalities exist among the autosomal trisomy syndromes than distinctions (Taylor, 1968) and that in subjects with poly X syndromes the phenotype is similar in many respects to that of DS (Thompson et al., 1991; Linden et al., 1995).

The question that remains, of course, is what is it about triple doses of many loci on one autosome that permits, most of the time, clinical distinction of one autosomal trisomy from another? Ultimately, the ability to distinguish different trisomies clinically must be referable to interactions of products of those chromosomes with developmental and physiological systems affected by the entire genome, pre- and postnatal environmental factors (Shapiro, 1983, 1994) and stochastic effects (Kurnit et al., 1987). The assumption persists that because a genetic accident (i.e. nondisjunction or unbalanced translocation) is involved that clinical findings in DS are a direct result of gene products. This is no more true of DS (where mutant genes are not at issue) than that an abnormal gene product in hypomelanotic conditions *causes* sun-induced skin carcinoma or that the severe hemolytic anemia following ingestion of fava beans is caused by an abnormal glucose-6-phosphate dehydrogenase molecule. As the foregoing review demonstrates, in the presence of different segments of chromosome 21 the ability to make a clinical diagnosis of DS has led investigators to attribute responsibility for the syndrome to particular loci on the segment studied. Unfortunately, the distinctions (and their clinical recognition) among the autosomal trisomies and between them and the general population cannot form the basis for understanding the development of aneuploid associated anomalies since the ability to diagnose a condition is far removed from understanding its pathogenesis.

Application of sophisticated molecular techniques to determine the mapping of a particular locus to chromosome 21 or a subsegment of it and a leap to the conclusion that the product of a particular locus causes a particular sign in DS is mistaken as well as nonproductive. For example, Fuentes et al. (1995) described a new human gene from 21q22.1–q22.2 that they named "Down syndrome candidate region 1 (DSCR1)". In the absence of any indication that the gene product in question was, in fact overexpressed in DS, they concluded that structural characteristics of its inferred product and its expression in brain and heart "encourage us to suggest that the overexpression of DSCR1 may be involved in the pathogenesis of Down syndrome, in particular mental retardation and/or cardiac defects." A more meaningful approach would have been to determine whether or not the product of a particular locus on 21 expresses a dosage effect in tissues thought to be affected in DS before speculating on its role. Greber-Platzer and her coworkers (1999) determined mRNA levels of the protooncogene ets-2 coded by a locus between D21S17 and DS21S55. They found no evidence of overexpression of this gene in cardiac tissue from children with Down syndrome and congenital heart disease (CHD) in comparison with chromosomally balanced children with CHD. Had they found increased levels of the gene product in subjects with DS they could have concluded that, even in the absence of direct evidence of its role in CHD pathogenesis, this increase was not inconsistent with a role in cardiac defects in those subjects with DS. However, they did not find its increase in tissues of subjects with DS and concluded, therfore, that their findings provided no evidence for the current gene dosage-effect hypothesis.

Without exception, the congenital and postnatal abnormalities often found in subjects with DS are each multifactorial conditions. No rationale exists for assuming that the pathogenesis of these abnormalities in DS is any different from that in the general population. Thus analysis of complex traits associated with DS should correspond to methods used to study multifactorial traits in general. Autosomal trisomic conditions appear to predispose affected individuals to susceptibility to genetic and (pre- and postnatal) environmental insults that organisms with balanced genomes usually buffer. Amplified instability of developmental and physiological pathways in DS because of genetic imbalance (Shapiro, 1983) offers a natural experimental model for obtaining insight into the pathogenesis of complex conditions in the general population as well as uncovering mechanisms of development of aneuploidy associated clinical findings.

The review presented here illustrates that speculations about a minimal critical DS region or that DS is a contiguous gene syndrome are oversimplified and incorrect explanations for a condition that warrants a different paradigm from the classical linear gene to phenotype model. The reductionist linear approach which appears to form the basis for the minimal critical region presumptions about DS has been rejected by me (Shapiro, 1975, 1983, 1989) and by Opitz and Gilbert (1982). Continued mapping of chromosome 21 is certainly worthwhile as part of the acquisition of complete knowledge of the human genome. Nevertheless, the expectation of a linear (i.e. gene locus to phenotype) explanation for the highly complex anomalies that comprise DS is unlikely just as complete knowledge of the genome will be insufficient to fully understand development of an organism. The competing models of gene locus — phenotypic specificity (Epstein, 1988) and generalized decreased homeostasis (Shapiro, 1983) seeking to explain DS pathogenesis were contrasted by Wilson (1990) who concluded that "concepts of colinearity and continuity derived from single gene action are. . . . likely to be false assumptions."

References

Aula PJ, Leisti J, von Kuskull H (1973) Partial trisomy 21. Clin Genet 4: 241–251
Baeteman MA, Baret A, Courtiere A, Rebuffel P, Mattei JF (1983) Immunoreactive Cu-SOD and Mn-SOD in lymphocytes sub-populations from normal and trisomy 21 subjects according to age. Life Sci 32: 895–902

Baltimore Conference (1975) Third International Workshop on Human Gene Mapping, vol XII. Birth defects: original article series. The National Foundation, New York, 1976

Cantu JM, Hernandez A, Plasencia L, Vaca G, Moller M, Rivera H (1980) Partial trisomy and monosomy 21 in an infant with an unusual de novo 21/21 translocation. Ann Genet 23: 183–186

Cervenka J, Gorlin RJ, Djavadi GR (1977) Down syndrome due to partial trisomy 21q. Clin Genet 11: 119–121

Cox DR, Shimizu N (1991) HGM 11: Report of the committee on the genetic constitution of chromosome 21. Cytogenet Cell Genet 58: 800–826

Craig JM, Bickmore WA (1993) Chromosome bands — flavours to savour. Bioessays 15: 349–354

Crosti N, Serra A, Rigo A, Viglino P (1976) Dosage effect of SOD-A gene in 21-trisomic cells. Hum Genet 31: 197–202

Daniel A (1979) Normal phenotype and partial trisomy for the G positive region of chromosome 21. J Med Genet 16: 227–229

Daumer-Haas C, Schuffenhauer S, Walther JU, Schipper RD, Porstmann T, Korenberg JR (1994) Tetrasomy 21 pter-q22.1 and Down syndrome. Am J Med Genet 53: 359–365

Dekaban AS, Zelson J (1968) Retardation in a child with an extra submetacentric chromosome fragment and partial mongolism. J Ment Defic Res 12: 21–225

Delabar JM, Sinet PM, Chadefaux B, Nicole A, Gegonne A, Stehelin D, Fridlansky F, Creau-Goldberg N, Turleau C, de Grouchy J (1987) Submicroscopic duplication of chromosome 21 and trisomy 21 phenotype (Down syndrome). Hum Genet 76: 225–229

Delabar JM, Theophile D, Rahmani Z, Chettouh JL, Blouin M, Preur B, Noel B, Sinet P-M (1993) Molecular mapping of twent-four features of Down syndrome on chromosome 21. Eur J Hum Genet 1: 114–124

de la Chapelle A, Kovisto M, Schroder J (1973) Segregating reciprocal (4;21)(q21;q21) translocation with proposita trisomic for parts of 4q and 21. J Med Genet 10: 384–389

de la Torres JA, Sumner T, Gosalvez J, Stuppia L (1992) The distribution of genes on human chromosomes as studied by in situ nick translation. Genome 35: 890–894

Dent T, Edwards JH, Delhanty JDA (1963) A partial mongol. Lancet ii: 484–487

Emberger JM, Lloret R, Rossi D (1980) Partial trisomy with 45 chromosomes due to translocation of two chromosomes 21 onto a chromosome 14: 45,XX-14,-21,+t(14q21q21q). Ann Genet 23: 179–180

Epstein CJ (1988) Specificity versus nonspecificity in the pathogenesis of aneuploid phenotypes. Am J Med Genet 29: 161–165

Feaster WW, Kwok LW, Epstein CJ (1977) Dosage effects for superoxide dismutase-1 in nucleated cells aneuploid for chromosome 21. Am J Hum Genet 29: 563–570

Frants RR, Eriksson AW, Jongbloet PH, Hamers AJ (1975) Superoxide dismutase in Down syndrome. Lancet ii: 42–43

Fuentes JJ, Pritchard MA, Planas AM, Bosch A, Ferrer I, Estivill X (1995) A new human gene from the Down syndrome critical region encodes a proline-rich protein highly expressed in fetal brain and heart. Hum Mol Genet 4: 1935–1944

Ganner E, Evans HJ (1971) The relationship between patterns of DNA replication and of quinacrine fluorescence in the human chromosome complement. Chromosoma 35: 326–341

Gilles L, Ferradini C, Foos J, Pucheault J, Allard D, Sinet PM, Jerome H (1976) The estimation of red cell superoxide dismutase activity by pulse radiolysis in normal and trisomic 21 subjects. FEBS Lett 69: 55–58

Greber-Platzer S, Schatzmann-Turhani D, Wollenek G, Lubec G (1999) Evidence against the current hypothesis of "gene dosage effects" of trisomy 21: ets-2, encoded on chromosome 21" is not overexpressed in hearts of patients with Down syndrome. Biochem Biophys Res Comm 254: 395–399

Groner Y, Elroy-Stein O, Avraham KB, Schickler M, Knobler H, Minc-Golomb D, Bar-Peled O, R Yaron, Rotshenker S (1994) Cell damage by excess CuZnSOD and Down's syndrome. Biomed Pharmacother 48: 231–240

Habedank M, Rodewald A (1982) Moderate Down's syndrome in three siblings having partial trisomy 21q22.2 → qter and therefore no SOD-1 excess. Hum Genet 60: 74–77

Hagemeijer A, Smit EME (1977) Partial trisomy 21. Further evidence that trisomy of band 21q22 is essential for Down's phenotype. Hum Genet 38: 15–23

Hernandez D, Pannett AA, Tybulewicz V, Fisher EM (1995) Highly polymorphic sequence at D21S1448 mapping close to D21S55, within the Down syndrome critical region. Hum Genet 95: 721–722

Hongell K, Airaksinen EA (1972) A Gg deletion in a girl with Down's syndrome. Hum Hered 22: 80–85

Huret JL, Delabar JM, Marlhens F, Aurias A, Nicole A, Berthier M, Tanzer J, Sinet PM (1987) Down syndrome with duplication of a region of chromosome 21 containing the CuZn superoxide dismutase gene without detectable karyotypic abnormality. Hum Genet 75: 251–257

Ilbery PLT, Lee CWG, Winn SM (1976) Incomplete trisomy in a mongoloid child exhibiting minimal stigmata. Med J Aust 2: 182–184

Jackson JF, North ER 3rd, Thomas JG (1976) Clinical diagnosis of Down syndrome. Clin Genet 9: 483–487

Jenkins EC, Duncan CJ, Wright CE, Giordano FM, Wilbur L, Wisniewski K, Sklower SL, French JH, Jones C (1983) Atypical Down syndrome and partial trisomy 21. Clin Genet 24: 9–102

Jeziorowska A, Jakubowsk Li, A Armatys A, Kaluzewski B (1982) Copper/zinc superoxide dismutase (SOD-1) activity in regular trisomy 21, trisomy 21 by translocation and mosaic trisomy 21. Clin Genet 22: 160–164

Jeziorowska A, Jakubowsk L, Lach J, Kaluzewski B (1988) Regular trisomy 21 not accompanied by increased copper-zinc superoxide dismutase (SOD1) activity. Clin Genet 33: 11–19

Kedziora J, Rozynkowa D, Kopff M, Jeske J (1976) Indophenol-oxidase in patients with Down's syndrome due to simple trisomy and to translocation 21/22. Hum Genet 34: 9–12

Kedziora J, Bartorz G, Leyko W, Rozynkawa D (1979) Dismutase activity in translocation trisomy. Lancet i: 105

Kerem BS, Goitein R, Diamond G, Cedar H, Marcu M (1984) Mapping of DNAase I sensitive regions on mitotic chromosomes. Cell 38: 493–499

Kirklionis AJ, Sergovich FR (1986) Down syndrome with apparently normal chromosomes: an update. J Pediat 108: 793–794

Kitsiou-Tzeli S, Hallett JJ, Atkins L, Latt SA, L B Holmes LB (1984) Familial translocation 4;21 (q2.4;q2.2) leading to unbalanced offspring with partial duplication 4q and 21q without manifestations of the Down syndrome. Am J Med Genet 18: 725–729

Korenberg JR, Kawashima H, Pulst S-M, Ikeuchi T, Ogasawara N, Yamamoto K, Schonberg SA, West R, Allen L, Magenis E, Ikawa K, Taniguchi N, Epstein CJ (1990) Molecular definition of a region of chromosome 21 that cause features of the Down syndrome phenotype. Am J Hum Genet 47: 236–246

Korenberg JR, Bradley C, Disteche CM (1992) Down syndrome: molecular mapping of the congenital heart disease and duodenal stenosis. Am J Hum Genet 50: 294–302

Korenberg JR, Chen XN, Schipper R, Sun Z, Gonsky R, Gehwehr S, Carpenter N, Daumer C, Dignan P, Disteche C et al (1994) Down syndrome phenotypes: the consequences of chromosomal imbalance. Proc Natl Acad Sci USA 91: 4997–5001

Kurnit DM, Layton WM, Matthysse S (1987) Genetics, chance and morphogenesis. Am J Hum Genet 42: 979–955

Lejeune J, Gautier M, Turpin R (1959) Etude des chromosome somatique des neufs enfants mongoliens. CR Acad Sci Paris 248: 1721–1722

Leonard C, Gautier M, Sinet PM, Selva J, Huret JL (1986) Two Down syndrome patients with rec (21),dup q,inv(21)(p11;q2109) from a familial pericentric inversion." Ann Genet 29: 181–183

Leschot NJ, Slater RM, Joenje H, Becker-Bloemkolk MJ, Nef JJ (1981) SOD-A and chromosome 21. Conflicting findings in a familial translocation (9p24;21q21.4). Hum Genet 57: 220–223

Lewontin RC (1992) Biology as ideology: the doctrine of DNA. Harper, New York, p 26

Linden MG, Bender BG, Robinson A (1995) Sex chromosome tetrasomy and pentasomy. Pediatrics 96: 672–682

Mattei JF, Mattei MG, Baeteman MA, Giraud F (1981) Trisomy 21 for the region 21q22.3: Identification by high resolution R-banding patterns. Hum Genet 56: 409–411

McCormick MK, Schinzel A, Petersen M, Stetten G, D Driscoll D, Tranebjaerg L, Mikkelsen M, Watkins P, Antonarkis S (1989) Molecular genetic approach to the characterization of the "Down syndrome region" of chromosome 21. Genomics 5: 325–331

Miyazaki K, Yamanaka T, Ogasawara N (1987) A boy with Down's syndrome having recombinant chromosome 21 but no SOD-1 excess. Clin Genet 32: 383–387

Neu RL, Voorhess ML, Gardner LI (1971) A case of 47,XX,(21q-)+ with some stigmata of Down's syndrome and an IQ of 77. J Med Genet 8: 528–529

Niebuhr E (1974) Down's syndrome — the possibility of a pathogeneic segment on chromosome no. 21. Humangenetik 21: 99–101

OMIM (TM) (1995) Online Mendelian Inheritance in Man. The Human Genome Data Base Project, World Wide Web ed. <URL: http://gdbwww.gdb.org/omim/docs/omimtop.html>. Johns Hopkins University, Baltimore

OMIM (1998) Online Mendelian Inheritance in Man, OMIM (TM). Johns Hopkins University, Baltimore, MD. MIM Number: {190685}: {8/21/98}:. World Wide Web URL: http://www.ncbi.nlm.nih.gov/omim/

Paris Conference (1971) Standardization in human genetics, vol 8. Birth defects. Orig Art Ser 1972

Park JP, Wurster-Hill D, Andrews P, Cooley W, Graham J (1987) Free proximal trisomy 21 without the Down syndrome. Clin Genet 32: 342–348

Pellissier MC, Laffage M, Philip N, Passage E, Mattei M-G, Mattei J-F (1988) Trisomy 21q223 and Down's phenotype correlation evidenced by in situ hybridization. Hum Genet 80: 277–281

Petersen MB, Tranebjaerg L, McCormick MK, Michelsen N, Mikkelsen M, Antonarakis SE (1990) Clinical, cytogenetic, and molecular characterization of two unrelated patients with different duplications of 21q. Am J Med Genet [Suppl] 7: 104–109

Pfeiffer RA, Kessel EK, Soer K-H (1977) Partial trisomies of chromosome 21 in man. Two new observations due to translocations 19;21 and 4;21. Clin Genet 1: 207–213

Poissonnier M, Saint-Paul B, Dutrillaux B, Chaissaigne M, Gruyer P, de Blignieres-Strouk G (1976) Trisomie 21 partielle (21q21→ q22.2). Ann Genet (Paris) 19: 69–70

Polani PE, Briggs JH, Ford CE, Clarke CM, Berg JM (1960) A mongol child with 46 chromosomes. Lancet i: 721–724

Priscu R, Sichitui S (1975) Types of enzymatic overdosing in trisomy 21: erythrocytic dismutase-AJ and phosphoglucomutase. Humangenetik 29: 79–83

Pueschel SM, Padre-Mendoza T, Ellenbogen R (1980) Partial trisomy 21. Clin Genet 18: 392–395

Rahmani Z, Blouin JL, Creau-Goldberg N, Watkins PC, Mattei JF, Poissonnier M, Prieur M, Chettouh Z, Nicole A, Aurias A et al (1989) Critical role of the D21S55 region on chromosome 21 in the pathogenesis of Down syndrome. Proc Natl Acad Sci USA 86: 5958–5962

Rahmani Z, Blouin JL, Creau-Goldberg N, Watkins PC, Mattei JF, Poissonnier M, Prieu M, Chettouh Z, Nicole A, Aurias A et al (1990) Down syndrome critical region around D21S55 on proximal 21q22.3. Am J Med Gyenet 7 [Suppl]: 98–103

Raoul O, Carpentier S, Dutrillaux B, Mallet R, Lejeune J (1976) Trisomies partielles du chromosome 21 par translocation maternelle t (15;21) (q262;q21). Ann Genet Paris 19: 187–190

Sanchez O, Mamunes P, Yunis JJ (1977) Partial trisomy 20 (20q13) and partial trisomy 21 (21pter → 21q21.3). J Med Genet 14: 459–462

Shabtai FS (1990) Free proximal trisomy 21 in the mother and malformation syndrome in the son. Am J Med Genet [Suppl] 7: 182–185

Shapiro BL (1975) Amplified developmental instability in Down syndrome. Ann Hum Genet 38: 429–437

Shapiro BL (1983) Down syndrome — a disruption of homeostasis. Am J Med Genet 14: 241–269

Shapiro BL (1989) The pathogenesis of aneuploid phenotypes — the fallacy of explanatory reductionism. Am J Med Genet 33: 146–150

Shapiro BL (1994) The environmental basis of the Down syndrome phenotype. Dev Med Child Neurol 36: 84–90

Shapiro BL (1997) Whither Down syndrome critical regions? Hum Genet 99: 4211–423

Sichitui S, Sinet PM, Lejeune J, Frezal J (1974) Surdosage de la forme dimerique de l'indophenoloxydase dans la trisomie 21, secondaire au surdosage genique. Humangenetik 23: 65–72

Sinet PM, Allard D, Lejeune J, Jerome H (1974) Augmentation d'activite de la superoxide dismutase erythrocytaire dans la trisomie pour le chromosome 21. CR Acad Sci Paris 278: 3267–3270

Sinet PM, LaVelle F, Michelson AM, Jerome H (1975) Superoxide dismutase activities of blood platelets in trisomy 21. Biochem Biophys Res Comm 67: 904–909

Sinet P-M, Couturier J, Dutrillaux B, Poissonnier M, Raoul O, Rethore M-O, Allard D, Lejeune J (1976) Trisomie 21 et superoxyde dismutase-1 (IPO-A) — Tentative de localisation sur la sous bande 21q22.1. Exp Cell Res 97: 47–55

Smith GF, Berg JM (1976) Down's anomaly. Churchill Livingstone, Edinburgh

Summer AT (1994) Functional aspects of the longitudinal differentiation of chromosomes. Eur J Histochem 38: 91–109

Surbt I, Prchlikova H (1970) An extra chromosomal centric fragment in an infant with stigmata of Down's syndrome. J Med Genet 7: 407–409

Tan YH, Tischfield J, Ruddle FM (1973) The linkage of genes for the human interferon-induced antiviral protein and indophenol oxidase-B traits to chromosome G-21." J Exp Med 137: 317–330

Taylor AI (1968) Autosomal trisomy sndromes: a detailed study of 27 cases of Edwards syndrome and 27 cases of Patau syndrome. J Med Genet 5: 227–252

Thompson MW, McInnes RR, Willard HF (1991) Thompson & Thompson: Genetics in medicine, 5th edn. W. B. Saunders, Philadelphia, pp 240–241

Verma RS, Peakman DC, Robinson A, Lubs HA (1977) Two cases of Down syndrome with unusual de novo translocation. Clin Genet 11: 227–234

Vogel F, Motulsky AG (1986) Human genetics: problems and approaches, 2nd edn. Springer, Berlin Heidelberg New York Tokyo

Wahrman J, Goitein R, Richler C, Goldman B, Ackstein E, Chaki R (1976) The mongoloid phenotype in man is due to trisomy of the distal pale G-band of chromosome 21. In: Pearson PL, Lewis KR (eds) Chromosomes today, vol 5. Halstead Press, New York

Williams JD, Summitt RL, Martens PR, Kimbrell RA (1975) Familial Down syndrome due to t(10;21) translocation: evidence that the Down syndrome phenotype is related to trisomy of a specific segment of chromosome 21. Am J Hum Genet 27: 478–485

Williams CA, Frias JL, McCormick MK, Antonarakis SE, Cantu ES (1990) Clinical, cytogenetic, and molecular evaluation of a patient with partial trisomy 21 (21q11-q22) lacking the classical Down syndrome phenotype. Am J Med Genet [Suppl] 7: 110–114

Wilson GN (1990) Karyotype/phenotype controversy: genetic and molecular implications of alternative hypotheses. Am J Med Genet 36: 500–505

Wulfsberg EA, Carrel RE, Klisak IJ, O'Brien TJ, Sykes JA, Sparkes RS (1983) Normal superoxide dismutase-1 (SOD-1) activity with deletion of chromosome band 21q21 supports localization of SOD-1 locus to 21q22. Hum Genet 64: 271–272

Yunis JJ, Kuo MT, Saunders GF (1977) Localization of sequences specifying messenger RNA to light-staining G-bands of human chromosomes. Chromosoma 61: 335–344

Author's address: Dr. B. L. Shapiro, Departments of Oral Science and Laboratory Medicine and Pathology and Institute of Human Genetics, University of Minnesota, 17-220 Moos Tower, 515 Delaware Street SE, Minneapolis, MN 55455, U.S.A., e-mail: burt@mailbox.mail.umn.edu

J Neural Transm (1999) [Suppl] 57: 61–74

Neuropathology

N. J. Cairns

Department of Neuropathology, Institute of Psychiatry, King's College,
London, United Kingdom

Introduction

The more effective treatment of heart disease and infections during infancy
and young adulthood has led to a remarkable increase in life expectancy of
patients with Down's syndrome (DS) from about 10 years at the beginning of
the twentieth century to more than 70% surviving to their sixth decade at its
end (Baird and Sadovnick, 1987). Unfortunately, this increased longevity has
been associated with dementia in a population already mentally handicapped.
The discovery of Alzheimer's disease (AD) in the brains of elderly Down's
syndrome patients has provided a unique model for studying the pathogenesis
of the disease. Conversely, new insights that have been gained from the
molecular genetics and biology of Alzheimer's disease may be helpful in
leading to novel treatments for sporadic and familial forms of Alzheimer's
disease and Alzheimer's disease in Down's syndrome.

Macroscopic changes

The naked eye appearance of the Down's syndrome brain reveals an
abnormally short and rounded brain with an almost vertical occipital
convexity. The brain weight is usually about 1,000g and rarely exceeds 1,200g.
The pattern of gyri is similar to that of a normal brain although the superior
temporal gyrus may be somewhat narrow bilaterally in about half the cases;
this is not specific to Down's syndrome as this anomaly occurs in other
disorders (Esiri and Morris, 1997; Graham and Lantos, 1997). The brain
stem and cerebellum may be small in relation to the size of the cerebral hemispheres (Raz et al., 1995). Benda (1947) has suggested that many
of the abnormalities of shape have been impressed on the developing brain
by retardation in the growth of the skull. The abnormalities of brain
development that result in learning disability and intellectual impairment are
compounded by the neuropathological changes that lead to dementia in
middle-aged Down's syndrome patients. In one large series, Alzheimer-type
changes were seen in the brains of all those dying over the age of 40 years
(Malamud and Gaitz, 1972). The onset of Alzheimer's disease is accom-

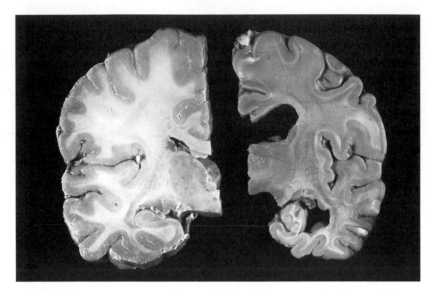

Fig. 1. On the right, a coronal slice of the hemi-brain of a patient aged 47 years with Down's syndrome showing atrophy: the lateral ventricle is enlarged with rounding of its angle and there is additional space between the hippocampus and the inferior horn of the lateral ventricle. Several gyri are moderately narrowed and the lateral fissure is widened. On the left, a slice of the right hemi-brain of a normal subject

panied by a number of neurodegenerative changes. Morphometric studies have shown that when the underdevelopment of the frontal and temporal lobes in younger DS patients is taken into account, the additional decrease in brain size that occurs in DS in later life is brought about, as in Alzheimer's disease, by neurofibrillary tangle and neuritic plaque formation and associated neuronal loss (De la Monte and Hedley-White, 1990). This neurodegeneration leads to loss of grey and white matter. The ventricles become enlarged and there is narrowing of the gyri (Fig. 1). In DS individuals, a larger parahippocampal gyrus and smaller hippocampus and neocortex are reported relative to age-matched controls using magnetic resonance imaging (Kesslak et al., 1994).

Microscopic changes

The neurohistological features of DS are complex and variable. The two hallmark lesions, senile plaques and neurofibrillary tangles, are complemented by granulovacuolar degeneration, Hirano bodies, neuronal loss, abnormalities of neuronal processes and synapses, astrocytic and microglial response, and vascular changes. Although these changes are also found in Alzheimer's disease and to a lesser degree in ageing, there are subtle differences in the morphology and distribution of lesions in DS (Mann, 1997).

Neuritic plaques

The senile or neuritic plaque (SP) is one of the major lesions found in the AD and DS brain and is present to a lesser extent in the ageing brain (Fig. 2). The SP was first described by Blocq and Marinesco in 1892. These structures ranging in size from 50 to 200 μm (Terry, 1985) can be readily demonstrated by silver impregnation methods. The lesion consists of an amyloid core with a corona of argyrophilic axonal and dendritic processes, amyloid fibrils, glial cell processes and microglial cells. Neuritic processes in the periphery of the SP are frequently dystrophic and contain PHFs which are composed largely of abnormally phosphorylated tau protein (Gonatas et al., 1967; Hanger et al., 1991). In 1927 Divry demonstrated that the core of the SP contained an amyloid-like congophilic substance which gave an apple green colour under polarised light. The amyloid is composed of 5 to 10 nm filaments made up of a 40–43 amino acid (4 kDa) protein which was sequenced by Glenner and Wong (1984), and now referred to as Aβ protein according to its proposed secondary structure of β-pleated sheets. Aβ is the amyloid plaque core protein found in AD and DS (Masters et al., 1985).

Studies using antibodies raised against Aβ protein have revealed the presence of a much more widespread deposition of amyloid than is visualised by traditional staining methods (Mann et al., 1989; Spargo et al., 1992; Armstrong et al., 1996). Aβ immunoreactivity has been detected throughout the central nervous system including the neocortex, hippocampus, thalamus, amygdala, caudate nucleus, putamen, nucleus basalis of Meynert, midbrain, pons, medulla oblongata, the cerebellum and spinal cord in DS and AD (Joachim et al., 1989b; Ogomori et al., 1989; Mann et al., 1990; Murphy et al., 1992). These deposits take a variety of forms and include: subpial, vascular,

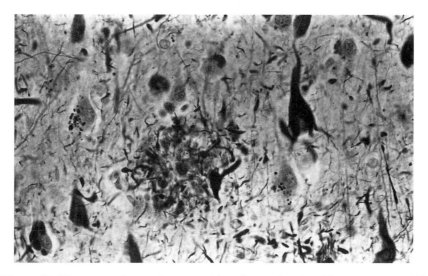

Fig. 2. Neurofibrillary tangles and a neuritic plaque in the hippocampus. Glees and Marsland silver impregnation

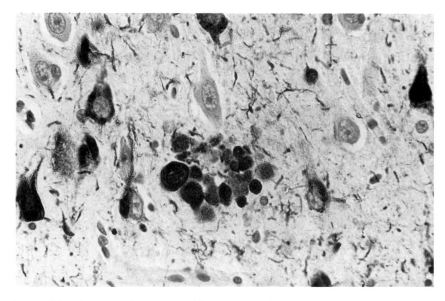

Fig. 3. A neuritic plaque and neurofibrillary tangles in the hippocampus demonstrated by
anti-tau immunohistochemistry

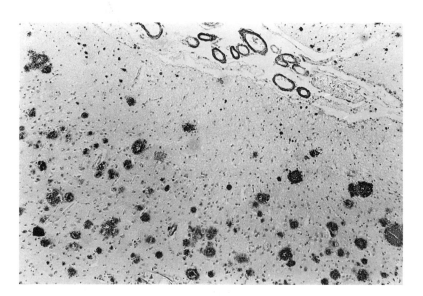

Fig. 4. β-Amyloid protein deposits in the neocortex of the temporal lobe of a patient aged
47 years with Down's syndrome. Several deposit morphologies can be seen. β-Amyloid
immunohistochemistry

dyshoric, punctate or granular, diffuse, stellar, ring-with-core, and compact
deposits (Ogomori et al., 1989; Fig. 4). A laminar pattern of Aβ deposits in the
neocortex has been described with concentrations in layers II, III and V
(Majocha et al., 1988), and spatial pattern analysis reveals clustering of Aβ
deposits in the cortex in DS (Armstrong, 1994a,b).

Although all cells have the potential for producing APP it seems likely that the major source of APP and Aβ is the neuron (Allsop et al., 1989). Nonetheless, astrocytes and microglial cells have also been implicated. The hypothesis that Aβ deposition is an early event in the pathogenesis of AD has been given support by the early detection of soluble $Aβ_{42}$ at 21 gestational weeks before any $Aβ_{42}$ plaques are observed in the DS brain (Teller et al., 1996). Also, extracellular APP and Aβ have been described in diffuse plaques not associated with any neuritic change, or astrocytic involvement (Mann et al., 1989; Yamaguchi et al., 1991). More condensed Aβ is, however, associated with a neuritic change, reactive astrocytosis and microglial infiltration and phagocytosis (Mann et al., 1992; Murphy et al., 1992). The Aβ fragment may act as a growth-promoting factor causing neuritic attraction and outgrowth leading to promotion of the SP about an Aβ core. Most species of Aβ have been detected by immunohistochemistry using antibodies that recognise a full length $Aβ_{1-42}$ or a truncated $Aβ_{1-40}$. The predominant species in DS and AD is $Aβ_{42(43)}$ and this is found in plaques of all morphological types. Diffuse plaques contain only $Aβ_{42(43)}$ (Mann et al., 1996a,b). However, a greater proportion of diffuse plaques is found in DS and a higher proportion of more compact and ring-with-core deposits is observed in AD indicating that different factors may be operating in DS to modify the amyloidogenic pathway (Armstrong, 1994b).

A number of proteins co-localise with Aβ including apolipoprotein E (ApoE), a plasma protein involved in the transport of cholesterol. There are three common isoforms of ApoE which are encoded by three alleles ε2, ε3, and ε4 and there is a strong association between the presence of the ε4 allele and AD. ApoE has been shown to bind to fibrous Aβ and a high proportion of Aβ and ApoE deposits tend to be co-localised. Thus, both diffuse and compact Aβ plaques may be immunolabelled by anti-Aβ and anti-ApoE antibodies (Cairns et al., 1997; Armstrong et al., 1998). The presence of one or more ε4 alleles leads to both an earlier onset and a more severe Aβ amyloidosis in AD (Roses, 1994) and an extra copy of the APP gene in trisomy 21 leads to increased plaque size and Aβ deposition (Hyman et al., 1995).

SPs are also immunohistochemically positive for a number of neuro-active substances. Acetylcholinesterase, a marker of the cholinergic system, can be demonstrated in some of the neuritic fibres of the corona. Several other neurotransmitters have been demonstrated in the SP including: substance P, neuropeptide Y, neurotensin, cholecystokinin, 5-hydroxytryptamine (5-HT) and catecholamine (Walker et al., 1988; Hauw et al., 1991). Markers of the cholinergic, noradrenergic, dopaminergic, and serotonergic systems are all reduced in DS and AD (Mann et al., 1985; Schapiro and Rapoport, 1988). Ubiquitin, a protein acting as a signal for degradation of abnormal proteins, is also present in SPs and NFTs (Perry et al., 1987). The presence of complement proteins and inflammatory cytokines suggest the involvement of inflammatory processes (McGeer and McGeer, 1998).

The distribution of SPs varies widely not only within architectonic units but also from one individual to another. SPs are more numerous in associative regions of neocortex than in sensory areas and SPs are also largest in those laminae characterised by large pyramidal neurons (Rafalowska et al., 1988). The primary focus of SPs and NFTs in DS is the limbic region including the entorhinal cortex, hippocampus and amygdala, with a spreading out to subsequently involve all areas of cortex, certain subcortical regions and the olfactory bulbs and tracts (Mann et al., 1986).

Neurofibrillary tangles

The second histological hallmark of AD is the neurofibrillary tangle (NFT; Fig. 2). These are not specific to AD, since they occur in ageing and other neurodegenerative disorders, including Down's syndrome, postencephalitic Parkinsonism, dementia pugilistica, in the amyotrophic lateral sclerosis-Parkinsonism-dementia complex of Guam, subacute sclerosing panencephalitis, dementia with tangles with and without calcification, and in myotonic dystrophy (Kiuchi et al., 1991). NFTs also develop in progressive supranuclear palsy and sporadic motor neuron disease, but their ultrastructure and immunocytochemistry differ from those seen in ageing, DS and AD.

In AD and DS, NFTs are not only more numerous than in normal ageing, but they also affect the brain more extensively. Like in ageing, they are common in the medial temporal structures, in the hippocampus, amygdala and parahippocampal gyrus, but also occur throughout the neocortex and the deep grey matter. NFTs are neuronal inclusions composed of abnormal cytoskeletal elements. Their configuration is determined by the shape of neurons in which they develop. NFTs are flame-shaped in pyramidal neurons, whilst in the neurons of the brainstem they assume more complex, globose forms. They can be easily discerned in histological sections stained with haematoxylin and eosin, but are best demonstrated by one of the silver impregnation techniques or with Congo red which renders them birefringent under polarised light. They can be demonstrated by immunocytochemistry using antibodies against various neurofilament proteins, tau protein and ubiquitin. Electron microscopy reveals their finer structural details: NFTs are mainly composed of paired helical filaments (Kidd, 1963) which in turn are formed by two filaments wound around each other. Each filament has a diameter of 10nm with crossover points at every 80nm resulting in the typical periodicity of a double helix (Crowther and Wischik, 1985). NFTs also contain straight filaments with an average diameter of 15 to 30nm (Crowther, 1991; Itoh and Yagishita, 1998). In the adult human brain there are six isoforms of tau and all of these are hyperphosphorylated in DS and AD (Goedert et al., 1991; Hanger et al., 1991). Antibodies that recognise specific phosphorylation-dependent anti-tau antibodies may be used to identify the three sites of tau pathology: the NFT, dystrophic neurites of SPs and neuropil threads (Fig. 3).

Neuronal loss

Neuronal loss is far more difficult to assess than the histological abnormalities previously described. There is destruction of nerve cells associated with anoxia and oedema during infancy in DS (Benda, 1947). Whilst well-defined, neuronal groups, like the locus coeruleus, are easier to count, the cerebral cortex presents considerable technical problems. Nevertheless, quantitative studies reveal neuronal loss up to 50% compared with controls, from birth onwards, but not in fetal life (Colon, 1972; Wisniewski, 1984). Neurons in the granular layers are particularly affected (Ross, 1984). In qualitative terms, all areas of the brain found to be damaged in AD are also damaged in DS at middle age, although there are differences in relative neuron number and mean nucleolar volume (Mann et al., 1987). Neuronal markers are decreased more in DS than AD and this neuronal loss is marked by a reactive astrocytosis (Jørgensen et al., 1990). There is also a decrease in the number of dendritic spines and alterations in the pattern of arborization in neonatal and infantile brains but not in the fetus (Takashima, 1981). The mechanism causing neuronal loss is not fully understood. However, Aβ fragments have been shown to have toxic effects in vitro (Yanker et al., 1989) and apoptotic processes have been demonstrated in AD and DS brains in those regions where neuronal loss is greatest (Sawa et al., 1997). Alterations in the expression of DNA excision-repair-cross-complementing proteins in AD and DS may indicate an impaired cellular response to oxidative damage (Hermon et al., 1998).

Neuropil threads

Neuropil threads or "curly" fibres and dystrophic neurites have been well-documented in silver impregnated preparations developed by Gallyas (1971), but they attracted attention when it was realised that they could substantially contribute to dementia (Braak et al., 1986; Braak and Braak, 1988, 1981). Neuropil threads, or curly fibres, are slender, short structures found in the neuropil of the cortex. Dystrophic neurites, some of which contribute to the formation of SPs, also appear as somewhat contorted, thread-like structures in these preparations. These abnormal neurites are likely to be a heterogeneous population of dendrites and axons (Tourtellotte and Van Hoesen, 1991). Their ultrastructural and immunocytochemical features are strikingly similar to those of NFTs, indicating that they also originate from the altered neuronal cytoskeleton. Neuropil threads immunostain with antibodies to tau (Kowall and Kosik, 1987; Probst et al., 1989) and to ubiquitin (Joachim et al., 1987). A higher density of neuropil threads has been observed in AD in comparison with elderly DS patients (de la Monte and Wands, 1992). Electron microscopy has revealed that they are often in dendrites and contain straight filaments of 14–16nm in diameter in addition to paired helical filaments (Tabaton et al., 1989; Yamaguchi et al., 1990b). Considerable decrease of synapses also occurs and this loss is likely to contribute to the development of the disease (Davis et al., 1987). Synaptophysin, a protein localised in presynaptic terminals, is reduced by 50% in the parietal, temporal and midfrontal cortex in AD and similar deficits may be expected in DS (Masliah et al., 1989).

Neurofibrillary tangles and SPs have been associated with patterns of gliosis in AD and ageing brain (Beach et al., 1989; Itagakai et al., 1989; Cairns et al., 1992). In AD and ageing, perivascular gliosis is prominent throughout the cerebrum. Both AD and the ageing brain show extensive gliosis which has a laminar pattern in the neocortex in AD, but there does not appear to be an AD-specific pattern of subcortical gliosis (Beach and McGeer, 1988; Beach et al., 1989). In addition to NFTs, SPs and dystrophic neurites, reactive astrocytosis is associated with degenerating neurons in AD. In the patho-genesis of the SP in DS and AD, it appears as though Aβ deposition is an early event preceding astrocytosis (Mann et al., 1992) and electron microscopic studies indicate that unaggregated amyloid fibrils are not found in proximity to astrocytes (Yamaguchi et al., 1990a).

Amyloid angiopathy

Cerebral blood vessels containing amyloid, when stained with Congo red, appear apple green under polarised light. Pantelakis (1954) called this deposition congophilic angiopathy. These congophilic changes affect lepto-meningeal and cortical arterioles, and occasionally intracortical capillaries and venules. The parieto-occipital cortex is usually more affected than that in the frontal and temporal lobes and the change is most easily identified in the striate cortex in the occipital lobe and cerebellum (Mann, 1988). Vascular amyloid is now known to be, basically, the same protein, Aβ, as the amyloid found in SPs and other deposits in AD. Occasionally, Aβ appears to extend from the vessel wall into the surrounding cerebral tissue; this is referred to as dyshoric angiopathy.

White matter changes

The white matter is also affected. The myelination of the grey and white matter is poor with a striking loss of myelin in the U-fibres (Brenda, 1947). Meyer and Jones (1939) were the first to demonstrate a fibrillary gliosis of the central white matter in most, but not all, cases. There is a frequent association between heart disease and brain pathology in Down's syndrome. Myelination was not delayed in the fetal brain, but in 129 cases, from birth to 6 years, myelination was delayed in 23% compared with normal controls. In Down's syndrome infants with congenital heart disease, this figure was increased to 48% (Wisniewski and Schmidt-Sidor, 1989).

Conclusions

Alzheimer's disease patients with amyloid precursor protein (APP) gene and two related genes, presenilin-1 (PS1) and presenilin-2 (PS2) mutations may have combined pathology. Cortical and brainstem Lewy bodies may co-exist with AD pathology in APP (Lantos et al., 1992) and PS mutation patients. It is now known that the Lewy body and dystrophic neurites of Parkinson's disease and dementia with Lewy bodies contain the protein α-synuclein (Spillantini et al., 1997). Lewy bodies have also been reported in Down's syndrome and this observation suggests that common pathogenic mechanisms may underlie aspects of neuronal degeneration in Parkinson's disease, Alzheimer's disease and Down's syndrome (Raghavan et al., 1993).

AD in DS, like autosomal dominant familial AD, is a disease primarily of adulthood to middle age, unlike the majority of sporadic cases of AD where the age at onset is old age. Although familial forms of AD with multiple affected individuals are rare and account for probably fewer than 5% of AD cases, the discovery of mutations in APP and PS genes linked to the disease has opened up new avenues for research into pathogenesis. Mutations in APP, PS1 and PS2 account for the majority of early-onset autosomal dominant cases of familial AD (Goate et al., 1991; Lamb, 1997). These mutations modulate the processing of APP and by means of mechanisms not fully understood they lead to increased production of the amyloidogenic fragment Aβ. Also, the presence of one or more ε4 alleles leads to both an earlier onset and a more severe Aβ amyloidosis in AD (Roses, 1994) and an extra copy of the APP gene in trisomy 21 leads to both increased plaque size and Aβ deposition (Hyman et al., 1995). Insight gained from the study of amyloidogenesis may be applicable to DS and vice versa. Anti-Aβ pharmacological agents may offer the prospect of inhibiting the amyloidogenic pathway that leads, ultimately, to dementia. Secondly, the identification of enzymes that phosphorylate tau abnormally may lead to the discovery of drugs which may interfere with cytoskeletal pathology in DS and AD.

References

Allsop D, Haga S-I, Haga C, Ikeda S-I, Mann DMA, Ishii T (1989) Early senile plaques in Down's syndrome brains show a close relationship with cell bodies of neurons. Neuropathol Appl Neurobiol 15: 531–542
Alzheimer A (1907) Über eine eigenartige Erkrankung der Hirnrinde. Allg Z Psychiat 64: 146–148
Armstrong RA (1994a) The spatial patterns of β/A4 deposit subtypes in Down's syndrome. Acta Neuropathol 88: 353–358
Armstrong RA (1994b) Differences in β-amyloid (β/A4) deposition in human patients with Down's syndrome and sporadic Alzheimer's disease. Neurosci Lett 169: 133–136
Armstrong RA, Cairns NJ, Myers D, Smith CUM, Lantos PL, Rossor MN (1996) A comparison of β-amyloid deposition in the medial temporal lobe in sporadic Alzheimer's disease, Down's syndrome and normal elderly brains. Neurodegeneration 5: 35–41
Armstrong RA, Cairns NJ, Lantos PL (1998) The spatial pattern of β-amyloid (Aβ) deposits in Alzheimer's disease patients is related to apolipoprotein genotype. Neurosci Res Commun 22: 99–106
Baird PA, Sadovnick AD (1987) Life expectancy in Down's syndrome. J Paediat 110: 849–854

Beach TG, McGeer EG (1988) Lamina-specific arrangement of astrocytic gliosis and senile plaques in Alzheimer's disease visual cortex. Brain Res 473: 357–361

Beach TG, Walker R, McGeer EG (1989) Patterns of gliosis in Alzheimer's disease and aging cerebrum. Glia 2: 420–436

Benda CE (1947) Mongolism and cretinism. Grune and Stratton, New York

Blocq P, Marinesco G (1892) Sur les lesions et la pathogénie de l'epilepsie dite essentielle. Semaine Medicale 12: 445–446

Braak H, Braak E (1988) Neuropil threads occur in dendrites of tangle-bearing nerve cells. Neuropathol Appl Neurobiol 14: 39–44

Braak H, Braak E (1991) Neuropathological staging of Alzheimer-related changes. Acta Neuropathol 82: 239–259

Braak H, Braak E, Grundke-Iqbal I, Iqbal K (1986) Occurrence of neuropil thread, in the senile human brain and in Alzheimer's disease: a third location of paired helical filaments outside of neurofibrillary tangles and neuritic plaques. Neurosci Lett 65: 351–355

Cairns NJ, Chadwick A, Luthert PJ, Lantos PL (1992) Astrocytosis, βA4-protein deposition and paired helical filament formation in Alzheimer's disease. J Neurol Sci 112: 68–75

Cairns NJ, Fukutani Y, Chadwick A, et al (1997) Apolipoprotein E, β-amyloid (Aβ), phosphorylated tau and apolipoprotein E genotype in Alzheimer's disease. Alzheimer's Res 3: 109–114

Colon EY (1972) The structure of the cerebral cortex in Down's syndrome: quantitative analysis. Neuropediatrics 3: 362–376

Crowther RA (1991) Straight and paired helical filaments in Alzheimer's have a common structural unit. Proc Natl Acad Sci USA 88: 2288–2292

Crowther RA, Wischik CM (1985) Image reconstruction of the Alzheimer paired helical filament. EMBO J 4: 3661–3665

Davis CA, Mann DMA, Sumpter PQ, Yates PO (1987) A quantitative morphometric analysis of the neuronal and synaptic content of the frontal and temporal cortex in patients with Alzheimer's disease. J Neurol Sci 78: 151–164

de la Monte SM, Hedley-White ET (1990) Small cerebral hemispheres in adults with Down's syndrome. Contributions of developmental arrest and lesions of Alzheimer's disease. J Neuropathol Exp Neurol 49: 509–520

de la Monte SM, Wands JR (1992) Neuronal thread protein over-expression in brains with Alzheimer's disease lesions. J Neurol Sci 113: 152–164

Divry P (1927) Etude histo-chimique des plaques séniles. J Belge Neurol Psychiat 27: 643–657

Esiri MM, Morris JH (1997) The neuropathology of dementia. Cambridge University Press, Cambridge

Gallyas F (1971) Silver staining of Alzheimer's neurofibrillary changes by means of physical development. Acta Morphol Acad Sci Hung 19: 1–8

Glenner GG, Wong CW (1984) Alzheimer's disease and Down's syndrome: sharing of a unique cerebrovascular amyloid fibril. Biochem Biophys Res Commun 122: 1131–1135

Goate A, Chartier-Harlin MC, Mullan M, et al (1991) Segregation of a missense mutation in the amyloid precursor protein gene with familial Alzheimer's disease. Nature 349: 704–706

Goedert M, Spillantini MG, Cairns NJ, Crowther RA (1992) Tau proteins of Alzheimer paired helical filaments; abnormal phosphorylation of all six isoforms. Neuron 8: 159–168

Gonatas NK, Anderson A, Vongelista I (1967) The contribution of altered synapses in the senile plaque: an electronmicroscopic study in Alzheimer's dementia. J Neuropathol Exp Neurol 26: 25–39

Graham DI, Lantos PL (eds) (1997) Greenfield's Neuropathology, 6th ed. Edward Arnold, London

Hanger DP, Brion J-P, Gallo J-M, Cairns NJ, Luthert PJ, Anderton BH (1991) Tau in Alzheimer's disease and Down's syndrome is insoluble and abnormally phosphorylated. Biochem J 275: 99–104

Hauw J-J, Duyckaerts C, Delaère P (1991) Alzheimer's disease. In: Duckett S (ed) The pathology of the aging nervous system. Lea and Febiger, London, pp 113–47

Hyman BT, West HL, Rebeck GW, Buldyrev SV, Mantegna RN, Ukleja M, et al (1995) Quantitative analysis of senile plaques in Alzheimer's disease: observation of log-normal size distribution and molecular epidemiology of differences associated with apolipoprotein E genotype and trisomy 21 (Down syndrome). Proc Natl Acad Sci USA 92: 3586–3590

Itagakai S, McGeer PL, Akiyama H, et al (1989) Relationship of microglia and astrocytes to amyloid deposits of Alzheimer's disease. J Neuroimmunol 24: 173–182

Itoh Y, Yagishita S (1998) Scanning electron microscopical study of neurofibrillary tangles in a presenile patient with Down's syndrome. Acta Neuropathol 96:179–184

Joachim CL, Morris JH, Selkoe DJ, Kosik KS (1987) Tau epitopes are incorporated into a wide range of lesions in Alzheimer's disease. J Neuropathol Exp Neurol 46: 611–622

Joachim CL, Morris JH, Selkoe DJ (1989) Diffuse senile plaques occur commonly in the cerebellum in Alzheimer's disease. Am J Pathol 135: 309–319

Jørgensen OS, Brooksbank BWL, Balázs R (1990) Neuronal plasticity and astrocytic reaction in Down syndrome and Alzheimer disease. J Neurol Sci 98: 63–79

Kesslak JP, Nagata SF, Lott I, Nalcioglu O (1994) Magnetic resonance imaging analysis of age-related changes in the brains of individuals with Down's syndrome. Neurology 44: 1039–1045

Kidd M (1963) Paired helical filaments in electron microscopy of Alzheimer's disease. Nature 197: 192–193

Kiuchi A, Otsuka N, Namba Y, et al (1991) Presenile appearance of abundant neurofibrillary tangles without senile plaques in the brain in myotonic dystrophy. Acta Neuropathol 82: 1–5

Kowall NW, Kosik KS (1987) Axonal disruption and aberrant localisation of tau protein characterise the neuropil pathology of Alzheimer's disease. Ann Neurol 22: 639–643

Lamb BT (1997) Presenilins, amyloid-β and Alzheimer's disease. Nature 3: 28–29

Lantos PL, Luthert PJ, Hanger D, Anderton BH, Rossor M (1992) Familial Alzheimer's disease with the amyloid precursor protein position 717 mutation and sporadic Alzheimer's disease have the same cytoskeletal pathology. Neurosci Lett 137: 221–224

Majocha RE, Benes FM, Reifel JL, et al (1988) Laminar-specific distribution and infrastructural detail of amyloid in the Alzheimer's disease cortex visualised by computer-enhanced imaging of epitopes recognised by monoclonal antibodies. Proc Natl Acad Sci USA 85: 6182–6186

Malamud N, Gaitz CM (eds) Aging and the brain. Plenum Press, New York

Mann DMA (1988) The pathological association between Down's syndrome and Alzheimer's disease. Mech Ageing 43: 99–136

Mann DMA (1997) Neuropathological changes of Alzheimer's disease in persons with Down's syndrome. In: Esiri MM, Morris JM (eds) The neuropathology of dementia. Cambridge University Press, Cambridge, pp 122–136

Mann DMA, Esiri MM (1989) The pattern of acquisition of plaques and tangles in the brains of patients under 50 years of age with Down's syndrome. J Neurol Sci 89: 169–179

Mann DMA, Yates PO, Marcynink B (1985a) Some morphometric observations in the cerebral cortex and hippocampus in presenile Alzheimer's disease, senile dementia of Alzheimer type and Down's syndrome in middle age. J Neurol Sci 69: 139–159

Mann DMA, Yates PO, Marcynink B, Ravindra CR (1985b) Pathological evidence for neurotransmitter deficits in Down's syndrome of middle age. J Ment Defic Res 29: 125–135

Mann DMA, Yates PO, Marcynink B, Ravindra CR (1986) The topography of plaques and tangles in Down's syndrome patients of different ages. Neuropathol Appl Biol 12: 447–457

Mann DMA, Yates PO, Marcynink B, Ravindra CR (1987) Loss of neurons from cortical and subcortical areas in Down's syndrome patients at middle age: quantitative comparisons with younger Down's patients and patients with Alzheimer's disease. J Neurol Sci 80: 79–89

Mann DMA, Brown A, Prinja D, Davies CA, Landon M, Masters CL, et al (1989) An analysis of the morphology of senile plaques in Down's syndrome patients of different ages using immunocytochemical and lectin histochemical techniques. Neuropathol Appl Neurobiol 15: 317–329

Mann DMA, Jones D, Prinja D, Purkiss MS (1990) The prevalence of amyloid (A4) protein deposits within the cerebral and cerebellar cortex in Down's syndrome and Alzheimer's disease. Acta Neuropathol 80: 318–327

Mann DMA, Younis N, Jones D, Stoddart RW (1992) The time course of pathological events in Down's syndrome with particular reference to the involvement of microglial cells and deposits of β/A4. Neurodegeneration 1: 210–215

Mann DMA, Iwatsubo T, Cairns NJ, et al (1996a) Amyloid β protein (Aβ) deposition in chromosome 14-linked Alzheimer's disease: predominance of $A\beta_{42(43)}$. Ann Neurol 40: 149–156

Mann DMA, Iwatsubo T, Ihara Y et al (1996b) Predominant deposition of amyloid-$\beta_{42(43)}$ in plaques in cases of Alzheimer's disease and hereditary cerebral hemorrhage associated with mutations in the amyloid precursor protein gene. Am J Pathol 148: 1257–1266

Masliah E, Terry RD, DeTeresa RM, Hansen LA (1989) Immunohistochemical quantification of the synapse-related protein synaptophysin in Alzheimer disease. Neurosci Lett 103: 234–239

Masters CL, Simms G, Weinman NA, Multhaup G, McDonald BL, Beyreuther K (1985) Amyloid plaque core protein in Alzheimer disease and Down syndrome. Proc Natl Acad Sci USA 82: 4245–4249

McGeer EG, McGeer PL (1998) The importance of inflammatory mechanisms in Alzheimer's disease. Exp Geront 33: 371–378

Meyer A, Jones TB (1939) Histological changes in brain in mongolism. J Ment Sci 85: 206–221

Murphy GM, Ellis WG, Lee Y-L, Stultz KE, Shrivastava R, Ticklenberg JR, et al (1992) Astrocytic gliosis in the amygdala in Down's syndrome and Alzheimer's disease. Prog Brain Res 94: 475–483

Ogomori K, Kitamoto T, Tateishi J, et al (1989) β-protein amyloid is widely distributed in the central nervous system of patients with Alzheimer's disease. Am J Pathol 134: 243–251

Pantelakis S (1954) Un type particulier d'angiopathie sénile du système nervaux central: l'angiopathie congophile. Topographie et fréquence. Monatsschr Psychiat Neurol 128: 219–256

Perry G, Friedman R, Shaw G, Chau V (1987) Ubiquitin is detected in neurofibrillary tangles and senile plaque neurites of Alzheimer's disease brains. Proc Natl Acad Sci USA 84: 3033–3036

Probst A, Anderton BH, Brion J-P, Ulrich J (1989) Senile plaque neurites fail to demonstrate anti-paired helical filament and anti-microtubule associated protein-tau immunoreactive proteins in the absence of neurofibrillary tangles in the neocortex. Acta Neuropathol 77: 430–436

Rafalowska J, Barcikowska M, Wen GY, Wisniewski HM (1988) Laminar distribution of neuritic plaques in normal aging, Alzheimer's disease and Down's syndrome. Acta Neuropathol 77: 21–25

Raghavan R, Khin-Nu C, Brown A, Irving D, Ince PG, Day K, et al (1993) Detection of Lewy bodies in Trisomy 21 (Down's syndrome). Can J Neurol Sci 20: 48–51

Raz N, Torres IJ, Brggs SD, Spencer WD, Thornton AE, Loken WJ, et al (1995) Selective neuroanatomic abnormalities in Down's syndrome and their cognitive correlates: evidence from MRI morphometry. Neurology 45: 356–366

Roses AD (1994) Apolipoprotein E affects the rate of Alzheimer's disease expression: β amyloid burden is a secondary consequence dependent on APOE genotype and duration of disease. J Neuropathol Exp Neurol 53: 429–437

Ross MH, Galaburda AM, Kemper TL (1984) Down's syndrome: is there a decreased population of neurons? Neurology 34: 909–916

Sawa A, Oyama F, Cairns NJ, Amano N, Matsushita M (1997) Aberrant expression of bcl-2 gene family in Down's syndrome brains. Mol Brain Res 48: 53–59

Schapiro MB, Rapoport SI (1988) Alzheimer's disease in premorbidly normal and Down's syndrome individuals: selective involvement of hippocampus and neocortical associative brain regions. Brain Dysfunct 1: 2–11

Seidl R, Greber S, Schuller E, Bernert G, Cairns N, Lubec G (1997) Evidence against increased oxidative DNA-damage in Down syndrome. Neurosci Lett 235: 137–140

Spargo E, Luthert PJ, Janota I, Lantos PL (1992) βA4 deposition in the temporal cortex of adults with Down's syndrome. J Neurol Sci 111: 26–32

Spillantini MG, Schmidt ML, Lee VM-Y, Trojanowski JQ, Jakes R, Goedert M (1997) α-Synuclein in Lewy bodies. Nature 388: 839–840

Tabaton M, Mandybur TI, Perry G, et al (1989) The widespread alteration of neurites in Alzheimer's disease may be unrelated to amyloid deposition. Ann Neurol 26: 771–778

Takashima S, Becker LE, Armstrong DL, Chan F (1981) Abnormal neuronal development in the visual cortex of the human fetus and infant with Down's syndrome: a quantitative and qualitative Golgi study. Brain Res 225: 1–21

Teller JK, Russo C, DeBusk LM, Angellini G, Zaccheo D, Dagna-Bricarelli F, et al (1996) Presence of soluble amyloid β-peptide precedes amyloid plaque formation in Down's syndrome. Nature Med 2: 93–95

Terry RD (1985) Alzheimer's disease. In: Davis RL, Robertson DM (eds) Textbook of neuropathology. Williams and Wilkins, Baltimore, pp 824–841

Tourtellotte WG, Van Hoesen GW (1991) The axonal origin of a subpopulation of dystrophic neurites in Alzheimer's disease. Neurosci Lett 129: 11–16

Walker LC, Kitt CA, Cork LC, et al (1988) Multiple transmitter systems contribute neurites to individual senile plaques. J Neuropathol Exp Neurol 47: 138–144

Wisniewski KE, Schmidt-Sidor B (1989) Postnatal delay of myelin formation in brains from Down syndrome infants and children. Clin Neuropathol 8: 55–62

Wisniewski KE, Laure-Kamionowska M, Wisniweski HM (1984) Evidence of arrest of neurogenesis and synaptogenesis in Down syndrome brains. N Engl J Med 311: 1187–1188

Yamaguchi H, Nakazato Y, Hirai S, Shoji M (1990a) Immunoelectron microscopic localization of amyloid β protein in the diffuse plaques of Alzheimer-type dementia. Brain Res 508: 320–324

Yamaguchi H, Nakazato Y, Shoji M, et al (1990b) Ultrastructure of the neuropil threads in the Alzheimer brain: their dendritic origin and accumulation in the senile plaques. Acta Neuropathol 80: 368–374

Yamaguchi H, Nakazato Y, Shoji M, et al (1991) Ultrastructure of diffuse plaques in senile dementia of the Alzheimer type: comparison with primitive plaques. Acta Neuropathol 82: 13–20

Yanker BA, Dawes LR, Fisher S, et al (1989) Neurotoxicity of a fragment of the amyloid precursor associated with Alzheimer's disease. Science 245: 417–420

Authors' address: Dr. N. J. Cairns, Department of Neuropathology, Institute of Psychiatry, King's College, De Crespigny Park, London SE5 8AF, United Kingdom, e-mail: n.cairns@iop.kcl.ac.uk

J Neural Transm (1999) [Suppl] 57: 75–85

c-fos expression in brains of patients with Down Syndrome

S. Greber-Platzer[1], **B. Balcz**[1], **N. Cairns**[2], and **G. Lubec**[1]

[1] Department of Pediatrics, University of Vienna, Austria
[2] Department of Neuropathology, Institute of Psychiatry, University of London, United Kingdom

Summary. c-fos is a protooncogene serving in multiple physiological processes in brain from signalling to proliferation and synaptic plasticity. We therefore decided to determine this transcription factor in control and Down Syndrome (DS) brain with the Rationale that c-fos may be linked to brain damage in DS.

We determined mRNA steady state levels in frontal, parietal, occipital, temporal cortex and cerebellum of 9 patients with DS and 9 controls using RT-PCT.

Significantly increased levels of mRNA c-fos normalized versus the housekeeping gene beta-actin mRNA were found in frontal, parietal and temporal cortex of DS brain. c-fox mRNA levels comparable to controls were found in occipital cortex and cerebellum.

Deteriorated c-fos expression in the individual brain regions may be linked to increased apoptosis and neurodegeneration, overexcitation by excitatory amino acids or reactive oxygen species.

Introduction

C-fos is a protooncogene and transcription factor with putatively pivotal functions in the brain. c-fos induction has been shown to occur following stimulation with excitatory amino acids (EAAs): Cole et al. (1989) reported that synaptic NMDA receptor activation led to a rapid increase of immediate early gene messenger RNAs in hippocampal neurons and Giovanelli and coworkers (1998) demonstrated that c-fos was induced by quisqualate excitotoxic lesions in the rat. Contestabile and coworkers (1998) showed induction of c-fos following kainic acid application to rats, a fact which has been already addressed by Schreiber et al. (1993).

Elevated c-Fos expression followed rapidly after the addition of glutamate to pyramidal neurons of the hippocampus (Figiel, 1997). Hasegawa and coworkers (1998) showed recently that activation of NMDA receptors induced c-fos and hsp 70 mRNA in neonatal rat cortical slices thus confirming that NMDA activates c-fos expression (Yee, 1993).

Overexcitation by EAAs is known to lead to neuronal apoptosis and the role for c-fos in programmed cell death is now widely accepted (Marcus et al., 1998; Walton et al., 1998; Hasegawa et al., 1998; Smeyne et al., 1993).

The induction of c-fos by active oxygen species has been shown by Richter-Landsberg and Vollgraf recently (1998): Brief exposure to H2O2 led to apoptotic cell death accompanied by increased transcriptional activity of c-fos and c-jun.

Furthermore, the involvement of c-fos for neuronal plasticity and long term potentiation (LTP), the theoretical and experimental basis for memory and learning, is still a matter of controversy (Lanahan et al., 1997; Roberts et al., 1996; Worley et al., 1993; Jefferey et al., 1990; Dragunow et al., 1989).

The postulated presence of apoptosis, overexcitation, oxidative stress and impaired neuronal plasticity in brain of patients with Down Syndrome (DS), all linked to c-fos expression, formed the Rationale for our study.

It was the aim of our investigation to determine mRNA steady state levels of c-fos in five individual brain regions of patients with DS in comparison to control brain.

Materials and methods

Patients

5 different brain regions including the frontal lobe, parietal lobe, temporal lobe, occipital lobe and the cerebellum of 9 adult Down Syndrome individuals (n = 9; 3 females, 6 male; 56,1 ± 7, 1 years) and 9 adult control persons without any known neurological and psychiatric disorder (n = 9; 5 females, 4 males; 72,6 ± 9, 6 years) were analysed (see Table 1).

Datas on agonal state, postmortem and storage condition are described in Seidl (1997).

Total RNA was isolated by using RNAzol B following the protocol by Chomczynski and Sacchi (1987) using an acid guanidinium thiocyanate-phenol-chloroform-extraction method and precipitation with isopropanol. Total RNA was treated with DNAse I for 15min at room temperature, reextracted, air-dried and resuspended in RNAse free water. Quantification of total RNA was done by UV absorbance spectrrophotometry at 260/280.

Reagents

— RNAzol B: Molecular Research Center, Inc., Cincinnati, OH
— pSP64 Poly(A) Vector: Promega Corporation, Madison, WI, USA
— TA Cloning Kit: INFaF' from Invitrogen Corporation, San Diego, CA, USA
— restriction enzymes and Rapid DNA Ligation Kit: Boehringer Mannheim GmbH, Mannheim, Germany
— Oligo(dt) Cellulose Columns, DNA and RNA molecular weight standards, yeast tRNA and cRNA SP6 transcription kit — SP6 Polymerase, DDT, buffer from Life Technologies, Gibco BRL, Maryland, USA

— cDNA transcription reagents — RNase Inhibitor, Oligo d(T)$_{16}$, dNTPs, MuLV Reverse Transcriptase, buffer and PCR reagents — dNTPs, primers, MgCl$_2$, AmpliTaq DNA Polymerase, buffer: PE Applied Biosystems, Foster City, CA, USA and Roche Molecular System, Inc, Branchburg, NeW Jersey, USA
— GELase TM Agarose Gel-Digestion Preparation Epicentre Technologies, Madison, WI, USA, PCR purification kit: Qiagen Inc, Santa Clarita, CA, USA
— oligosynthesis: Applied Biosystems GmbH, Weiterstadt, Germany
— Chemicals as phenol, chloroform, isopropanol, different salts, agarose were purchased from Sigma Chemical Co
— Reagents as Formamide, POP4, capillaries, 10xbuffer with EDTA and TAMRA 500 for the ABI Prism 310 Genetic Analyzer: PE Applied Biosystems, Foster City, CA, USA

Instruments

PCR was performed on a Perkin — Elmer GeneAmp PCR System 9,600 and Perkin — Elmer CeneAmp PCR System 2,400.

The PCR products labeled with FAM were analysed on the ABI-310 Genetic Analyzer from Applied Biosystem containing a 47 cm × 50 μm long capillary with Performance Optimized Polymer 4 (POP4) and as matrix standard TAMRA 500.

Syntheis of RNA internal standard for the protooncogene c-fos and the house keeping gene beta-actin

We performed a synthetic RNA internal standard using a Competitor DNA — cloning strategy. Primer sequences were selected with the primer express program (Primer Express, PE Applied Biosystems, Foster City, CA, USA).

The internal standards for the human c-fos and beta-actin cDNAs are synthetic RNAs with lacking sequences of 13 and 41 bp respectively was compared to the wild-type sequences. Therefore serial PCR amplification and purification steps were performed.

For the synthesis of the c-fos internal standard following PCR amplification procedures were performed. Using the primerpair A and the primerpair B (see Table 2) and 1 μg of reverse transcribed human heart cDNA as template two DNA fragments of 530 bp and 620 bp length could be produced. Both DNA fragments, each with an overlapping sequence of 60 nucleotides were purified by extraction from a low melting agarose gel and thereafter linked by a PCR amplification procedure using the primerpair 1 (see Table 2). The forward primer was linked by a 5′ SP6 promotor sequence and the reverse primer by a 5′ poly T tail sequence to perform in vitro RNA synthesis. After gel extraction the 1,169 bp fragment was used as template for in vitro synthetic RNA transcription.

For the synthesis of the beta-actin internal standard following PCR amplification procedures were performed. The first PCR amplification reaction produced with primerpair 1 (see Table 3 — primerpair 1) and 1 μg of reverse transcribed human heart cDNA as template a 1,600 bp fragment of the house keeping gene beta-actin cDNA. After treatment with the Qiagen PCR purification kit 100 ng of the 1,006 bp fragment were used to perform in two different PCR amplification reactions a 975 bp and a 624 bp fragment, each with an overlapping sequence of 39 nucleotides by using a reverse primer with a 20 bp linkage and a forward primer with a 19 bp linkage respectively (see Table 3 — primerpair A and primerpair B). The third step PCR amplification reaction was performed with primerpair 2 (see Table 3) by 5′ linkage of the SP6 promotor and poly T

Table 1

Patients	Sex	Age (years)	Postmortem Interval (h)	Agonal State	cfos mRNA fg./10 ng total RNA Frontal Lobe	cfos mRNA fg./100 pg beta-Actin Frontal Lobe	cfos mRNA fg./10 ng total RNA Parietal Lobe
Morbus Down							
DS1	F	47 years	24 h	Bronchopneumonia			
DS2	M	46 years	72 h	Bronchopneumonia	3,7	160,9	3,2
DS3	M	57 years	31 h	Bronchopneumonia	5,6	88,9	
DS4	M	56 years	36 h	Bronchopneumonia	25,2	66,3	24,3
DS5	M	69 years	30 h	Bronchopneumonia	18,8	13,6	
DS6	F	54 years	24 h	Pulmonary Embolus			10,8
DS7	F	53 years	48 h	N.A.	10,6	30	5,8
DS8	M	65 years	9 h	Bronchopneumonia	8,2	105,1	5,1
DS9	M	54 years	9 h	Bronchopneumonia			35,3
Controls							
C1	F	53 years	83 h	Ischaemic Heart Disease	12,7	7,1	9,7
C2	M	48 years	59 h	Ruptured Aortic Aneurysm	10,1	25,3	8,5
C3	F	64 years	58 h	Coronary Artery Occlusion			23,8
C4	M	64 years	48 h	Acute Pumonary Oedema	58,1	29,5	47,8
C5	M	63 years	26 h	Coronary Artery Occlusion			189
C6	M	51 years	15 h	Chron. Obst. Airways Disease	5,6	12,8	18,6
C7	F	32 years	22 h	Pulmonary Embolus	9,6	11,8	8,4
C8	M	55 years	31 h	Acute Myocardial Infarction	17,5	14,6	8,9
C9	M	58 years	22 h	Myocardial Infaction	52,2	17,2	5,6

tail sequences to produce a 1,277 bp fragment. After gel extraction the 1,277 bp fragment was used as template for in vitro synthetic RNA transcription.

Extraction of the DNA fragments was performed with the GELase Agarose Gel-Digesting Preparation Kit^R from Epicentre Technologies after separation on a 1% low melting agarose gel and quantification by uv absorbance spectrophotometry.

Following PCR reagents and conditions were used:

10x reaction buffer II from Perkin Elmer, 2 mM MgCl$_2$, 150 μM each dNTP, 300 nM each primer and 1 U Taq Polymerase in a 50 μl reaction volume. The amplification procedure was performed in a Perkin Elmer thermocycler 9600 and 2400 starting with a 3 min. denaturing step at 94°C, followed by a 35 cycle step with denaturing step for 30 sec. at 94°C, annealing step for 30 sec at 56°C and elongation step for 30 or 60 sec. at 72°C and finished by an elongation step for 7 min. at 72°C.

We confirmed the correct sequences of the ets-2 and beta-actin internal standards by direct sequencing with fluorescent labeled nucleotides using the ABI Prism 310 Genetic Analyzer from Perkin Elmer.

In vitro RNA-synthesis of the c-fos internal standard and beta-actin internal standard

For in vitro RNA synthesis 5x Gibco buffer (200 mM Tris-HCl, 30 mM MgCl$_2$, 10 mM spermidine hydrochloride), 10 mM DTT, 20 μM each rNTP, 10 U human placental RNase

Table 1. *Continued*

cfos mRNA fg./100 pg beta-Actin Parietal Lobe	cfos mRNA fg./10 ng total RNA Temporal Lobe	cfos mRNA fg./100 pg beta-Actin Temporal Lobe	cfos mRNA fg./10 ng total RNA Occipital Lobe	cfos mRNA fg./100 pg beta-Actin Occipital Lobe	cfos mRNA fg./10 ng total RNA Cerebellum	cfos mRNA fg./100 pg beta-Actin Cerebellum
	0,16	10,9	4,2	123,5		
200			4,5	13	3,6	12,8
			6,4	108,5		
168,8						
	3,9	22,3			35,9	114,3
36,7	13,3	19,9	15,1	129,1	25,3	12,5
15,9	9,2	12,7				
53,7	1,6	108,1			8	25,3
91,2	45,3	30,7	65,8	55		
6,7	6,6	5,9	53,1	13,9	7,4	6
21,9					4,1	27,5
16,5			27,2	38,9	201,7	123
22,7	11,1	12,2				
98,9	0,35	0,2	58,8	69		
15,3	1,1	5,1	29,2	40,3		
8,2	0,05	0,15	3,7	14,1	180,2	189,7
14,5						
9,3	2,7	1,2	15,5	68,6	37,1	24,6

Inhibitor and 35 U SP6 polymerase were mixed with 50–100 ng of the 1,169 bp fragment for the c-fos standard and 100 ng of the 1,277 bp fragment for the beta-actin standard and incubated in a 25 µl volume at 40°C for 1 hour. After treatment with RNase-free DNase I (Gibco BRL) for 15 min the in vitro synthesized RNA fragments were recovered by ethanol precipitation, resolved in DNase and RNase free water and aliquots were visualized on an 1% agarose gel electrophoresis by ethidiumbromide staining (fragment size was confirmed by comparison with a bp RNA ladder). Thereafter the synthetic RNA fragments were purified by oligo(dt) affinity chromatography using Oligo(dt) Cellulose Columns (Gibco BRL). Quantity and purity of the synthetic RNAs were determined by OD 260/280 absorbance spectrophotometry.

Reverse transcription PCR (RT-PCR) amplification from the synthetic RNAs were confirmed by PCR amplification reaction with an RNAse-treated sample used as a negative control to ensure no contamination of any DNA template.

Several synthetic RNA aliquots ranging from 222 to 0,02 femtogram/µl for the c-fos internal standard and from 500 to 5 picogram/µl for the beta-actin internal standard were diluted in yeast tRNA (1 mg/ml) in silanised tubes (silanization solution SERVA) to prevent RNA adhesion to plasticware. Several PCR reactions performed with yeast tRNA as template were negative and therefore interferences between the synthetic RNA and yeast tRNA during transcription and amplification could be excluded. Storage was performed at −80°C for 3 months maximum.

S. Greber-Platzer et al.

Table 2. Human c-FOS primerpairs

Primerpair		Pimersequence	cDNA position of the Primer	Length of the amplified segment (bp)
Primerpair A	forward	5′GGCTTCAACGCAGACTACGA 3′	166 bp	
	reverse	5′TTGGCAATCTCGGTCTGCAAAG CAGACTTCTCATCTTCTCCGCT TGGAGTGTATCAGTCA 3′	655 bp	530 bp
Primerpair B	forward	5′TGACTGATACACTCCAAGCGGA GAAGATGAGAAGTCTGCTTTGC AGACCGAGATTGCCAA 3′	689 bp	620 bp
	reverse	5′AGCGAGTCAGAGGAAGGCTCAT 3′	1,268 bp	
Primerpair 1	forward	5′linked SP6 promotor 5′GTATCATACACATACGATTTAGG TGACACTATAGAAGGCTTCAACG CAGACTACGA 3′	166 bp	wildtype 1,182 bp/ internal standard 1,169 bp
	reverse	5′linked poly T 5′GAATTCGGTTTTTTTTTTTTTTTT TTTTTTTTTTTGGGAGAGCGAG TCAGAGGAAGGCTCAT 3′	1,268 bp	
Primerpair 2	forward	5′CTGGCGTTGTGAAGACCATGA 3′	479 bp	wildtype 235 bp/ Internal standard 222 bp
	reverse	5′AGCAGGTTGGCAATCTCGGT 3′	713 bp	

Table 3. Human beta-actin primerpairs

Primerpair		Primersequence	cDNA position of the primer	Length of amplified segment (bp)
Primerpair 1	forward	5′-GCCAGCTCACCATGGATGAT-3′	31 bp	
	reverse	5′-GCACGAAGGCTCATCATTCA-3′	1,612 bp	1,601 bp
primerpair A	forward	5′-GCCAGCTCACCATGGATGAT-3′	31 bp	
	reverse	5′-CGCTCAGGAGGAGCAATGAT TCTGCATCCTGTCGGCAAT-3′	966 bp	974 bp
primerpair B	forward	5′-ATTGCCGACAGGATGCAGA ATCATTGCTCCTCCTGAGCG-3′	1,026 bp	625 bp
	reverse	5′-GCACGAAGGCTCATCATTCA-3′	1,612 bp	
primerpair 2	forward	5′linked SP6 promotor 5′-GTATCATACACATACGATTTA GGTGACACTATAGAACCATGTA CGTTGCTATCCAGGC-3′	433 bp	wildtype 1,277 bp/ internal standard 1,236
	reverse	5′linked poly T 5′-GAATTCGGTTTTTTTTTTTTTTT TTTTTTTTTTTTTTTGGGAGGC ACGAAGGCTCATCATTC-3′	1,612 bp	
pimerpair 3 — FAM labeled				
	forward	5′-CCATCATGAAGTGTGACGTG-3′	883 bp	wildtype 225 bp/ internal standard 184 bp
	reverse	5′-ACATCTGCTGGAAGGTGGAC-3′	1,087 bp	

Competitive RT-PCR

Serial dilutions of the synthetic RNAs for c-fos internal standard (22,2 to 0,02 femtgram/μl) and beta-actin internal standard (500 picogram/ml to 5 picogram/μl) were added to equal concentrations of total RNA (10 ng) extracted from 5 different brain regions of DS patients and control persons in series of RT-PCR reaction mixtures. We could confirm linearity for all samples. Reverse transcription was performed in a 20 μl volume containing 4 mM MgCl₂, PCR Buffer II from Perkin Elmer, 1 mM dGTP, 1 mM dATP, 1 mM dTTP, 1 mM dCTP, 1 U/μL RNase Inhibitor, 2.5 μM Oligo d(T) 16 and 2.5 U/μL Moloney murine leukemia virus reverse transcriptase (all components from Perkin-Elmer). The temperature profile started at room temperature for 10 minutes, followed by 15 minutes at 42°C, heat inactivation of the enzyme at 99°C for 5 minutes and cooled to 5°C for 5 minutes. Immediately 5 μl of the produced cDNA were used for direct labeled PCR reaction. 15 μl were stored at −80°C.

PCR amplification reaction was performed for c-fos quantification in a 50 μl volume. The PCR reation mixture contained PCR buffer II from Perkin Elmer, 2 mM MgCl₂, 150 μM each dNTP, 200 nM each primer (see Table 2 — primerpair 2: forward FAM-labeled 5′CTGGCGTTGTGAAGACCATGA 3′ and reverse 5′AGCAGGTTGG-CAATCTCGGT 3′) and 1 U AmpliTaq DNA polymerase and produced a 235 bp fragment for the wild-type c-fos and a 222 bp fragment for the internal standard. The amplification temperature profile started with denaturing for 3 min. at 94°C, followed by 25,30 and 35 cycles with a denaturing step for 30 sec at 94°C, annealling step for 30 sec at 56°C and elongation step for 30 sec at 72°C and finished by an elongation step for 7 min. at 72°C.

For the beta-actin PCR amplification reaction the mixture contained buffer II from Perkin Elmer, 2 mM MgCl₂, 150 μM each dNTP and 600 nM each primer (see table 3 — primerpair 3: forward primer 5′ FAM-labeled 5′CCATCATGAAGTGTGACGTG 3′ and reverse primer 5′ACATCTGCTGGAAGGTGGAC 3′) and 1 U AmpliTaq DNA polymerase and produced a 225 bp fragment for the wild-type beta-actin cDNA and 1 184 bp fragment for the synthetic cDNA standard. Temperature profile consisted of 25 cycles and followed the above mentioned protocol for the c-fos ampliciation procedure except the annealing temperature time was increased to 60°C instead of 56°C.

1 μl of each nested PCR product was solved in 20 μl formamide containing 0,5 μl TAMRA standard solution from Perkin Elmer for quantification on the ABI PRISM 310 genetic analyzer of Perkin Elmer.

Results

We have determined c-fos and beta-actin mRNA levels in frontal, parietal, temporal, occipital lobe and cerebellum of 9 adult DS and 9 control persons. Mean c-fos mRNA levels (SD) are shown in Fig. 1 as values in the femtogram range per 10 ng total RNA and in Fig. 2 as normalized values versus 100 picogram beta-actin mRNA, a house keeping gene.

We could find the lowest cfos mRNA levels in the temporal lobe of the control group (3,65 +/−4,37 fg cfos/10 ng total RNA; 4,13 +/−4,67 fg cfos/100 pg beta-actin mRNA).

To analyze statistical significances we used the Mann Whitney test between the DS group and control group for all 5 brain regions.

In frontal (p = 0,014), parietal (p = 0,0176) and temporal (p = 0,023) lobe significant increases of cfos mRNA/100 pg beta-actin mRNA were shown in

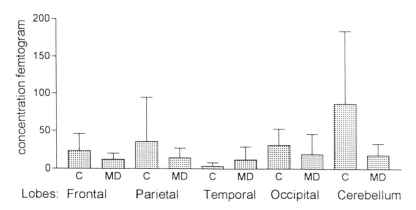

Fig. 1. Mean c-fos mRNA level ± SD in 10 ng total RNA of 5 different regions in DS patients (MD) and the control group (C). The Mann Whitney test has shown no statistically significant difference in all brain regions

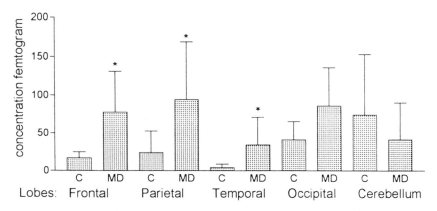

Fig. 2. Mean c-fos mRNA level ± SD in 100 pg beta-actin of 5 different brain regions in DS patients (MD) and the control group (C). The Mann Whitney test has shown statistically significant increases of c-fos mRNA levels in the frontal lobe (p = 0,014), parietal lobe (p = 0,0176) and temporal lobe (p = 0,0023) in DS patients (*). There are no statistically significant differences in the occipital lobe and cerebellum

the DS group. There were no significant differences in these brain lobes, when c-fos levels were related to 10 ng total RNA.

Comparable c-fos levels in the DS and control group were found in occipital lobe and cerebellum, when either related to 10 ng total RNA or to 100 pg beta-actin mRNA.

Discussion

As shown in the Results c-fos mRNA normalized versus mRNA of the housekeeping gene beta-actin showed variable expression in the individual brain regions of controls. Comparing control brain regions with DS brain regions, c-fos expression was significantly increased in temporal, parietal and frontal cortex of DS brain, comparable in occipital cortex and cerebellum.

As values were normalized versus the housekeeping gene beta-actin, cell loss in brain of patients with DS should have been respected and compensated. However, when c-fos mRNA steady state levels were expressed versus total RNA, no significant changes between DS and control brain regions could be detected. The expression of c-fos versus beta-actin is a more realistic reflection as that versus total RNA. Methodologically, total RNA would not reflect cell loss but would be highly dependent on total RNA composition, handling and metabolism by the cell, a factor wholly unknown in both groups.

We cannot assign altered c-fos levels to DS pathology specifically as c-fos changes may be very well reflecting neurodegeneration per se: Marcus and coworkers (1998) studied c-fos protein expression in hippocampal areas of Alzheimer's disease (AD) brain and demonstrated an increase in c-fos and c-jun density, although MacGibbon and coworkers did not find any significant changes of a series of transcription factors including c-fos at the protein level (1997), whereas Zhang and coworkers (1992) found increased c-fos staining in hippocampus of AD brain.

Our finding of increased c-fos mRNA steady state levels of temporal, parietal and frontal cortex may be in line with observations in AD: All our adult DS patients examined in this study show AD — neuropathological changes and c-fos overexpression may be due to abundant amyloid beta-peptide. Mouse hippocampal cell lines exposed to amyloid beta-peptide fragment caused a rapid and sustained increase in c-Fos immunoreactivity (Gillardon et al., 1996). On the other hand, there was an even distribution of amyloid pathology in our brain samples of patients with DS and this finding is in contradiction with c-fos in occipital cortex and cerebellum.

Although the active oxygen species — hypothesis for the development of neurodegeneration in DS would be helpful for the interpretation of increased temporal, parietal and frontal cortex — c-fos expression, we have shown in a comparable cohort of DS patients that in postmortem brain no parameters for increased oxidative stress were found (Hayn et al., 1996; Seidl et al., 1997).

Apoptosis, well-known and well-documented to occur in DS brain, could contribute to increased c-fos cortex levels in DS but the identical arguments as given above are not supporting this mechanism. Increased excitatory amino acids have been repeatedly reported in several regions and areas of DS brain (Risser et al., 1997) and this in turn could be responsible for the observed increase of cortical c-fos expression in DS.

In the brain c-fos mainly reflects neuronal activity and there is some support for that: Changes in synaptic plasticity of DS brain may be the major source for activation of c-fos expression and synaptic plasticity of the brain is unequivocally deteriorated in DS. This would be in agreement with observations by Ryabinin (1998) revealing that deterioration of learning, i.e. synaptic plasticity, in behavioral paradigms is associated with the induction of c-fos and other immediate early genes.

Studies referenced in a review by Herrera (1996) on the activation of c-fos in the brain would be supportive for the hypothesis that c-fos in the brain is neurone-derived, in addition.

Here, we provided data revealing altered c-fos gene expression in DS brain. The underlying mechanisms for increased c-fos expression in temporal, parietal and frontal cortex range from neurodegeneration, apoptosis to increased EAA levels. The reason for comparable levels of c-fos in cerebellum and occipital lobe remain unclear, as they show neuropathological lesions comparable to other regions in our cohort.

We are currently in the process of investigating matched brain samples of AD in order to rule out "AD-neurodegeneration" found in our DS brain samples, per se.

Acknowledgement

We are highly indebted to the Red Bull Company, Salzburg, Austria, for generous financial support.

References

Chomczynski P, Sacchi N (1987) Single — step method of RNA isolation by acid guanidinium thiocyanate-phenol-chloroform extraction. Anal Biochem 162: 156–159

Cole AJ, Saffen DW, Baraban JM, Worley PF (1989) Rapid increase of an immediate early gene messenger RNA in hippocampal neurons by synaptic NMDA receptor activation. Nature 340: 474–476

Contestabile A, Ciani E, Sparapani M, Guarnieri T, Dell'Erba G, Bologna F, Cicognani C (1998) Activation of the ornithin decarboxylase-polyamine system and induction of c-fos and p53 expression in relation to excitotoxic neuronal apoptosis in normal and microencephalic rats. Exp Brain Res 120: 519–526

Dragunow M Abraham WC, Goulding M, Mason SE, Robertson HA, Faull RL (1989) Long-term potentiation and the induction of c-fos mRNA and proteins in the dentate gyrus of unanaesthetized rats. Neurosci Lett 101: 274–280

Figiel I, Kacmarek L (1997) Cellular and molecular correlates of glutamate-evoked neuronal programmed cell death in the in vitro cultures of rat hippocampal dentate gyrus. Neurochem Int 31: 229–240

Gillardon F, Skutella T, Uhlmann E, Holsboer F, Zimmermannn M, Behl C (1996) Activation of c-fos contributes to amyloid beta-peptide — induced neurotoxicity. Brain Res 706: 169–172

Giovanelli L, Casamenti F, Pepeu G (1998) C-fos expression in the rat nucleus basalis upon excitotoxic lesion with quisqualic acid: a study in adult and aged animals. J Neural Transm 105: 935–948

Hasegawa K, Litt L, Espanol MT, Sharp FR, Chan PH (1998) Expression of c-fos and hsp70 mRNA in neonatal rat cerebrocortical slices during NMDA-induced necrosis and apoptosis. Brain Res 785: 262–278

Hayn M, Kremser K, Singewald N, Cairns N, Nemethova M, Lubec B, Lubec G (1996) Evidence against the involvement of reactive oxygen species in the pathogenesis of neuronal death in Down's syndrome and Alzheimer's disease. Life Sci 59: 537–544

Herrera DG, Robertson HA (1996) Activation of c-fos in the brain. Progr Neurobiol 50: 83–107

Jefferey KJ, Abraham WC, Dragunow M, Mason SE (1990) Induction of Fos-like immunoreactivity and the maintenance of long-term potentiation in the dentate gyrus of unanaesthetized rats. Brain Res Mol Brain Res 8: 267–274

Lanahan A, Lyford G, Stevenson GS, Worley PF, Barnes CA (1997) Selective alteration of long term potentiation-induced transcripütional response in hippocampus of aged, memory — impaired rats. J Neurosci 17: 2876–2885

MacGibbon GA, Lawlor PA, Walton M, Sirimanne E, Faull RL, Synek B, Mee E, Connor B, Dragunow M (1997) Expression of Fos, Jun, and Krox family proteins in Alzheimer's disease. Exp Neurol 147: 316–332

Marcus DL, Strafaci JA, Miller DC, Masia S, Thomas CG, Rosman J, Hussain S, Freedman ML (1998) Quantitative neuronal c-fos and c-jun expression in Alzheimer's disease. Neurobiol Aging 19: 393–400

Richter-Landsberg C, Vollgraf U (1998) Mode of cell injury and death after hydrogen peroxide exposure in cultured oligodendroglia cells. Exp Cell Res 244: 218–219

Risser D, Lubec G, Cairns N, Herrera-Marschitz M (1997) Excitatory amino acids and monoamines in parahippocampal gyrus and frontal cortical pole of adults with Down syndrome. Life Sci 60: 1231–1237

Roberts LA, Higgins MJ, O'Shaugnessy CT, Stone TW, Morris BJ (1996) Changes in hippocampal gene expression associated with the induction of long-term potentiation. Brain Res-Mol Brain Res 42: 123–127

Ryabinin AE (1998) Role of hippocampus in alcohol-induced memory impairment: implications from behavioral and immediate early gene studies. Psychopharmacology 139: 34–43

Schreiber SS, Najm I, Tocco G, Baudry M (1993) Co-expression of HSP72 and c-fos in rat brain following kainic acid treatment. Neuroreport 5: 269–272

Seidl R, Greber S, Schuller E, Bernert G, Cairns N, Lubec G (1997) Evidence against increased oxidative DNA-damage in Down Syndrome. Neurosci Lett 235: 137–140

Smeyne RJ, Vendrell M, Hayward M, Baker SJ, Miao GG, Schilling K, Robertson LM, Curran T, Morgan JI (1993) Continuous c-fos expression precedes programmed cell death in vivo. Nature 363: 166–169

Walton M, MacGibbon G, Young D, Sirimanne E, Williams C, Gluckmann P, Dragunow M (1998) Do c-jun, c-Fos and amyloid precursor protein play a role in neuronal death or survival? J Neurosci Res 53: 330–342

Worley PF, Bhat RV, Baraban JM, Erickson CA, McNaughton BL, Barnes CA (1993) Thresholds for synaptic activation of transcription factors in hippocampus: correlation with long-term enhancement. J Neurosci 13: 4776–4786

Yee WM, Frim DM, Isacson O (1993) Relationship between stress protein induction and NMDA-mediated neuronal death in the entorhinal cortex. Exp Brain Res 94: 193–202

Zhang P, Hirsch EC, Damier P, Duyckaerts C, Javoy-Agid F (1992) c-fos protein-like immunoreactivity: distribution in the human brain and over-expression in the hippocampus of patients with Alzheimer's disease. Neuroscience 46: 9–21

Authors' address: Prof. Dr. G. Lubec, University of Vienna, Department of Pediatrics, Währinger Gürtel 18, A-1090 Vienna, Austria e-mail: gert.lubec@akh-wien.ac.at

J Neural Transm (1999) [Suppl] 57: 87–97

Neuronal cell death in Down's syndrome

A. Sawa

Department of Neuroscience, Johns Hopkins University School of Medicine,
Baltimore, MD, USA

Summary. Down's syndrome (DS), occurring in 0.8 out of 1,000 live births, is a genetic disorder in which an extra portion of chromosome 21 leads to several abnormalities. With respect to the nervous system, it causes mental retardation. It is conceived that abnormal neuronal cell death in development is involved, but there is no direct evidence yet. In addition to developmental brain abnormalities, almost all DS brains over 40 years old manifest a similar pathology to Alzheimer's disease (AD), including the presence of senile plaques (SP) and neurofibrillary tangles (NFT). Although there was a debate to segregate dementia from underlying mental retardation, at least some portion of DS patients exhibit deteriorated mental status with aging. The mechanism underlying these abnormalities at the molecular level remains to be elucidated. Recently there have been several reports suggesting abnormalities reflecting increased risk to apoptosis in DS brains. Increased expression of several apoptosis-related genes (p53, fas, ratio of bax to bcl-2, GAPDH) in DS brains has been reported. Cultured neurons from both patients and model animals are reportedly more vulnerable to apoptosis. Overproduction of reactive oxygen species and its causative roles for increased apoptosis in DS tissues are suggested. One possible hypothesis is an increased susceptibility to apoptosis due to p53 overactivation in DS brains. $A\beta42$, a critical peptide for AD pathology from amyloid precursor protein (APP), can be detected in DS brains. $A\beta42$ is deposited in SP from an early stage, suggesting common molecular mechanisms in DS and AD. Animal models for DS are important in the search of molecular mechanisms. Several types of models are now available. Future DS studies are expected to integrate information from animal models and human tissues.

Abbreviations

AD Alzheimer's disease, *APP* amyloid precursor protein, *DS* Down's syndrome, *GAPDH* glyceraldehyde-3-phosphate dehydrogenase, *HD* Huntington's disease, *NFT* neurofibrillary tangle, *ROS* reactive oxygen species, *RT-PCR* reverse trascriptase-coupled polymerase chain reaction,

SAGE serial analysis of gene expression, *SOD* superoxide dismutase, *SP* senile plaque.

Introduction

DS patients display several abnormalities in various tissues (Epstein, 1986). They manifest typical facial features such as a round face and a short nose with a flat nasal bridge. Skeletal abnormalities include a short stature, as well as stubby fingers and toes. Increased incidence of congenital heart diseases, such as septal and endocardial cushion defects, and increased risk of leukemia are clinically important. Above all, abnormalities of the nervous system are most prominent, including mental retardation and dementia (Coyle et al., 1986; Lai and Williams, 1989; Schapiro et al., 1992). The clinical difficulty in separating dementia from underlying metal retardation has caused debate in the past. It is currently believed that at least some portion (40%) of DS patients show dementia (Lai and Williams, 1989; Schapiro et al., 1992). Regardless of the existence of dementia, macroscopic analysis utilizing autopsy samples, CT or MRI studies (Pearlson et al., 1998) has clarified that DS brains are smaller than normal subjects in various areas. The difference in cerebellar volume between DS and normal brains is most prominent, but this volume reduction in the DS cerebellum is not age-related (Aylward et al., 1997). Comparison of DS with dementia to DS without dementia revealed that additional atrophy related to dementia also occurs macroscopically (Lai and Williams, 1989; de la Monte and Hedley-Whyte, 1990; Schapiro et al., 1992).

Evidence that DS brains show AD-like pathology at molecular level

Two lines of evidence strongly suggest some common pathological mechanisms between DS and AD. First of all, most DS brains over 40 years old show senile plaques (SP) and neurofibrillary tangles (NFT), hallmarks of AD. Wisniewski and her colleagues (Wisniewski et al., 1985) conducted a detailed histopathological analysis of the brains of one hundred DS cases of various ages. According to their report, all aged DS brains exhibit AD pathology (SP, NFT). This data has been recapitulated by several groups (Mann and Esiri, 1988; Rumble et al., 1989). The main component of SP is the Aβ peptide, which is cleaved from amyloid precursor protein (APP) (Goldgaber et al., 1987; Kang et al., 1987; Tanzi et al., 1987). In DS brains, Aβ42, a longer form of Aβ crucial to AD pathology, is detected (Iwatsubo et al., 1995; Mann and Iwatsubo, 1996). It deposits in SP from an early stage of its pathology. Secondly, some types of early-onset familial AD have mutations in the APP gene, which is located on chromosome 21(Goate et al., 1991). Taken together, the derangement of the APP gene result in abnormal metabolisms of the Aβ peptide both in DS and AD.

AD has multiple etiologies. DS can be used as a good pathological model for AD with its advantage of homogeneity. One of the most suggestive findings using autopsy brains of various ages is that SP formation precedes NFT in DS brains (Mann and Esiri, 1989). This pathological sequence might be a good model for AD studies. Although DS and AD seem to share pathological mechanisms at least in SP formation, it is not yet clear whether or not they share common mechanisms of neuronal cell death.

Evidence that DS brains show abnormalities reflecting increased risks of apoptosis at molecular level

DS brains are small in several regions (Coyle et al., 1986). There is no direct evidence that these changes are the outcome of increased neuronal cell death in development and aging. Several recent reports suggest that there are abnormalities linked to increased risks of apoptosis in DS brains.

The first approach utilizes primary neurons from DS fetal brains (Busciglio and Yankner, 1995). Cortical neurons from fetal DS and age-matched normal brains differentiate normally in culture, but DS neurons subsequently degenerate and undergo apoptosis. DS neurons exhibit a three- to fourfold increase in intracellular reactive oxygen species (ROS) and elevated levels of lipid peroxidation that preceed neuronal death.

The second approach is utilizing autopsy brains from DS patients (Brooksbank and Balazs, 1984; Hayn et al., 1996; Sawa et al., 1997a; de la Monte et al., 1998; Labudova et al., 1998; Lubec et al., 1999; Seidl et al., 1999). Our group evaluated bcl-2 gene family expression (Sawa et al., 1997a). Currently, mitochondria are recognized as the site of regulation for the execution of cell death initiation, especially apoptosis (Green and Reed, 1998). Bcl-2 family gene products exist on mitochondria membranes (Adams and Cory, 1998). Bax promotes apoptosis by regulating the release of cytochrome c from mitochondria. Bcl-2 and Bcl-X protect against apoptosis. Using quantitative reverse trascriptase-coupled polymerase chain reaction (RT-PCR), these three gene expressions are evaluated in frontal, temporal, and occipital lobes, respectively. Seven DS samples are compared with 9 controls. We concluded that bax expression is constant, but that bcl-2 and bcl-X are decreased in DS. We standardized bax, bcl-2, and bcl-X expression by glyceraldehyde-3-phosphate dehydrogenase (GAPDH) expression, which had been assumed to be a marker of constant expression. However, recently we have found GAPDH expression is dramatically increased in some types of apoptosis in cultured cells (Sawa et al., 1997b). In fact, increased GAPDH expression is reported in DS brains this year (Lubec et al., 1999). Thus our original results can be interpreted that bax expression is elevated and bcl-2 and bcl-X are constant or decreased. We also examined the ratios of bax to bcl-2 and bax to bcl-X, which were not influenced by GAPDH expression in our original study (Fig. 1). These ratios can be regarded as apoptotic indices. Both ratios were increased in DS brains, especially in the temporal lobes. Two groups reported that P53 and Fas/APO-1 expressions are elevated in DS brains (de la Monte et al., 1998; Seidl et al., 1999). These data from postmortem brains can be interpreted rather consistently as an overactivation of p53 in DS brains. P53 can activate apoptosis-related genes. The expression

bax /bcl-2

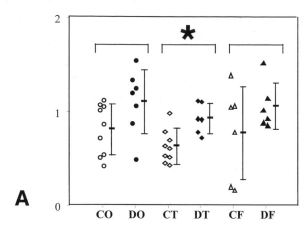

A

bax / bcl-x

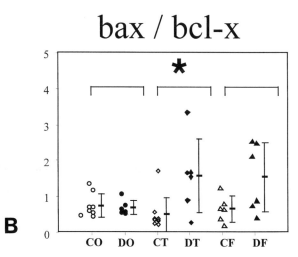

B

Fig. 1. Apoptotic indices, bax / bcl-2 (**A**) and bax/bcl-x (**B**), in various brain areas from DS and control brains. The expression levels of these genes were quantified by quantitative reverse transcriptase-coupled polymerase chain reaction (quantitative RT-PCR). The apoptotic indices tend to be elevated in DS brains. The increases in the values were the most prominent in the DS temporal lobes, that is, bax / bcl-2, *P < 0.01 (significant); bax / bcl-x, #P = 0.593 (highly suggestive). The data were subjected to statistical analysis using Mann-Whitney's U test. The bars indicate the mean with standard deviation. *CO* control occipital; *DO* DS occipital; *CT* control temporal; *DT* DS temporal; *CF* control frontal; *DF* DS frontal (from Sawa et al., 1997a, with permission)

of Fas is augmented by p53 in some systems (Bennett et al., 1998; Muller et al., 1998). Bax is under transcriptional control of p53 (Miyashita and Reed, 1995). Polyak and her colleagues (Polyak et al., 1997) screened genes transactivated by p53 induction through their original differential expression assay named serial analysis of gene expression (SAGE). Most of the genes (called PIGs, for p53-induced genes) are related to oxidative stress. Among them, they particularly examined PIG-3, and found ROS production follows PIG-3 overexpression. These observations coincide with the overproduction of ROS and lipid peroxidation in primary neurons from fetal DS brains. Biochemical measurements of lipoperoxidation in fetal brains of DS support this finding (Brooksbank and Balazs, 1984). But Hayn and his colleagues (Hayn et al., 1996) show no elevation in lipid peroxidation and SOD1 activity in aged DS patients. Using postmortem DS brains, decreased expression of transcription factor junD was recently reported (Labudova et al., 1998). Although Jun N-terminal kinase (JNK) has possible signaling linkages between junD and p53, it is not so clear if junD downregulation can be interpreted in the p53 paradigm or not.

If an apoptotic process is involved in DS brains, what is the cause(s) of this? The genetic abnormality in DS is the existence of an extra portion of chromosome 21. Which gene(s) of overdosage on chromosome 21 contributes to this process? To address this question, our group evaluated the correlation of the apoptotic indices to APP gene expression (Oyama et al., 1994; Sawa et al., 1997a). There was a significant correlation between APP expression and the apoptotic indices; The Spearman's correlation coefficients were 0.426 for APP and bax / bcl-2 ($P < 0.01$), and 0.404 for bax / bcl-2 ($P < 0.05$) (Fig. 2). No correlation was found between the indices and other genes in the brain including tau and MAP2 (data not shown). These results suggest that at least APP can be a candidate gene to promote an apoptotic cascade. The effect of APP overexpression is controversial. Some groups insist that increased expression of APP can cause apoptosis, especially with mutations found in familial AD (Yoshikawa et al., 1992; Yamatsuji et al., 1996). Based on the results obtained from DS patient tissues, the relationship between APP and p53 can be examined in future study.

If there is an increased risk of apoptosis in DS tissues, the next question is whether these changes are linked to a developmental abnormality or degeneration, probably related to AD pathology. In brain development, neuronal apoptosis plays a critical role in forming a proper network. In neurodegenerative diseases, apoptotic processes seem to be involved. Through application of in situ end labelling of DNA to autopsy brain sections (the TUNEL method), several groups have suggested the involvement of apoptosis in AD and Huntington's disease (HD) (Su et al., 1994; Dragunow et al., 1995; Lassmann et al., 1995). Utilizing peripheral tissues from patients, increased vulnerability to apoptosis was reported in HD (Sawa et al., 1999) and Friedreich's ataxia (Wong et al., 1999). Bcl-2 protein expression in AD brains has been examined by several groups, but without consensus (Satou et al., 1995; O'Barr et al., 1996; Su et al., 1996; Nagy and Esiri, 1997). The apoptotic indices (the ratios of bax to bcl-2 and bax to bcl-X at messenger RNA level) were not changed with respect to aging, suggesting weak relationship of the indices to AD pathology (Sawa et al., unpublished data). P53 and Fas/APO-1 expression were also examined in AD, and the results were similar to those of DS (de la Monte et al., 1997). Although the use of biochemical approaches in the search for pathogenesis in DS brains can be of great importance, some of the data so far might overlook that point if the changes are due to neuron or glial cells. Research trying to identify cell types in which the changes occur is expected.

A. Sawa

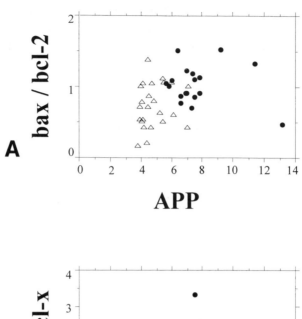

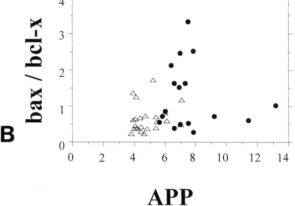

Fig. 2. Significant correlation between the APP expression and apoptotic indices. The expression level of APP in each sample was calculated by nuclease S1 protection analysis in our previous study. There was a significant correlation between APP expression and apoptotic indices; The Spearman's correlation coefficient were 0.426 for APP and bax / bcl-2 ($P < 0.01$), and 0.404 for bax / bcl-2 ($P < 0.05$). No correlation was found between the indices and other genes in brain including tau and MAP2 (data not shown) (from Sawa et al., 1997a, with permission)

Animal models for DS

Although we can get very important suggestive data from DS patient tissues, animal models are mandatory to clarify its molecular mechanisms.

The trisomy 16 (Ts16) mouse can be a potential genetic model of DS (Coyle et al., 1988). Mouse chromosome 16 (MMU-16) contains a cluster of genes and loci, including APP, CuZn-superoxide dismutase (SOD-1), which are also located on the long arm of human chromosome 21 (HSA21). Ts16 mice and DS patients share several abnormalities (Holtzman et al., 1992). Both have endocardial cushion defects, similar craniofacial malformations, and hematological abnormalities. Reduced size is a striking feature of the

Ts16 brain. How much this abnormality in Ts16 reflects DS pathology is not so clear. Although the Ts16 mice can't survive birth, analysis of several fruitful results have been reported by utilizing cultured neurons from a fetus, implantation of Ts16 mouse derived neurons into normal mice, and production of chimera mice. Altered Ca2+ signaling and mitochondrial deficiencies are found in hippocampal neurons of Ts 16 mice. This may be involved in a process of neurodegeneration (Schuchmann et al., 1998). Cultured hippocampal neurons from Ts 16 mouse exhibit decreased glutathione (reduced state) levels and augmented cell death when compared to diploid cells (Stabel-Burow et al., 1997). Cortical neurons from Ts 16 show increased apoptosis. This is prevented by treatment with anti-γ-interferon (Hallam and Maroun, 1998). The increased susceptibility to apoptosis in Ts16 neurons is parallel to the data from cultured neurons originally from fetal DS patients (Busciglio and Yankner, 1995). When Ts16 or control basal forebrain is transplanted into the hippocampus of young adult mice, transplanted neurons survive and grow neurites in all grafts. It is only in cholinergic neurons of Ts16 grafts, however, that selective atrophy is observed six months after transplantation (Holtzman et al., 1992). This atrophy is prevented by the addition of nerve growth factor (NGF) (Holtzman et al., 1993).

Ts65Dn was developed in 1990 (Davisson et al., 1990). It was generated through a reciprocal translocation, T(16C3–4; 17A2)65Dn, which can be used to produce segmental trisomy for distal chromosome 16. This mouse can survive after birth, enabling researchers to carry out detailed behavioral, morphological, and molecular analyses in both developing and mature subjects. The mice do not show early-onset of AD pathology. Ts65Dn mice do, however, demonstrate impaired performance in a complex learning task requiring the integration of visual and spatial information (Reeves et al., 1995). Ts65Dn brains show age-related degeneration of septohippocampal cholinergic neurons and astrocytic hypertrophy, markers of the AD pathology that is present in elderly DS individuals (Holtzman et al., 1996). Although the changes are not gross, Ts65Dn mice show developmental delay during the postnatal period as well as abnormal behaviors in both young and adult animals that may be analogous to mental retardation. The atrophy in the cerebellum is prominent in both Ts65Dn brains and in DS patients.

A newer DS model mouse, Ts1Cje, carries an extra copy of the MMU16 segment, smaller than that of Ts65D (Sago et al., 1998). The segment does not contain APP and SOD1 gene. The mouse exhibits learning and behavioral abnormalities, but no degeneration. Degeneration of basal forebrain cholinergic neurons, which was observed in Ts65Dn, was absent in Ts1Cje. This can be a good model for some clinical aspects of DS, but it may be inappropriate to seek for possible mechanisms of neuronal cell death in DS.

Recently yeast artificial chromosome (YAC) transgenic mice containing human DNA from 21q22.2 have been produced and exhibit impaired learning (Smith et al., 1997).

SOD1 and APP are located on chromosome 21, and they are believed to play important roles in cell death. Transgenic animals with SOD1 or APP gene can be good models for DS in some aspects (Avraham et al., 1988;

Games et al., 1995; Nalbantoglu et al., 1997). As apoptotic indices of bax to bcl-2 or bax to bcl-X correlate to APP in DS brains, transgenic mice with the APP gene may be of upmost importance. Depending on constructs, phenotypes of mice are different. Transgenic mice expressing the 104 amino acid carboxyl-terminal of APP show cell loss in the CA1 region of the hippocampus, and impaired learning and LTP (Nalbantoglu et al., 1997). Mice overexpressing the APP gene with a disease mutation show SP in the brain (Games et al., 1995).

Future DS studies are expected to integrate information from animal models and human tissues.

Acknowledgements

I thank Miss L. J. Hanle and Miss A. K. Kodaira for the preparation of the manuscript.

References

Adams JM, Cory S (1998) The Bcl-2 protein family: arbiters of cell survival. Science 281: 1322–1326

Avraham KB, Schickler M, Sapoznikov D, Yarom R, Groner Y (1988) Down's syndrome: abnormal neuromuscular junction in tongue of transgenic mice with elevated levels of human Cu/Zn-superoxide dismutase. Cell 54: 823–829

Aylward EH, Habbak R, Warren AC, Pulsifer MB, Barta PE, Jerram M, Pearlson GD (1997) Cerebellar volume in adults with Down syndrome. Arch Neurol 54: 209–212

Bennett M, Macdonald K, Chan SW, Luzio JP, Simari R, Weissberg P (1998) Cell surface trafficking of Fas: a rapid mechanism of p53-mediated apoptosis. Science 282: 290–293

Brooksbank BW, Balazs R (1984) Superoxide dismutase, glutathione peroxidase and lipoperoxidation in Down's syndrome fetal brain. Brain Res 318: 37–44

Busciglio J, Yankner BA (1995) Apoptosis and increased generation of reactive oxygen species in Down's syndrome neurons in vitro. Nature 378: 776–779

Coyle JT, Oster-Granite ML, Gearhart JD (1986) The neurobiologic consequences of Down syndrome. Brain Res Bull 16: 773–787

Coyle JT, Oster-Granite ML, Reeves RH, Gearhart JD (1988) Down syndrome, Alzheimer's disease and the trisomy 16 mouse. Trends Neurosci 11: 390–394

Davisson MT, Schmidt C, Akeson EC (1990) Segmental trisomy of murine chromosome 16: a new model system for studying Down syndrome. Prog Clin Biol Res 360: 263–280

de la Monte SM, Hedley-Whyte ET (1990) Small cerebral hemispheres in adults with Down's syndrome: contributions of developmental arrest and lesions of Alzheimer's disease. J Neuropathol Exp Neurol 49: 509–520

de la Monte SM, Sohn YK, Wands JR (1997) Correlates of p53- and Fas (CD95)-mediated apoptosis in Alzheimer's disease. J Neurol Sci 152: 73–83

de la Monte SM, Sohn YK, Ganju N, Wands JR (1998) P53- and CD95-associated apoptosis in neurodegenerative diseases. Lab Invest 78: 401–411

Dragunow M, Faull RL, Lawlor P, Beilharz EJ, Singleton K, Walker EB, Mee E (1995) In situ evidence for DNA fragmentation in Huntington's disease striatum and Alzheimer's disease temporal lobes. Neuroreport 6: 1053–1057

Epstein CJ (1986) The consequences of chromosomal imbalance. Cambridge University Press, Cambridge

Games D, Adams D, Alessandrini R, Barbour R, Berthelette P, Blackwell C, Carr T, Clemens J, Donaldson T, Gillespie F, Guido T, Hagopias S, Johnson-Wood K, Khan K, Lee M, Leibowits P, Lieberburg I, Little S, Masliah E, McConlogue L, Montoya-Zavala M, Mucke L, Paganini L, Penniman E, Power M, Schenk D, Seubert P, Snyder B, Soriano F, Tan H, Vitale J, Wadsworth S, Wolozin B, Zhao J (1995) Alzheimer-type neuropathology in transgenic mice overexpressing V717F beta-amyloid precursor protein. Nature 373: 523–527

Goate A, Chartier-Harlin MC, Mullan M, Brown J, Crawford F, Fidani L, Giuffra L, Haynes A, Irving N, James L, Mant R, Newton P, Rooke K, Roques P, Talbot C, Pericak-Vance M, Roses A, Williamson R, Rossor M, Owen M, Hardy J (1991) Segregation of a missense mutation in the amyloid precursor protein gene with familial Alzheimer's disease. Nature 349: 704–706

Green DR, Reed JC (1998) Mitochondria and apoptosis. Science 281: 1309–1312

Goldgaber D, Lerman MI, McBride OW, Saffiotti U, Gajdusek DC (1987) Characterization and chromosomal localization of a cDNA encoding brain amyloid of Alzheimer's disease. Science 235: 877–880

Hallam DM, Maroun LE (1998) Anti-gamma interferon can prevent the premature death of trisomy 16 mouse cortical neurons in culture. Neurosci Lett 252: 17–20

Hayn M, Kremser K, Singewald N, Cairns N, Nemethova M, Lubec B, Lubec G (1996) Evidence against the involvement of reactive oxygen species in the pathogenesis of neuronal death in Down's syndrome and Alzheimer's disease. Life Sci 59: 537–544

Holtzman DM, Li YW, DeArmond SJ, McKinley MP, Gage FH, Epstein CJ, Mobley WC (1992) Mouse model of neurodegeneration: atrophy of basal forebrain cholinergic neurons in trisomy 16 transplants. Proc Natl Acad Sci USA 89: 1383–1387

Holtzman DM, Li Y, Chen K, Gage FH, Epstein CJ, Mobley WC (1993) Nerve growth factor reverses neuronal atrophy in a Down syndrome model of age-related neurodegeneration. Neurology 43: 2668–2673

Holtzman DM, Santucci D, Kilbridge J, Chua-Couzens J, Fontana DJ, Daniels SE, Johnson RM, Chen K, Sun Y, Carlson E, Alleva E, Epstein CJ, Mobley WC (1996) Developmental abnormalities and age-related neurodegeneration in a mouse model of Down syndrome. Proc Natl Acad Sci USA 93: 13333–13338

Iwatsubo T, Mann DM, Odaka A, Suzuki N, Ihara Y (1995) Amyloid beta protein (A beta) deposition: a beta 42(43) precedes A beta 40 in Down syndrome. Ann Neurol 37: 294–299

Kang J, Lemaire HG, Unterbeck A, Salbaum JM, Masters CL, Grzeschik KH, Multhaup G, Beyreuther K, Muller-Hill B (1987) The precursor of Alzheimer's disease amyloid A4 protein resembles a cell-surface receptor. Nature 325: 733–736

Labudova O, Krapfenbauer K, Moenkemann H, Rink H, Kitzmuller E, Cairns N, Lubec G (1998) Decreased transcription factor junD in brains of patients with Down syndrome. Neurosci Lett 252: 159–162

Lai F, Williams RS (1989) A prospective study of Alzheimer disease in Down syndrome. Arch Neurol 46: 849–853

Lassmann H, Bancher C, Breitschopf H, Wegiel J, Bobinski M, Jellinger K, Wisniewski HM (1995) Cell death in Alzheimer's disease evaluated by DNA fragmentation in situ. Acta Neuropathol 89: 35–41

Lubec G, Labudova O, Cairns N, Fountoulakis M (1999) Increased glyceraldehyde 3-phosphate dehydrogenase levels in the brain of patients with Down's syndrome. Neurosci Lett 260: 141–145

Mann DM, Esiri MM (1988) The site of the earliest lesions of Alzheimer's disease. N Engl J Med 318: 789–790

Mann DM, Esiri MM (1989) The pattern of acquisition of plaques and tangles in the brains of patients under 50 years of age with Down's syndrome. J Neurol Sci 89: 169–179

Mann DM, Iwatsubo T (1996) Diffuse plaques in the cerebellum and corpus striatum in Down's syndrome contain amyloid beta protein (A beta) only in the form of A beta 42(43). Neurodegeneration 5: 115–120

Miyashita T, Reed JC (1995) Tumor suppressor p53 is a direct trascriptional activator of the human bax gene. Cell 80: 293–299

Muller M, Wilder S, Bannasch D, Israeli D, Lehlbach K, Li-Weber M, Friedman SL, Galle PR, Stremmel W, Oren M, Krammer PH (1998) p53 activates the CD95 (APO-1/Fas) gene in response to DNA damage by anticancer drugs. J Exp Med 188: 2033–2045

Nagy ZS, Esiri MM (1997) Apoptosis-related protein expression in the hippocampus in Alzheimer's disease. Neurobiol Aging 18: 565–571

Nalbantoglu J, Tirado-Santiago G, Lahsaini A, Poirier J, Goncalves O, Verge G, Momoli F, Welner SA, Massicotte G, Julien JP, Shapiro ML (1997) Impaired learning and LTP in mice expressing the carboxy terminus of the Alzheimer amyloid precursor protein. Nature 387: 500–505

O'Barr S, Schultz J, Rogers J (1996) Expression of the protooncogene bcl-2 in Alzheimer's disease brain. Neurobiol Aging 17: 131–136

Oyama F, Carins NJ, Shimada H, Oyama R, Titani K, Ihara Y (1994) Down's syndrome: up-regulation of beta amyloid protein precursor and tau mRNAs and their defective coordination. J Neurochem 62:1062–1066

Pearlson GD, Breiter SN, Aylward EH, Warren AC, Grygorcewicz M, Frangou S, Barta PE, Pulsifer MB (1998) MRI brain changes in subjects with Down syndrome with and without dementia. Dev Med Child Neurol 40: 326–334

Polyak K, Xia Y, Zweier JL, Kinzler KW, Vogelstein B (1997) A model for p53-induced apoptosis. Nature 389: 300–305

Reeves RH, Irving NG, Moran TH, Wohn A, Kitt C, Sisodia SS, Schmidt C, Bronson RT, Davisson MT (1995) A mouse model for Down syndrome exhibits learning and behaviour deficits. Nature Genet 11: 177–184

Rumble B, Retallack R, Hilbich C, Simms G, Multhaup G, Martins R, Hockey A, Montgomery P, Beyreuther K, Masters CL (1989) Amyloid A4 protein and its precursor in Down's syndrome and Alzheimer's disease. N Engl J Med 320: 1446–1452

Sago H, Carlson EJ, Smith DJ, Kilbridge J, Rubin EM, Mobley WC, Epstein CJ, Huang TT (1998) Ts1Cje, a partial trisomy 16 mouse model for Down syndrome, exhibits learning and behavioral abnormalities. Proc Natl Acad Sci USA 95: 6256–6261

Satou T, Cummings BJ, Cotman CW (1995) Immunoreactivity for Bcl-2 protein within neurons in the Alzheimer's disease brain increases with disease severity. Brain Res 697: 35–43

Sawa A, Oyama F, Cairns NJ, Amano N, Matsushita M (1997a) Aberrant expression of bcl-2 gene family in Down's syndrome brains. Brain Res Mol Brain Res 48: 53–59

Sawa A, Khan AA, Hester LD, Snyder SH (1997b) Glyceraldehyde-3-phosphate dehydrogenase: nuclear translocation participates in neuronal and nonneuronal cell death. Proc Natl Acad Sci USA 94: 11669–11674

Sawa A, Wiegand GW, Cooper J, Margolis RL, Sharp AH, Lawler Jr JF, Greenamyer JT, Snyder SH, Ross CA (1999) Increased apoptosis of Huntington's disease lymphoblasts associated with repeat lengh-dependent mitochondrial depolarization. Nature Med 5: 1194–1198

Schapiro MB, Haxby JV, Grady CL (1992) Nature of mental retardation and dementia in Down syndrome: study with PET, CT, and neuropsychology. Neurobiol Aging 13: 723–734

Schuchmann S, Muller W, Heinemann U (1998) Altered Ca2+ signaling and mitochondrial deficiencies in hippocampal neurons of trisomy 16 mice: a model of Down's syndrome. J Neurosci 18: 7216–7231

Seidl R, Fang-Kircher S, Bidmon B, Cairns N, Lubec G (1999) Apoptosis-associated proteins p53 and APO-1/Fas (CD95) in brains of adult patients with Down syndrome. Neurosci Lett 260: 9–12

Smith DJ, Stevens ME, Sudanagunta SP, Bronson RT, Makhinson M, Watabe AM, O'Dell TJ, Fung J, Weier HU, Cheng JF, Rubin EM (1997) Functional screening of 2 Mb of human chromosome 21q22.2 in transgenic mice implicates minibrain in learning defects associated with Down syndrome. Nature Genet 16: 28–36

Stabel-Burow J, Kleu A, Schuchmann S, Heinemann U (1997) Glutathione levels and nerve cell loss in hippocampal cultures from trisomy 16 mouse — a model of Down syndrome. Brain Res 765: 313–318

Su JH, Anderson AJ, Cummings BJ, Cotman CW (1994) Immunohistochemical evidence for apoptosis in Alzheimer's disease. Neuroreport 5: 2529–2533

Su JH, Satou T, Anderson AJ, Cotman CW (1996) Up-regulation of Bcl-2 is associated with neuronal DNA damage in Alzheimer's disease. Neuroreport 7: 437–440

Tanzi RE, Gusella JF, Watkins PC, Bruns GA, St George-Hyslop P, Van Keuren ML, Patterson D, Pagan S, Kurnit DM, Neve RL (1987) Amyloid beta protein gene: cDNA, mRNA distribution, and genetic linkage near the Alzheimer locus. Science 235: 880–884

Wisniewski KE, Wisniewski HM, Wen GY (1985) Occurrence of neuropathological changes and dementia of Alzheimer's disease in Down's syndrome. Ann Neurol 17: 278–282

Wong A, Yang J, Cavadini P, Gellera C, Lonnerdal B, Taroni F, Cortopassi G (1999) The Friedrich's ataxia mutation confers cellular sensitivity to oxidant stress which is rescued by chelators of iron and calcium and inhibitors of apoptosis. Hum Mol Genet 8: 425–430

Yamatsuji T, Matsui T, Okamoto T, Komatsuzaki K, Takeda S, Fukumoto H, Iwatsubo T, Suzuki N, Asami-Odaka A, Ireland S, Kinane TB, Giambarella U, Nishimoto I (1996) G protein-mediated neuronal DNA fragmentation induced by familial Alzheimer's disease-associated mutants of APP. Science 272: 1349–1352

Yoshikawa K, Aizawa T, Hayashi Y (1992) Degeneration in vitro of post-mitotic neurons overexpressing the Alzheimer amyloid protein precursor. Nature 359: 64–67

Author's address: A. Sawa, MD, PhD, Department of Neuroscience, Johns Hopkins University School of Medicine, 725 N. Wolfe Street, Baltimore, MD 21205, U.S.A., e-mail: akira@welchlink.welch.jhu.edu

J Neural Transm (1999) [Suppl] 57: 99–124

Altered gene expression in fetal Down Syndrome brain as revealed by the gene hunting technique of subtractive hybridization

E. Kitzmueller[1], **O. Labudova**[1], **H. Rink**[2], **N. Cairns**[3], and **G. Lubec**[1]

[1] Department of Pediatrics, University of Vienna, Austria
[2] Department of Radiobiology, University of Bonn, Federal Republic of Germany
[3] Brain Bank, Institute of Psychiatry, University of London, United Kingdom

Summary.   Information on gene expression in brain of patients with Down Syndrome (DS, trisomy 21) is limited and molecular biological research is focussing on mapping and sequencing chromosome 21. The information on gene expression in DS available follows the current concept of a gene dosage effect due to a third copy of chromosome 21 claiming overexpression of genes encoded on this chromosome.

Based upon the availability of fetal brain and recent technology of gene hunting, we decided to use subtractive hybridization to evaluate differences in gene expression between DS and control brains.

Subtractive hybridization was applied on two fetal brains with DS and two age and sex matched controls, 23rd week of gestation, and mRNA steady state levels were evaluated generating a subtractive library. Subtracted sequences were identified by gene bank and assigned by alignments to individual genes.

We found a series of up- and downregulated sequences consisting of chromosomal transcripts, enzymes of intermediary metabolism, hormones, transporters / channels and transcription factors (TFs).

We show that trisomy 21 or aneuploidy leads to the deterioration of gene expression and the derangement of transcripts describes the impairment of transport, carriers, channels, signaling, known metabolic and hormone imbalances. The dys-coordinated expression of transcription factors including homeobox genes, POU-domain TFs, helix-loop-helix-motifs, LIM domain containing TFs, leucine zippers, forkhead genes, maybe of pathophysiological significance for abnormal brain development and wiring found in patients with DS. This is the first description of the concomitant expression of a large series of sequences indicating disruption of the concerted action of genes in this disorder.

Introduction

Molecular biological approaches are focussing on mapping and gene hunting of chromosome 21, but data showing differences between gene expression in normal and Down Syndrome (DS) brain are limited. This maybe due to the fact that a major current concept incriminates gene dosage effects of a third copy of chromosome 21 as responsible for the development of the DS phenotype (Epstein, 1992).

Among the few publications of gene expression in the brain of patients with DS Pash and coworkers showed the overexpression of chromosomal protein HMG-14, which is mapped to chromosome 21 (Pash et al., 1991) and the authors propose a pathogenetic role for this upregulated structural chromatin — related protein.

Goodison and coworkers described that in frontal cortex of patients with DS chromosome-21 encoded mRNAs, neurofilament light subunit and amyloid precursor transcription were increased (Goodison, 1993). The upregulation of beta-amyloid precursor was confirmed by Oyama and coworkers and the authors showed the disruption of coordinated expression between the amyloid precursor APP-751 and four-repeat tau (Oyama, 1994).

Not all genes present in the critical "Down Syndrome region" of chromosome 21 are, however, overexpressed in the brain: Marks and coworkers reported comparable mRNA and protein levels of S 100, a calcium-binding protein encoded on chromosome 21 and accumulating during CNS maturation in mammals, in controls and DS cerebellum (Marks, 1996).

De la Monte et al. demonstrated that early in life neuronal thread protein was overexpressed in cerebral cortex of patients with DS, a tentative candidate for the development of Alzheimer's disease pathology which occurs from the fourth decade in DS brain (de la Monte, 1996).

Transient overexpression of a trifunctional protein with glycinamide ribonucleotide synthase, aminoimidazole ribonucleotide synthase and glycinamide ribonucleotide formyltransferase activity, key enzymes for purine de novo biosynthesis, in cerebellum of infants with DS has been published by Brodsky and coworkers (1997).

No gene hunting approach is known to us in peer reviewed accessible literature, testing the differences in gene expression between DS and control brain. However important the known reports on different expression on individual genes are, no pattern or derangement of the concerted action of genes involved in DS pathology has emerged yet. The availability of age and sex matched fetal brain samples of DS and controls on the one and the gene hunting technique of subtractive hybridization on the other hand made us create a subtractive library which enabled us to reveal up-and downregulated sequences which were subsequently assigned to genes by gene bank work.

In this paper we describe expressional differences in terms of mRNA steady state levels of transcription factors, channels, transporters, chromosomal transcripts, which may well serve to explain structural and functional abnormalities in DS brain pathology.

Methods

Subtractive hybridization was selected as a gene hunting method and carried out as given previously (Labudova and Lubec, 1998; Labudova, 1998):

Fetal brain samples of two fetuses with DS and two age and sex matched controls, 23rd week of gestation, were obtained from the Brain Bank of the Institute of Psychiatry (Denmark Hill, London, UK). Hippocampus (gyrus parahippocampalis) was taken into liquid nitrogen and ground for the isolation of mRNA. Isolation of mRNA was performed using the Quick Prep Micro mRNA purification kit (Pharmacia Biotech Inc., Uppsala, Sweden). One microgram of mRNA from each preparation was quality — checked by cDNA cloning kit (Gibco, Life Technologies, Eggenstein, Germany, cat. 18248-013) using the incorporation of (^{32}P]dATP (Amersham, Buckinghamshire, UK) with subsequent electrophoresis on 1% agarose followed by autoradiography. The Reflection film (Dupont, Germany) was exposed to the gel for a period of two hours at room temperature. Construction of the substractive library: 10 micrograms each of mRNA from brain of DS and control were biotinilated by UV irradiation at 360 nm according to the instructions supplied in the subtractor kit (Invitrogen, Leek, Netherlands). One microgram of mRNA-pools each from the DS brain sample was subject to reverse transcriptase reaction (subtractor kit, Invitrogen) and the cDNA-pools were hybridized with the corresponding biotinilated mRNAs from controls. The substractive hybridization mixture was incubated with streptavidin according to the subtractor kit given above and thus the biotinilated molecules (non-induced biotinilated mRNAs and the hybrid (biotinilated mRNAs/cDNAs]) complexed. The streptavidine complexes were removed by repeated phenol-chloroform extraction and subtracted cDNAs were separated from the aqueous phase by alcohol precipitation (subtractor kit). In order to amplify and clone subtracted cDNAs, they were ligated with Not I linkers followed by Not I -digestion. The Not I-linked cDNAs were ligated to Not I site of sPORT 1 cloning vector (cDNA cloning kit, Gibco). To enable visualization of subtracted cDNAs, the cloned cDNAs were amplified using universal primers I 5′-GTAAAACGACGGCCAGT-3′II 5′-ACAGCTATGACCATG-3′from multiple cloning site of the sPORT-1 vector (cDNA cloning kit, Gibco). Amplified cDNAs were analyzed by electrophoresis on 1% agarose. Cloning of subtracted Not I-linked cDNAs cDNAs ligated with sPORT 1 vector were used for the transformation of highly competent INFalpha F E. coli cells (Invitrogen, Leek, Netherlands) and plated clones were analysed by plasmid isolation kit (Quiagen, Hilden, Germany) and digestion with Eco RI/Hind III. Recombinant clones were sequenced by K. Granderath, MWG — Biotech (Ebersberg, Germany). Homologies were determined by computer assisted comparison of data from the genebank sequence library: fastA@ebi.ac.uk (GBALL, EMBL, Gen Bank Heidelberg, Germany). Subtractive hybridization was performed cross-wise, i.e. DS sample mRNA subtraction from control and vice versa at the 1:3 level (DSmRNAs:control mRNAs).

Results

Tables 1 shows the sequences overexpressed in fetal DS brain and Table 2 lists the sequences which were downregulated in fetal DS. The alignments of the indicated genes are given in Figs. 1 to 15.

Discussion

Subtractive hybridization, i.e. the generation of a subtractive library of DS and control brain, led to the discovery of a series of sequences and many of them could be assigned to corresponding genes by gene bank work. In this paper we have omitted the non-identified sequences but present those with a homology higher than 50%. Some of them could not be assigned to human

Table 1. Genes with higher transcriptional level in DS

Fig.	No of clone	% Homology	Name of gene	EMBL acces. number	Organism	Length of cDNA (bp)
1	9	70	chromosome 21 Down Syndrome region (from 75,588 nt)	Ac000002	hs	149
2	10	70	chromosome 21 Down Syndrome region (from 1,395 nt)	U44740	hs	149
3	11	70	chromoseme 21 Down Syndrome region (from 406 nt)	U44743	hs	149
9	5–15	100	XRCC1	M36089	hs	211
10	103	56	transducer Htc-protein	U75437	Halobacterium salinarium	515
11	101	55	LysR (mptA/R) (metalloproteinase and LysR-type transcriptional activator)	Z11929	Streptomyces coelicolor	527
12	78	59	dipeptidyl peptidase IV	D42121	Flavobacterium meningosepticum	545
13	241	58	dynamin	L07807	hs	267
14	102	100	apolipoprotein E	E08423	hs	359

Table 2. Genes with higher transcriptional level in controls

Fig.	No of clone	% Homology	Name of gene	EMBL acces. number	Organism	Length of cDNA (bp)
4	18	73	Neuropeptide GFAD precursor (acidic repeat)	P91952	Helix lucorum	150
5	269	55	Endothelin-2	U59510	Rat	405
6	317	59	OMP32 gene (outer membrane porin protein 32 precursor)	P24305	Comamonas acidovorans	761
7	318	59	TRX (thioredoxin)	X54539	Hs	72
8	263	100	U1-70K SnRNP	Y00886	hs	357
15	266	55	SOX-4 (DNA-binding, HMG box, transcriptional factor)	X70298	Mouse	386
16	141	100	TAX-helper protein (transcriptional factor, transcriptional suppressor, TRE-2S-binding-protein)	D14827	hs	187
17	180	53	Hsp70 (heat shoch protein, ATPase)	M84006	Halobacterium marismortui	408

```
Ac000002 Human Down Syndrome region of chromosome 21,  9


9                                        GCCTCCCAAGTATGTGGGACCACAGGT27
                                         ::::::  :  ::  :::::  ::::  :
HSAC00  TGCCTCCTGGGCTCAAACAATTCTCCTGCCTCAGCCTCCTGAATAGCTGGGATTACAGAT75617

9       GTGTGCTACCACTCTCAGCTCCAAGCTAATTTTGGGTTTTGGTTTTTTTGTTGTTTTTGT87
        :::  :  :: :  :  : :  :::   :  :  ::::  :::  :  :::   :::  :::  :::::
HSAC00  GTGCACCACTATGCCCGGCTAGTTGTT--TTTTTTGTTGTTGTTGCTTT-TTG-TTTTGT75673

9       TTTGTTTTGCTTCGTTTTTTTGTAGAGACAGGGTCCTACTATGTTGCCCAGGCTGATCTCC147
        ::::  ::::  ::   :::::  ::::  :::  ::::       ::  ::::::  ::::::::::  ::::
HSAC00  TTTGCTTTGTTTGTTTTTTTAGTAGTGACGGGGTTTCACCATGTTGGCCAGGCTGGTCTCA75730

9       AA149
        ::
HSAC00  AACTCCTGACCTCAGGTGATCCACCTGCCTCGGTCTCCCAAAGTGCTGGGATTATAGACA75793
```

Fig. 1

genes and were matching procaryotic or eucaryotic counterparts only, mainly as no human data were available. The high homology, however, makes the identification highly probable. Moreover, some sequences were never reported in the human or even in the mammalian system and we are the first to describe their presence in human brain. In this paper we are providing the alignments of the sequences to the gene bank data and more information is available on request from the corresponding author. A significant portion of genes expressed differently at least threefold (see the setting of 1:3 of mRNAs used for subtraction) could be assigned to chromosomal transcripts. The many genes up-or downregulated of the subtractive library indicate the multi-gene dysregulation expected in this disorder.

As suggested, the chromosomal transcripts of chromosome 21, DS region (accession numbers Ac0000002, U44740, U44743), were overexpressed, although not at the 1.5 fold but at the threefold level (Table 1, Figs. 1–3).

```
U44740 Human Down Syndrome region of chromosome 21,  10


10                                       CAGCCTCCCAAGTATGTGGGACCACAGGT29
                                         :::::::::  :  ::  :::::  ::::  :
HS4474  TGCCTCCTGGGCTCAAACAATTCTCCTGCCTCAGCCTCCTGAATAGCTGGGATTACAGAT1423

10      GTGTGCTACCACTCTCAGCTCCAAGCTAATTTTGGGTTTTGGTTTTTTTGTTGTTTTTGT89
        :::  :  :::  :  : :  :::    :  :  ::::  :::  :  :::   :::  :::  ::::::
HS4474  GTGCACCACTATGCCCGGCTAGTTGTT--TTTTTTGTTGTTGTTGCTTT-TTG-TTTTGT1479

10      TTTGTTTTGCTTCGTTTTTTTGTAGAGACAGGGTCCTACTATGTTGCCCAGGCTGATCTCC149
        ::::  ::::  ::   :::::  ::::  :::  ::::       ::  ::::::  ::::::::::  ::::
HS4474  TTTGCTTTGTTTGTTTTTTTAGTAGTGACGGGGTTTCACCATGTTGGCCAGGCTGGTCTCA1539
```

Fig. 2

```
U44743 Human Down Syndrome region of chromosome 21,  11


11                                      TCAGCCTCCCAAGTATGTGGGACCACAGGT³⁰
                                        ::::::::::  : ::  :::::  :::: :
HS4474  TGCCTCCTGGGCTCAAACAATTCTCCTGCCTCAGCCTCCTGAATAGCTGGGATTACAGAT⁴³⁵


11      GTGTGCTACCACTCTCAGCTCCAAGCTAATTTTGGGTTTTGGTTTTTTTGTTGTTTTTGT⁹⁰
        :::  : :: :  : : :::    : :  :::: ::: : :::  ::: ::: ::::::
HS4474  GTGCACCACTATGCCCGGCTAGTTGTT--TTTTTTGTTGTTGTTGCTTT-TTG-TTTTGT⁴⁹¹


11      TTTGTTTTGCTTCGTTTTTTGTAGAGACAGGGTCCTACTATGTTGCCCAGGCTGATCTC¹⁴⁹
        :::: :::: :: ::::: :::: ::: ::::   :: ::::::: :::::::: ::::
HS4474  TTTGCTTTGTTTGTTTTTTAGTAGTGACGGGGTTTCACCATGTTGGCCAGGCTGGTCTCA⁵⁵¹
```

Fig. 3

Poteryaev described a novel neuropeptide precursor gene in the nervous system of the terrestrial snail helix lucorum (Poteryaev, 1998). A homolog was found in our fetal brain samples and it was downregulated in DS brain (Table 2, Fig. 4). The function of this specific neuropeptide, which is expressed in several CNS neurons is the control of behaviour in the snail. The preprohormone is 108 amino acids long, contains a hydrophobic leader peptide and eight Lys-Arg recognition sites for endoproteolysis. The post-translational processing of the prohormone may lead to the generation of seven tetrapeptides, Gly-Phe-Ala-Asp-COOH (GFAD). The absence of this neuropeptide in brain of patients with DS may contribute to the behavioural changes seen in DS and extends previous work on neuropeptide dysregulation in DS brain (Risser, 1996; Labudova, 1998).

Endothelin-2 (ET-2) is another downregulated sequence in fetal DS brain (Table 2, Fig. 5).

As early as in 1990 Marsault and coworkers (1990) have shown that endothelins stimulate the hydrolysis of phosphatidylinositol by phospholipase C in primary cultures prepared from rat brain cortex. Endothelins including ET-2 induced the rapid mobilization of intracellular calcium stores and promote a more slowly developing influx of Ca^{2+}. Giaid and coworkers (1991) demonstrated the topographical localisation of endothelin mRNAs and

```
NEUROPEPTIDE GFAD PRECURSOR 18


P91952 (108aa)


          .           .           .           .           .
   1 DERDVGDERGVGDERGVGDERGVGDERGVGDERDVGDERGVDDERDVGHER 50
     ||:|    :||  :|:||  :|:||  :|:||  :|:|   :|:||   |:|   : :|
  37 DEEDFLGKRGFADKRGFADKRGFADKRGFADKRGFADKRGFADKRGFADKR 87
```

Fig. 4

U59510 Rattus norvegicus endothelin-2, 263

```
263                                                    ¹AGCAGGCCATGGGCCTGCAGGTAACCTGTG³⁰
                                                        ::  :::::  ::::::::::  ::  :  ::::
rEN   ⁴³⁵CGGGGAGCACTTGGGTTAACTTCGGAACTCAGAAGGCCTTGGGCCTGCTGGGAGCCTGGA⁴⁹⁵


263       GACTCCTGTCGGTGTCACGCCGAGGCGCATTGGCAATTGCAGGGCCTTGCTCCCCTGCTC⁹⁰
          ::::::::  :::::  :  ::  :::::  ::       ::   :::::  :  ::  ::   :
rEN       GACTCCTGAAGGTGTTCCAGCGCAGCGCAGAGGGTGGAGCCCAGCCTTCCACCTCTCTCC⁵⁵⁵


263       CAGGAGGCGCACCACCAGGGCGTGTTGGCAGCCAGTGGGCGCCGGTCCCTGGCTTTTGAC¹⁵⁰
          :        ::       ::::::::: :  :::: ::::  :: :  :::::::::::::::  :
rEN       CCCTCACCGGGAGGCCAGGGCGGGCAGGCACCCAGGAGGAGGCGGTCCCTGGCTTTCTGC⁶¹⁵


263       GCAAGGTGCCCCAGGTGAGCGAGGCGGCGCTGCACCACGGGCGATGGAACGCCCGGGCTT²¹⁰
          ::::::::   :::  :   ::  ::::::  :::::::::     :  :     :  ::::::  :  ::
rEN       CCAAGGTGCAGCAGCTAGGCAAGGCGGGGCTGCACCAGTAGGGGGCGGACGCCCTGCCTG⁶⁷⁵


263       GGGCGATGACGGCCCGGCGGTGTCCCTCTCGGATGACATCCCAGGGCGACCTCAGAGCAG²⁷⁰
          :     ::  :  :  :::::  :      ::::     ::::  ::::::  ::  :  :     :::::
rEN       GTTTAATAAGAGGCGGTCAGGGCACCTCGGCAATGAGCTCCCAGCTCGGCTTAGCAGCAG⁷³⁵


263       GCGGTGGGCAACCTGGATGGCCGCTTCATGAGCCCAACCAAGCCCGCAAGCCGTGCCTTC³³⁰
          :        :::  :::::  :     ::::  :  :::::: :      :::  :::::     :   ::::
rEN       GACACTGGCGACCTGCCTCCCCGCTCCCTTGGCCCAGCCAAGCACTC------TGCCGCT⁷⁸⁹


263       GCAATACACGCCCCACCAAGGTGTTCGCTTTC---GTGCATCCTTACGGCCCCACGCGCA³⁸⁷
          ::  ::    :  ::::   :  :::::::  ::  :    :  :  :  :::   :::::   :   :  :
rEN       GCTATGGTCTCCCCGGCCTGGTGTTCCATTGCTCTGGCCCTACTTTTGGCCCTGCATGAA⁸⁴⁹


263       CGCGAGCTGCCCTCCTTC⁴⁰⁵
          :  :::::::::::::::
rEN       GGTGAGCTGCCCTCCTTCGCATATTCGAATCCCAGGGTCTATAT⁸⁹⁰
```

Fig. 5

peptide immunoreactivity in neurones of the human brain of parietal, temporal and frontal cortex, hypothalamus, amygdala, hippocampus, cerebellum. ET-immunoreactivity was colocalized with other neuropeptides, neurophysin and galanin. ETs and their receptors are present in high levels in the brain and have been proposed to act as neuromodulators or

```
human trx,  318

318       ¹CAGGAAGCCTTGGACGCTGCAGCTGATAAACTTGTAGTAGTTGACTTCTCAGCCACGTGG
           ::::::::::::::::::::::::::::::::::::::::::::::::::::::::::::::::
hstrx    ¹⁴⁵CAGGAAGCCTTGGACGCTGCAGCTGATAAACTTGTAGTAGTTGACTTCTCAGCCACGTGG

318       TGTGGGCCTTGC⁷²
          ::::::::::::
hstrx     TGTGGGCCTTGC²¹⁶
```

Fig. 6

106   E. Kitzmueller et al.

hs U1 snRNP 70K, 263

```
263                      ¹GATCGTGACCGTGACCGTGACCGCGAGCACAAACGGGGGGA⁴²
                          ::::::::::::::::::::::::::::::::::::::::::
hsU1  ¹cGAGCGGCGCCGGGACCGGGATCGTGACCGTGACCGTGACCGCGAGCACAAACGGGGGGA⁶⁰
         E   R   R   R   D   R   D   R   D   R   D   R   D   R   E   H   K   R   G   E


263   GCGGGGCAGTGAGCGGGGCAGGGATGAGGCCCGAGGTGGGGGCGGTGGCCAGGACAACGG¹⁰²
      ::::::::::::::::::::::::::::::::::::::::::::::::::::::::::::::::
hsU1  GCGGGGCAGTGAGCGGGGCAGGGATGAGGCCCGAGGTGGGGGCGGTGGCCAGGACAACGG¹²⁰
        R   G   S   E   R   G   R   D   E   A   R   G   G   G   G   G   Q   D   N   G


263   GCTGGAGGGTCTGGGCAACGACAGCCGAGACATGTACATGGAGTCTGAGGGCGGCGACGG¹⁶²
      ::::::::::::::::::::::::::::::::::::::::::::::::::::::::::::::::
hsU1  GCTGGAGGGTCTGGGCAACGACAGCCGAGACATGTACATGGAGTCTGAGGGCGGCGACGG¹⁸⁰
          L   E   G   L   G   N   D   S   R   D   M   Y   M   E   S   E   G   G   D   G


263   CTACCTGGCTCCGGAGAATGGGTATTTGATGGAGGCTGCGCCGGAGTGAAGAGGTCGTCC²²²
      ::::::::::::::::::::::::::::::::::::::::::::::::::::::::::::::::
hsU1  CTACCTGGCTCCGGAGAATGGGTATTTGATGGAGGCTGCGCCGGAGtgaagaggtcgtcc²⁴⁰
        Y   L   A   P   E   N   G   Y   L   M   E   A   A   P   E   *


263   TCTCCATCTGCTGTGTTTGGACGCGTTCCTGCCCAGCCCCTTGCTGTCATCCCCTCCCCC²⁸²
      ::::::::::::::::::::::::::::::::::::::::::::::::::::::::::::::::
hsU1  tctccatctgctgtgtttggacgcgttcctgcccagccccttgctgtcatcccctccccc³⁰⁰


263   AACCTTGGCCACTTGAGTTTGTCCTCCAAGGGTAGGTGTCTCATTTGTTCTGGCCCCTTG³⁴²
      ::::::::::::::::::::::::::::::::::::::::::::::::::::::::::::::::
hsU1  aaccttggccacttgagtttgtcctccaagggtaggtgtctcatttgttctggcccccttg³⁶⁰


263   GATTTAAAAATAAAA³⁵⁷
      :::::::::::::::
hsU1  gatttaaaaataaaattaatttcctgtt³⁸⁸
```

Fig. 7

neurotransmitters. Sullivan and Morton (1996) showed that ETs induce Fos protein expression, a marker of neuronal activation, in both granule cells and glia in organotypic cultures of rat cerebellum. This finding supports previous evidence that ETS act directly on neurones and that they may play a neuromodulatory role in the cerebellum. Again, downregulation of ET-2 in DS brain may disturb the concerted action of neuropeptides.

A sequence was found to be downregulated that could be assigned to a bacterial outer membrane protein Omp32 (Table 2). It is an anion-selective porin (Gerbl-Rieger, 1992) consisting of 351 amino acid residues with a signal peptide of 19 amino acid residues. Nothing is known so far on the function of this membrane protein but the strong homology with other porins suggests a common function of transport and/or osmoregulation and sensing

hs XRCC1,   5-15

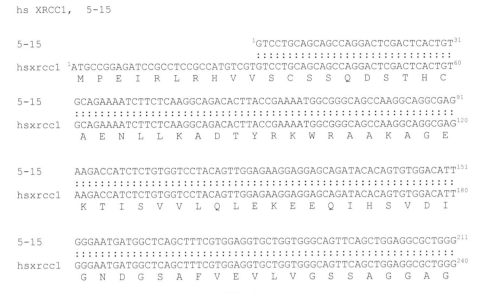

5-15 ^{1}GTCCTGCAGCAGCCAGGACTCGACTCACTGT^{31}
              :::::::::::::::::::::::::::::::::
hsxrcc1 ^{1}ATGCCGGAGATCCGCCTCCGCCATGTCGTGTCCTGCAGCAGCCAGGACTCGACTCACTGT^{60}
         M P E I R L R H V V S C S S Q D S T H C

5-15   GCAGAAAATCTTCTCAAGGCAGACACTTACCGAAAATGGCGGGCAGCCAAGGCAGGCGAG^{91}
              :::
hsxrcc1 GCAGAAAATCTTCTCAAGGCAGACACTTACCGAAAATGGCGGGCAGCCAAGGCAGGCGAG^{120}
         A E N L L K A D T Y R K W R A A K A G E

5-15   AAGACCATCTCTGTGGTCCTACAGTTGGAGAAGGAGGAGCAGATACACAGTGTGGACATT^{151}
              :::
hsxrcc1 AAGACCATCTCTGTGGTCCTACAGTTGGAGAAGGAGGAGCAGATACACAGTGTGGACATT^{180}
         K T I S V V L Q L E K E E Q I H S V D I

5-15   GGGAATGATGGCTCAGCTTTCGTGGAGGTGCTGGTGGGCAGTTCAGCTGGAGGCGCTGGG^{211}
              :::
hsxrcc1 GGGAATGATGGCTCAGCTTTCGTGGAGGTGCTGGTGGGCAGTTCAGCTGGAGGCGCTGGG^{240}
         G N D G S A F V E V L V G S S A G G A G

Fig. 8

(Gerbl-Rieger, 1991) No Omp gene has ever been described in the mammalian system and this is the first description in human brain. Its downregulation may be compatible with impaired membrane transport which may well be involved in abnormal neurotransmission taking place in DS brain.

The reduced expression of thioredoxin (TRX) (Table 2, Fig. 6) may have pronounced effects at several levels: TRX is a multifunctional protein with a redox-active site and is expressed in the brain where a neuroprotective action was proposed (Hori, 1994; Lippoldt, 1995). Apart from its potential redox-controlling -activity it directly interacts with Ref-1 (redox factor-1) to regulate AP-1 transcriptional activity (Hirota, 1997). As major components of the AP-1 transcription factor are members of the Jun family of TFs, and junD was recently described to be downregulated in brain of patients with DS, TRX mediated modulation of the AP-1 complex may be of relevance for the wiring and plasticity in fetal and adult brain (Labudova, 1998).

Decreased TRX may furthermore modulate apoptosis in DS brain: TRX is a direct inhibitor of apoptosis signal regulating kinase 1 (ASK1): the evidence that TRX is a negative regulator of ASK1 suggests possible pathomechanisms for redox-regulation of the apoptosis signal transduction pathway as well as the effects of antioxidants against cytokine- and stress-induced apoptosis (Saitoh, 1998).

The redox-related apoptotic pathway is of particular interest as there is evidence for the involvement of active oxygen species in the apoptosis of DS (Busciglio and Yankner, 1995) although not unequivocal (Seidl, 1997).

Other tentative roles for TRX in the pathomechanism of DS maybe impairment of recombination and DNA repair, which was suggested (Golubnitchaya-Labudova, 1998) and impaired assembly of microtubules (Khan, 1991).

```
HtC, 103

              *    .      *     *      *  *
              T  M  E  Q  D  R  R  I  A
103          1CACGATGGAACAGGATCGGAGGATCGCCA29
              :::: ::: ::::: ::: :::::::::
HtC      1457GTTCAACGACATGATGGGCGACGTGGAAGCCACGCTGGCACAGGTGCGGTCGATCGCCG1516
              A  T  L  A  Q  V  R  S  I  A

              .  .      .       *  *        .       *        .  .
              S  G  G  T  I  D  S  T  V  S  N  G  A  T  V  R  H  S  G  K
103          GCGGGGGCACCATCGACAGCACCGTCAGTAACGGCGCAACGGTGCGCCACAGCGGCAAGG89
              :: :: : : :: :::::::: :   ::: ::: ::: : :: ::::: :: :
HtC          ACGCGGTCGACGCCGCCAGCACCGACGTCCACGACGCCGCGGAGATCCGCAGCGCCAGCG1573
              D  A  V  D  A  A  S  T  D  V  H  D  A  A  E  I  R  S  A  S

              *  *  *     .  *       .      .  .       *     *  .  *
              V  P  V  S  E  E  A  Q  A  Q  A  D  E  G  A  E  Y  R  G  R
103          TCCCGGTCAGCGAATCTGCCCAGGCACAGGCCGACGAGGGCGCTGAGTACCGAGGCCGCA149
              :: :::::::::::: : :::::      ::: :: : :: ::: : :: : ::::
HtC          ACCAGGTCAGCGAATCCGTCCAGGACATCTCCGCGGACGCCGAGGAGCAACGCGACCGCC1633
              D  Q  V  S  E  S  V  Q  D  I  S  A  D  A  E  E  Q  R  D  R

              .     *  .  *  *     .      *  *  *  *  *        .  .  *
              T  A  T  G  G  D  R  L  D  A  L  S  A  T  D  I  F  T  G  A
103          TCGCCACGGGCGGCGACAGGCTCGATGCGCTGTCTGCCACAGACATATTCACCGGTGCGA209
              ::: :::: :::::::: : : :::::::::: :: :: : :   :: :: :: :
HtC          TCGGCACGGTCGGCGACGAGGTACATGCGCTGTCCGCGACCGTCGAGGACATCGCGGCCA1691
              L  G  T  V  G  D  E  V  H  A  L  S  A  T  V  E  D  I  A  A

              *  *  .       .  *       .  .  *       .  *  *  *  *
              R  P  S  A  L  K  A  T  F  H  R  V  A  H  L  C  E  R  G  Q
103          GGCCGTCCGCGTTGAAGGCCACGTTTCACCGCGTCGCCCATTTGTGCGAGCGCGGGCAGC269
              ::::: :  ::: : :  ::::: :  ::: : :::::     : :::::::::: :::
HtC          GGCCGACGACGTCGCCGAGCACGGTCAACCAGGCCGCCACCGAGAGCGAGCGCGGCCAGG1749
              R  P  T  T  S  P  S  T  V  N  Q  A  A  T  E  S  E  R  G  Q

              .  *  .     .       *  *       *  .  *        .  .  *
              Q  L  S  V  G  L  V  A  A  L  G  R  A  H  E  V  G  D  T  A
103          AGCTTGTCGTTGGGGTTGTCGCCGCGCTCGGCCGAGCCCACGAAGTAGGCGACACAGCAG329
              :::: : :: : :  ::::::: :::: :: : :     : ::::: :: :
HtC          AGCTCGGCGAGGACGCCGTCGCCGAACTCGAACGCATCGAGGCGACCGCCGACAGCGCCG1805
              E  L  G  E  D  A  V  A  E  L  E  R  I  E  A  T  A  D  S  A

              .  .  *     *     .  .  *     .  *  .     *  *  *  *  .
              A  G  R  R  A  A  Q  Q  Q  A  T  G  A  F  R  H  V  T  G  L
103          CAGGGCGAAGGGCAGCCCAACAGCAGGCGACTGGAGCGTTCCGGCACGTGACCGGGTTGA389
              : ::: : : :::: :: :::: : :: :: ::::::::::::: : :
HtC          TCGAGCGCGTGACCGCCCTGGAGGAGGCCGTCGACGCCATCGGGCACGTGACCGGCGTCA1861
              V  E  R  V  T  A  L  E  E  A  V  D  A  I  G  H  V  T  G  V

              *     .  *  .     .       .  .       *        .  *
              I  R  R  M  A  G  S  S  Y  S  V  P  S  A  A  I  M  H  V  A
103          TCAGGCGCATGGCCGGGTCCTCTTACAGCGTGCCGTCGGCGGCCATCATGCACGTAGCTG449
              ::: : ::: :::: :   : ::: :: ::: : :::: ::: : : ::
HtC          TCACGGACATCGCCGAGCAGACGAACATGCTGGCGTTGAACGCCAACATCGAGGCCGCCC1921
              I  T  D  I  A  E  Q  T  N  M  L  A  L  N  A  N  I  E  A  A
```

Fig. 9

snRNP 70K (small nuclear ribonucleoprotein) is essential for splicing of precursor mRNA, an activity depending upon both, the RNA and protein components of the U1 particle. We found that mRNA steady state levels were reduced in fetal DS brain (Table 2, Fig. 7). Casciola-Rosen and coworkers

```
              .   *   *    .           .  *   *   *   *   .   *   .    *   *
          G   V   D   K   C   S   L   R   L   A   V   V   A   D   G   V   T   P   L   A
   103    GGGTCGACAAATGCAGCTTGAGGTTGGCGGTGGTCGCCGACGGGGTCACGCCGTTGGCAA509
          : :  ::::::  :: ::     :  :: :::::::::::::::: :::::  :    :  ::
   HtC    GCGCCGACAAGAGCGGCGACGGCTTCGCGGTGGTCGCCGACGAGGTCAAGGACCTCGCCG1981
          R   A   D   K   S   G   D   G   F   A   V   V   A   D   E   V   K   D   L   A


              .
          S   H
   103    GTCATG515
          :  :  :
   HtC    ATGAGGTCAAAGAATCGGCCACGGAGATCGAGACGCTGGTCGACG2041
          D   E   V
```

PROTEIN SIMILARITY /112aa from 171aa/ 65%

Fig. 9. *Continued*

(1994) reported that the snRNP 70K protein is specifically cleaved in apoptotic cells resulting in the generation of a 40kDa fragment. The authors proposed interleukin 1 beta-converting enzyme (ICE) as the protease but ICE did not cleave snRNP 70K in vitro and a specific ICE inhibitor did not prevent cleavage of snRNP 70K.

Tewari and coworkers (1995) made a major step forward: Fas and tumor necrosis factor receptor (TNF-R) are two cell surface proteins that, when stimulated with ligand, trigger apoptotic cell death. The CrmA protein is a serpin family protease inhibitor inhibiting ICE and ICE-like proteases. CrmA has been shown to inhibit Fas or TNF-R induced apoptosis. The authors showed that snRNP 70K is rapidly cleaved during both apoptosis inducing pathways and that this cleavage was inhibited by expression of CrmA but not by expression of an inactive point mutant of CrmA; these data identify snRNP 70K as a death substrate cleaved during Fas and TNF-R mediated apoptosis. Casciola-Rosen (1996) added more nuclear repair proteins as DNA-dependent protein kinase (DNA-PK) and poly(ADP-ribose)polymerase (PARP) to the death substrate snRNP 70K extending knowledge on the impaired homeostasis of apoptotic pathways.

Greidinger and coworkers (1996) showed the sequential activation of distinct ICE activities with cleavage of fodrin in the first phase at 30min, snRNP 70K along with PARP, DNA-PK in the second phase of apoptotic cleavage at 50min and a third phase of lamin B cleavage at 90min.

Casiano (1996) demonstrated cleavage of snRNP 70K, lamin B, nuclear mitotic apparatus protein NuMa, DNA topoisomerase I and II, RNA polymerase I upstream binding factor UBF in CD95 (Fas/APO-1)mediated apoptosis. This observation fits own observations of deranged Fas /APO-1 in adult DS brain (Seidl, 1999) and the potential deterioration of the nuclear mitotic apparatus protein may play a role in the pathogenesis of DS/trisomy independent of its role as death substrate.

Choi and coworkers (1998) reported that snRNP 70K cleavage is also taking place in transforming growth factor beta1 — induced apoptosis, a pathway involving caspase-2.

LysR, 101

```
                                             .             .         *  *       .
                          A   C   P   R   G   D   T   A   N
101               ¹ CGGCGTGTCC-CCGCGG-CGACACCGCCAA²⁸
                          :::: :: :: :: :: : ::::::::
LysR     ²¹¹GCCGCTCTTCGACGCGACCACCGGCGGGCTCGGCCTGACCGGCGTGGTCCTCACCGCCAC²⁷⁰
                          L   G   L   T   G   V   V   L   T   A   T

            *   *   *   .           .       .       .           .
          L   Q   L   H   L   D   P   S   Y   V   Q   M   A   A   D   G   T   L   Y   E
101       CCTCAAGCTGCATTTCGACCCCAGCTACGTGCAGATGGCCGCCGACGGCACGCTGTATGA⁸⁸
          :::: :::: ::    :    ::: :: : ::: : ::: :    ::: :::  .
LysR      CCTCCAGCTCCAGCCGGTGGCCACCTCCCTGCTGCTGACCGGC-TCGGCCCGCGCCA--C³²⁷
          L   Q   L   Q   P   V   A   T   S   L   L   L   T   G   S   A   R   A       T

          D   P   A   M   R   L   I   N   R   H   P   E   R   S   S   R   L   L   F   V
101       GGACCCGGCCATGCGCCTGATCAACCGTCACCCGGAACGCTCCAGTCGCCTGCTGTTTGT¹⁴⁸
          :::: :: :       ::::::      :: :    ::   ::    :::: .   :  :
LysR      CGACCTGGAC---GACCTGATGGCGCGGCTCTGCGACGCCGACCTCCGCCACACCTACGC³⁸⁴
          D   L   D       D   L   M   A   R   L   C   D   A   D   L   R   H   T   Y   A

                          *   *               *               .           .
          L   L   P   F   A   L   L       L   F   A   Y   F   V   G   S   A   E   R   L
101       GCTGCTGCCCTTCGCCCTGCTG---CTGTTCGCCTACTTCGTGGGCTCGGCCGAGCGCCT²⁰⁵
          :           : : :::  :::::::     : :  :::: :  ::   :    ::: :: ::
LysR      GAGCGCCCGCGTCGACCTGCTGGCCCCGGGGCGCCGCCACCGGCCGGGGCACCGTGCTCCA⁴⁴⁴
          S   A   R   V   D   L   L   A   R   G   A   A   T   G   R   G   T   V   L   H

            *   *           .       .           *           .           .       .
          A   D   N       P   N   D   K   L   L   P   S               A   A   Q   M   G
101       GGCAGACAA--CCCCAAC-GACAAGCTGCTGCCCAG-------CGCCGCACAAATGGGCG²⁵⁵
          :: ::: :   :::: :  : ::    :   ::             ::::::: :    ::: ::
LysR      CGCCGACCACGCCCCGCCGGAGGCGCTCCCGGCCCGGCTCGCCCGCGC-CCGCTGGCCG⁵⁰³
          A   D   H   A   P   P   E   A   L   P   A   R   L   A   R   R       P   L   A

            *           .       *           .           *           .       *
          D   A   V   K   R   L   A   F   N   A   D   A   R   T   G   G   Y   V   L   W
101       ACGCGGTGAAACGCTTGGCCTTCAACGCGGACGCTCGCACCGGTGGATATGTGCTGTGGC³¹⁵
          : ::       ::: ::: : ::: : ::::  :::: : : :: :   ::
LysR      --CCCGTCCCCAGCT--GCCGGCCCCGCCGCAGCTCCCGCCGG-GCCT-CGTCCCCCAGC⁵⁵⁷
          A   R   P   Q   L       P   A   P   P   Q   L   P   P       G   L   V   P   Q

            .       .       .   *       *       *       *   *       .           *
          Q   D   S   A   S   S   L   R   R   L   A   I   G   L   G   I   S   A   L
101       AAGACAGCGCATCGAGC-CTGCGCCGCCTGGCGATCGGCCTCGGTATCAGCGCCCT----³⁷⁰
          ::    :: : ::   : ::: ::::::: :::: : :::::::        ::   :: :::
LysR      -TCACCCCGGACCGGCCGCTGGGCCGCCGGGCGCTGGGCCTC---CTCCACGACCTGCGC⁶¹³
          L   T   P   D   R   P   L   G   R   R   A   L   G   L       L   H   D   L   R
```

Fig. 10

Using a series of antibodies against snRNP 70K we found a decrease of snRNP 70K protein in adult DS and Alzheimer's disease brain which would match the findings given here and the well-documented ongoing apoptotic processes in these disorders. The finding of decreased snRNP 70K-mRNA steady state levels in fetal DS brain has to be evaluated in a series of fetal brain samples at the protein level to allow a conclusion.

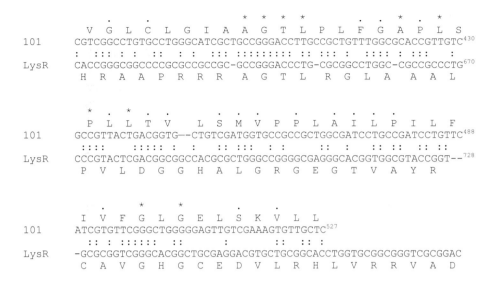

```
                                   *    *    *    *    .   .  *    .  *
            V  G  L  C  L  G  I  A  A  G  T  L  P  L  F  G  A  P  L  S
   101      CGTCGGCCTGTGCCTGGGCATCGCTGCCGGGACCTTGCCGCTGTTTGGCGCACCGTTGTC⁴³⁰
            :  ::: :  :: : : :::  :::::::::  ::: :  ::: :  : ::: :
   LysR     CACCGGGCGGCCCCGCGCCGCCGC-GCCGGGACCCTG-CGCGGCCTGGC-CGCCGCCCTG⁶⁷⁰
            H  R  A  A  P  R  R  A  G  T  L  R  G  L  A  A  A  L


                .    .  *          .          .          .          .
            P  L  L  T  V     L  S  M  V  P  P  L  A  I  L  P  I  L  F
   101      GCCGTTACTGACGGTG—-CTGTCGATGGTGCCGCCGCTGGCGATCCTGCCGATCCTGTTC⁴⁸⁸
            ::::     :::::: :  :  :: ::: : :    ::  :  :: :: :: ::
   LysR     CCCGTACTCGACGGCGGCCACGCGCTGGGCCGGGGCGAGGGCACGGTGGCGTACCGGT--⁷²⁸
            P  V  L  D  G  G  H  A  L  G  R  G  E  G  T  V  A  Y  R


                .       *       *             .          .
            I  V  F  G  L  G  E  L  S  K  V  L  L
   101      ATCGTGTTCGGGCTGGGGGAGTTGTCGAAAGTGTTGCTC⁵²⁷
            ::  :  : :::::::  ::      :         ::  :: :
   LysR     -GCGCGGTCGGGCACGGCTGCGAGGACGTGCTGCGGCACCTGGTGCGGCGGGTCGCGGAC
            C  A  V  G  H  G  C  E  D  V  L  R  H  L  V  R  R  V  A  D
```

protein similarity /89aa from 175aa/ 51%

Fig. 10. *Continued*

Although deficient DNA repair was proposed for neurodegenerative disorders including DS, repair genes for nucleotide excision repair or X-ray repair have not been studied in brain yet. As one of the hypotheses for the pathogenesis of brain damage in DS is oxidative stress and cells of patients with DS are more susceptible to ionizing irradiation, we decided to study ERCC2, ERCC3 and XRCC1, representatives of repair genes known to be involved in the repair of oxidative DNA damage. mRNA steady state levels of ERCC2, ERCC3, XRCC1 were determined and normalized versus the housekeeping gene, beta-actin in five individual brain regions of controls and adult DS patients. Although different in the individual regions, DNA repair genes were consistently higher in temporal, parietal and occipital lobes of patients with DS. Our results were the first to describe DNA repair gene patterns in human brain regions providing the basis for further studies in this area. We showed that DNA repair genes ERCC2 and ERCC3 (excision-repair-cross-complementing-) for nucleotide excision repair and XRCC1 (X-ray-repair-cross-complementing-) for X-ray repair, were increased at the transcriptional level with the possible biological meaning that this increase may be compatible with permanent (oxidative?) DNA damage (Fang-Kircher, 1999). The finding of overexpressed mRNA-XRCC1 in fetal DS brain may suggest an involvement in repair and recombination early in life (Table 1, Fig. 8).

The HtC transducer protein gene was described by Zhang and coworkers (1996) as one of the soluble or membrane-bound transducer proteins. They are methyl accepting proteins central to chemotaxis systems and serve in signal transduction in the archaeon halobacterium salinarium. This gene has

112   E. Kitzmueller et al.

```
dipeptidyl peptidase IV 78

                             *   *   *   *         *   *   .
                             D  M  L  A  A  D  G  Y  I  Q
78                          ¹TGATATGCTGGCTGCAGATGGCTACATACA²⁷
                            ::::::::::::    :  :  ::::::  :
DPIV¹⁸⁹⁶TTCCTGGGACGGCGGAAACGGAATCTGGTTTGATATGCTGGCACAAAAAGGCTACCTTGT¹⁹⁵⁵
                             F  D  M  L  A  Q  K  G  Y  L  V

            *    .    *   *   *   *          .       *       .   .   *   .
            V  S  L  D  G  R  G  T  P  A  P  K  G  A  A  W  R  K  A  I
78          AGTAAGTTTAGACGGAAGAGGAACACCCGCACCAAAAGGGGCAGCATGGCGTAAAGCTAT⁸⁷
            ::  ::  :  ::::::  :  ::::::  :        ::  :   ::     ::::  :
DPIV        CGTTTGTGTGGACGGGCGCGGAACA---GGTTTCCGTGGAACGAAATACAAAAAAGTAAC²⁰¹²
            V  C  V  D  G  R  G  T        G  F  R  G  T  K  Y  K  K  V  T

            *  *       *  *   .   *  *        *   *   *       .       *       *   .   *
            Y  K  Q  L  G  R  Y  E  C  E  D  Q  A  M  A  A  N  K  L  L
78          TTACAAACAATTAGGCAGATATGAATGTGAAGATCAGGCAATGGCTGCAAACAAATTATT¹⁴⁷
            ::  :::  :  ::  :  :::  ::    :::::::::::  ::   :::::   ::  :  ::
DPIV        CTATAAAAATCTGGGTAAATACGAGATTGAAGATCAG---ATTACTGCAGCGAAGTGGTT²⁰⁶⁹
            Y  K  N  L  G  K  Y  E  I  E  D  Q        I  T  A  A  K  W  L

               *   *   *   .   *       .   *    .   .   .   *   *   *      *   *
               Q  P  Q  S  Y  I  D  A  N  R  V  A  V  W  G  W  S  G  G  G
78             GCAAGGCCAGTCGTATATTGATGCAAACCGTGTAGCAGTTTGGGGTTGGAGTGGTGGCGG²⁰⁷
               :  :::::::::::  :  :::    :  :  ::  :  :  ::  ::::::    :::  ::
DPIV           AGGAAACCAGTCGTATGTGGATAAGTCCAGAATTGGTATCTTTGGTTGGTCGTATGGAGG²¹²⁹
               G  N  Q  S  Y  V  D  K  S  R  I  G  I  F  G  W  S  Y  G  G

            *          .    .    .        *   *    .   *    .   *      *   *   *    .
            Y  Q  T  T  F  T  I  S  I  G  P  D  I  Y  K  T  G  I  A  I
78          TTACAACACTACATTTACTATTTCAATAGGTCCTGATATTTATAAAACGGGTATTGCCAT²⁶⁷
            :::  :     ::  :   :     :  ::    :::  ::::  :::::  :  :::::  ::::::::  :
DPIV        TTATATGGCTTCTCTGGCGATGACAAAAGGTGCTGATGTATTTAAAATGGGTATTGCTGT²¹⁸⁹
            Y  M  A  S  L  A  M  T  K  G  A  D  V  F  K  M  G  I  A  V

            *   *   *       *          *   *    .   *   *       *   *    .       .
            A  P  V  A  N  Q  L  C  Y  D  N  I  Y  Q  E  R  Y  M  G  L
78          TGCACCGGTTGCAAACCAGTTATGTTATGATAATATTTATCAGGAAAGATACATGGGTTT³²⁷
            ::  :::::  :     ::  ::  ::  ::    ::::::::  :::::::::  :  ::
DPIV        AGCTCCGGTAACCAACTGGAGATTCTACGACAGTATTTATACGGAAAGATTCCTGCAGAC²²⁴⁹
            A  P  V  T  N  W  R  F  Y  D  S  I  Y  T  E  R  F  L  Q  T

            *   *   *   *   *   *    .    .        *   *    .   *   *   *   *      *
            P  Q  E  N  K  D  D  F  V  K  G  S  P  V  T  Y  A  K  N  L
78          GCCACAGGAAAATAAAGATGATTTTGTAAAAGGTTCGCCGGTTACGTATGCAAAAAATTT³⁸⁷
            ::::::::  ::::::::  :  ::  ::      ::::::    :::  :::::  :::  :
DPIV        ACCACAGGAGAATAAAGACGGTTATGACCTGAATTCGCCAACTACTTATGCTAAACTGCT²³⁰⁹
            P  Q  E  N  K  D  G  Y  D  L  N  S  P  T  T  Y  A  K  L  L

            *   *    .    .    *    .    *   *   *    .   *   *   *   *    .   *   *    .
            K  G  N  L  L  V  V  H  G  T  G  D  D  N  V  H  Y  Q  N  T
78          AAAAGGTAATTTATTGGTTGTTCATGGAACCGGTGATGATAATGTGCATTATCAAAATAC⁴⁴⁷
            ::::::  ::  ::  :   ::::::::  ::  :  :::::::::::::  ::  :   ::  ::  :
DPIV        AAAAGGCAAATTCCTGCTGATTCATGGTACAGCCGATGATAATGTACACTTCCAGAACTC²³⁶⁹
            K  G  K  F  L  L  I  H  G  T  A  D  D  N  V  H  F  Q  N  S
```

Fig. 11

never been described in the mammalian system and we now demonstrate its presence in an upregulated form in fetal DS brain (Table 1, Fig. 9). A tentative human homolog may well serve as a methyl-accepting protein or chemoattractant, possibly in recognition processes.

```
        *       .   .   *   *   *   .       *   *   *   .   .   *   .   *   *
        M   L   I   N   E   A   L   V   K   N   K   K   Q   F   Q   L   M   S   Y   P
78      AATGCTGATTAATGAAGCTTTGGTGAAAAATAATAAACAATTTCAGTTAATGAGTTACCC⁵⁰⁷
        :::   :   ::     :::::::::   :   ::   ::   ::   :::::   :   ::   :::   :::   ::
DPIV    TATGGAGTTTTCCGAAGCTTTGATCCAGAACAAAAAGCAATTTGACTTTATGGCTTATCC²⁴²⁹
        M   E   F   S   E   A   L   I   Q   N   K   K   Q   F   D   F   M   A   Y   P


        .       *   *   *   *       .   *   *   *   *
        N   R   T   H   S   I   S   E   G   N   T   R
78      CAACAGAACACACAGTATTAGTGAAGGAAATACACGTC⁵⁴⁵
        :::   ::     :::::   ::::     :   :::::::::::::::
DPIV    GGACAAAAACCACAGCATTATCGGAGGAAATACACGTCCGCAGTTATACGAAA²⁴⁸²
        D   K   N   H   S   I   I   G   G   N   T   R   P


protein similarity (148aa from 181aa) 82%
```

Fig. 11. *Continued*

```
L07807 Human dynamin,241


                                        *   *   .   *       *   .
                                        T   S   G   P   C   P   H   P   S
241                                     ¹ACCTCTGGCCCCTGTCCCCACCCCTCC²⁷
                                        ::   ::     :::::     ::   ::   :   :
hsdy²³⁵⁹GTGCAGAGCGTACCGGCCGGACGCAGGTCGCCCACGTCCAGCCCCACGCCGCAGCGCCGA²⁴¹⁹
                                        T   S   S   P   T   P   Q   R   R


        *       .   .   *       .   *   .   *       .       *       *   *   *
        A   L   S   G   A   P   G   G   P   E   S   P   S   T   A   H   I   P   P   P
241     GCCCTCTCCGGAGCCCCTGGCGGCCCCGAGTCACCCAGCACAGCTCACATTCCTCCGCCG⁸⁷
        ::::   :   :::     ::::   :   :   :::   :::   :     ::   :   ::::       :::::::
hsdy    GCCCCCGCCGTGCCCCCAGCCCGGCCCGGGTCGCGGGGCCCTGCTCCTGGGCCTCCGCCT²⁴⁷⁹
        A   P   A   V   P   P   A   R   P   G   S   R   G   P   A   P   G   P   P   P


        *   *       *   *   *   .       *       .       .       *       *   *   .
        A   G   L   A   L   G   P   E   P   L   G   K   A   P   G   E   R   A   S   A
241     GCCGGGTTGGCTCTCGGGCCAGAGCCCCTGGGCAAGGCCCCTGGGGAGAGGGCCTCAGCT¹⁴⁷
        ::   ::::     ::   ::   :::     :   :::::     :   :   ::   :     ::   :   ::::   ::   ::
hsdy    GCTGGGTCCGCCCTGGGGGGGGGCGCCCC---CCGTGCCCTCCAGGCCGGGGGCTTCCCCT²⁵³⁶
        A   G   S   A   L   G   G   A   P       P   V   P   S   R   P   G   A   S   P


        .   *   *       *   *   *   .       *       .       .       *   *   .
        G   P   F   W   P   P   P   K   A   S   S   R   A   N   P   D   P   P   V   S
241     GGGCCTTTCTGGCCTCCTCCGAAGGCCTCCTCGCGGGCCAACCCAGACCCTCCCGTGTCC²⁰⁷
        :     :::::::   :   :::::   ::     :::     :::::::     :::::::     :   :::   :::   :   :
hsdy    GACCCTTTCGGCCCTCCCCCTCAGGTGCCCTCGCGCCCCAACCGCGCCCCGCCCGGGGTC²⁵⁹⁶
        D   P   F   G   P   P   P   Q   V   P   S   R   P   N   R   A   P   P   G   V


        *   *       .   .   *   *   .       *           .   .   *   *       *   *
        P   S   P   A   A   Q   A   G   S   R   R   R   A   C   A   R   P   H   F   D
241     CCCAGCCCAGCAGCTCAGGCAGGCTCACGCCGACGGGCTTGCGCCAGGCCCCACTTCGAC²⁶⁷
        :::::::   :   :   :   :::::::   :     ::     :::   :   :       ::   :::::::::   :::::::
hsdy    CCCAGCCGATCGGGTCAGGCAAGTCCATCCCGTCCTGAGAGCCCCAGGCCCCCCCTTCGAC²⁶⁵⁶
        P   S   R   S   G   Q   A   S   P   S   R   P   E   S   P   R   P   P   F   D


Protein (89AA,similar 64AA):72% similarity
```

Fig. 12

A sequence with high homology to the metalloproteinase gene (mprA) and its transcriptional activator (mprR), a member of the LysR family, was found to be overexpressed in DS fetal brain (Table 1, Fig. 11). The protease secreted by the heterologous host Streptomyces lividans was characterised biochemically as a metalloproteinase with a M(r) of 20.000, which is in good agreement with data derived from DNA sequence analysis. The mprA gene is transcribed divergently from mprR, the deduced protein of which displays homology to prokaryotic transcriptional regulators of the LysR family. It was found to activate transcription in S. lividans and in E. coli (Dammann, 1992). The biological meaning of this finding may be deterioration of the concerted action of about 5,000 transcription factors, modulating and controling wiring and plasticity of the DS brain.

Upregulation of dipeptidyl peptidase IV (DPP-IV) m RNA steady state levels in DS fetal brain (Table 1, Fig. 11) may be responsible for deteriorated neuropeptide and -protein handling and metabolism in the brain: DPP-IV is ubiquitously expressed in glia and neurons in the rat brain, although activities are low in contrast to DPP-II (Mentlein, 1990). The major role of DPP-IV according to present knowledge is neuropeptide Y (NPY) converting enzyme activity (Zukowska-Grojec, 1998): NPY action is specific and is mediated by Y1 and Y2 receptors. The expression of both receptors is upregulated during cell growth; however, the Y2 appears to be the main NPY angiogenic receptor. Its upregulation parallels the NPY-induced capillary tube forming activity of NPY, whereas the Y2 antagonist blocks it. Brain endothelium contains NPY receptors, the peptide itself and the "NPY-converting enzyme" dipeptidyl peptidase IV, which terminates Y1 activity of NPY and cleaves the Tyr1-Pro2 from NPY to form an angiogenic Y2 agonist, NPY3-36. Upregulation may therefore implicate a developmental effect on sympathetic nerves and angiogenesis during tissue development and repair in adult brain. As DPP-IV is cleaving also glucagon-like peptides (Drucker, 1998), the upregulation could be involved in impaired brain glucose metabolism, a clinically and biochemically well-documented phenomenon in DS (Labudova, 1999).

Upregulation of the sequence assigned to dynamin (Table 1, Fig. 12), may be seen in the context of impaired synaptic functions in DS. Dynamin is a 100kDa GTPase, necessary for normal development and function of mammalian neural tissue. In neurons, it is necessary for the biogenesis of synaptic vesicles. The function mechanism entails the formation of a horseshoe-shaped dynamin polymer at the neck of the budding vesicle, followed by neck scission through a GTP hydrolysis dependent activity.

Dynamin is activated by stimulus coupled PKC phosphorylation in brain (Scaife, 1997) and this is a major step in the clathrin-mediated endocytosis involving cycles of assembly and disassembly of clathrin coat components; dephosphorylation of rat brain extract was shown to promote the assembly of dynamin 1, synaptojanin 1 and amphiphysin into complexes that also included clathrin and AP-2 (Slepnev, 1998).

Apart from endocytic processes dynamin(s) are thought to play roles in cell growth, cell spreading and neurite outgrowth (Urrutia, 1997), a function which was examined by antisense studies in rat hippocampal neurons (Torre,

```
apolipo E,  102

102    ¹GGAGCTGCGGGTGCGCCTCGCCTCCCACCTGCGCAAGCTGCGTAAGCGGCTCCTCCGCGA⁶⁰
       ::::::::::::::::::::::::::::::::::::::::::::::::::::::::::::::::
hsAP ⁴⁶⁰GGAGCTGCGGGTGCGCCTCGCCTCCCACCTGCGCAAGCTGCGTAAGCGGCTCCTCCGCGA⁵²⁰

102     TGCCCATGACCTGCAGAAGCGCCTGGCAGTGTACCAGGCCGGGGCCCGCGAGGGCGCCGA¹²⁰
        ::::::::::::::::::::::::::::::::::::::::::::::::::::::::::::::::
hsAP    TGCCCATGACCTGCAGAAGCGCCTGGCAGTGTACCAGGCCGGGGCCCGCGAGGGCGCCGA⁵⁸⁰

102     GCGCGGCCTCAGCGCCATCCGCGAGCGCCTGGGGCCCCTGGTGGAACAGGGCCGCGTGCG¹⁸⁰
        ::::::::::::::::::::::::::::::::::::::::::::::::::::::::::::::::
hsAP    GCGCGGCCTCAGCGCCATCCGCGAGCGCCTGGGGCCCCTGGTGGAACAGGGCCGCGTGCG⁶⁴⁰

102     GGCCGCCACTGTGGGCTCCCTGGCCGGCCAGCCGCTACAGGAGCGGGCCCAGGCCTGGGG²⁴⁰
        ::::::::::::::::::::::::::::::::::::::::::::::::::::::::::::::::
hsAP    GGCCGCCACTGTGGGCTCCCTGGCCGGCCAGCCGCTACAGGAGCGGGCCCAGGCCTGGGG⁷⁰⁰

102     CGAGCGGCTGCGCGCGCGCGGATGGAGGAGATGGGCAGCCGGACCCGCGACCGCCTGGACGA³⁰⁰
        ::::::::::::::::::::::::::::::::::::::::::::::::::::::::::::::::
hsAP    CGAGCGGCTGCGCGCGCGCGGATGGAGGAGATGGGCAGCCGGACCCGCGACCGCCTGGACGA⁷⁶⁰

102     GGTGAAGGAGCAGGTGGCCGAGGTGCGCGCCAAGCTGGAGGAGCAGGCCCAGCAGATAC³⁵⁹
        ::::::::::::::::::::::::::::::::::::::::::::::::::::::::::::::::
hsAP    GGTGAAGGAGCAGGTGGCCGAGGTGCGCGCCAAGCTGGAGGAGCAGGCCCAGCAGATAC⁸¹⁹
```

Fig. 13

1994). It is entirely unclear at the moment how increased dynamin mRNA steady state levels in DS brain may modulate neurite outgrowth or synaptogenesis or endo/exocytosis, functions impaired in DS brain.

The overexpression of a transcript with 100% homology to apolipoprotein E3 in fetal DS (Table 1, Fig. 13) supports considerations on a tentative role of apolipoproteins (APOs) in the pathogenesis of DS pathology.

Although most work on APOs in DS patients was performed epidemiologically in plasma samples (Tyrrell, 1998) and not in the brain, there are reports showing an association of APOs with senile plaques and amyloid deposition: Arai (1995) described a temporal profile of apo-E expression as revealed by immunehistochemistry in DS brain. The number of apo-E immunoreactive astrocytes in frontal cortex was larger than in controls and apo-E immunoreactive senile plaques were noted in DS from the age of 25 years while amyloid precursor protein immunoreactivity was first noted in senile plaques at the age of 32 years.

Hyman and coworkers (1995) performed a neuropathological study in the hippocampal formation of DS patients and concluded that the Apo E epsilon 4 genotype appears to be an additional, independent risk factor for developing higher levels of amyloid accumulation.

Mann and coworkers (1995) showed that the extent of amyloid deposition in brain of patients with DS does not depend upon the apolipoprotein E genotype; no significant differences in the amount of amyloid A beta deposited in the brain, either as A beta 42(43) or beta 40, were noted in patients possessing an ApoE4 allele, compared with those without. They conclude that APOs may influence the timing of onset, rate of progression of disease but without affecting the type or total amount of tangles or plaques.

SOX4, 266

```
266                                           ¹GAAGGCACGGCTTGCGGGCTTGTGGTTGGG³⁰
                                              :: ::    : :: : :: ::    :::::::
SOX4  ⁴⁷⁰gcccgcaggagcacttcagcggtgagggaggacggagggcctcggggactctaggttggc⁵²⁹
      5´UTR

266   CTCATGAAGCGGCCATCCGTTGCCCAC-CGCCGCTCTGAGGTCGCCCGGATGTTCCGAGA⁸⁹
      :  ::  ::::::  :: : :::: : :::::::: : :: : :  ::  :
SOX4  ggcgggaggcggcc-gccctggcccgcgcgccgctcaggggaccctggtctccgccccg-⁵⁸⁷

266   GGGACACCGCCGGGCCGTCATCGCCCAAGCCCTGGGCGTTCCATCGCCCGTGGTGAGCGC¹⁴⁹
      ::::  ::::  :: :::    :: ::   :::  : ::::   : ::: :::::    ::
SOX4  gggagcccgcaggcccg---gcgaccgcgccgcgagcgtgtgagcgcgcgtgggcgccgg⁶⁴⁴

              *       .      .       .
              M  G  L   R  Q  K  P  R  R  P  L          A  D
266   CGACCTCGCTCACATGGGGTTGC-GTCAAAAGCCCCGGCGCCCACTG-------GCTGAC²⁰¹
      : :  :: ::    :::::    :::::    : ::  : ::::    ::::  :
SOX4  CAAGC-CGGGGCCATGGTACAACAGACCAACAACGCGGAGAACACTGAGGCTCTGCTGGC⁷⁰³
              M  V  Q  Q  T  N  N  A  E  N  T  E  A  L  L  A

           .        *   *     *         .         .
           Q  H  A  L     V  V  R  L  L  E  Q  G  S  N  G  P  A  I  A
266   CAACACGCCCTGG---TGGTGCGCC-TCCTGGAGCAGGGGAGCAACGGCCCTGCAATTGC²⁵⁷
      :     :  :: :  :: : :::::  :::::::: ::: :  :: ::    : ::
SOX4  CGGGGAGAGCTCGGACTCGGGCGCCGGCCTGGAGCTGGGCATC-GCGTCCTCCCCGACGC⁷⁶²
           G  E  S  S  D  S  G  A  G  L  E  L  G  I  A  S  S  P  T

              .     *        .      .        .
              K  R  L  G  V  T  R  Q  A  V  H  G  Y  L  Q  A  H  G  L  L
266   CAAGCGCCTCGGCGTGACCCGACAGGCAGTCCACGGTTACCTGCAGGCCCATGGCCTGCT³¹⁷
      : :: :::: ::::   : :: :: :   :   :: :: :: ::    :: :: :::
SOX4  CTGGCTCCACCGCGTCGACGGGCGGCAAGGCGGACGACCCCAGCTGGTGCAAGAC--GCC⁸²⁰
           P  G  S  T  A  S  T  G  G  K  A  D  D  P  S  W  C  K  T     P

              .        *  *     .     *    .   *     .         *
              P  T  Q  M  E  D  P  M  T  V  F  L  V  V  V  L  F  E  L  A
266   CCCAACCCAGATGGAGGATCCCATGACCGTTTTTCTGGTGGTCGTGCTGTTTGAGCTCGC³⁷⁷
      :      ::: ::  ::   :::::::: ::  ::: :::     :: : :::
SOX4  CAGTGGCCACATCAAGCGGCCCATGAACGCCTTTATGGTGTGGTCGCAGATCGAGCGGCG⁸⁸⁰
           S  G  H  I  K  R  P  M  N  A  F  M  V  W  S  Q  I  E  R  R

              .
              M  F  A
266   CATGTTTGC³⁸⁶
      :: : :
SOX4  CAAGATCATGGAGCAGTCGCCCGACATGCACAACGCCGAGATCTCCAAGCGGCTAGGCAA⁹⁴⁰
           K  I  M  E  Q  S  P  D  M  H  N  A  E  I  S  K  R  L  G  K
```

protein similarity (37aa from 74aa) 50%

Fig. 14

Lemere and coworkers (1996) studied the sequence of deposition of amyloid beta-peptides and APO E in DS brain of patients 3–73 years old. A beta ending at amino acid 42 (A beta 42) was the earliest form of A beta deposition in DS cortex with earliest deposition age of 12 years. A beta ending

```
HSP 70,  180
                                                    .         *   *   .   *
                                            K   D  R  T  H  A  D  E  A
180                                       ¹CAAGAGACCGGACA-CATGCGGATGAGGCC²⁹
                                          ::  :::  :::  :::  ::  ::  ::  ::::  :
HSP  ¹⁴³⁸CTCTCCGACGACCAGATCGAGGAGATGCAACAGGAGGCCGAACAGCACGCCGAAGAGGAC¹⁴⁹⁷
                                            Q   E  A  E  Q  H  A  E  E  D


                      .           *       .     .       *       .   *            *  *
                  G   D   Q   R   Y   Q   T   K   P   S   N   Q   A   G   T   L   L   K   R   A
180               GGTGACCAGAGGTACCAGACCAAGCCATCCAACCAGGCCGGTACCTTGCTGAAGAGGGC⁸⁸
                  :  :  :  :  :  ::  :  :  :  :  ::::  ::::  :  :  :  :  :
HSP               GAGCAGCGCCGCGACGGCATCGAAGCGCGCAACGAGGCCGAGGCGTCCGTCCGCCGTGC¹⁵⁵⁶
                  E   Q   R   R   D   G   I   E   A   R   N   E   A   E   A   S   V   R   R   A


                      *           .       *           .       *               *
                  E   R   N   Q   Q   E   A   D   Q   D   A   N   D   A   V   G   Y   S   Y   V
180               CGAGAGGAATCATCACGAGGCCGATCAGGATGCGAACGATGCGGTCGGCTGTTCCTACGT¹⁴⁸
                  :::::  :  :  :  ::::::  ::    ::::  :  ::  :::::  ::  :   :::  :  :
HSP               CGAGACGCTCCTTGACGAGAACG---AGGAGGAGATCGATGAGGACCTCCAGTCCGACAT¹⁶¹³
                  E   T   L   L   D   E   N       E   E   E   I   D   E   D   L   Q   S   D   I


                      .       *           .       *               .           *       .   *
                  G   K   H   G   S   T   P   T   D   E   M   K   S   G   A   R   P   K   R
180               --TGGGAAACATGGCAGCACTCCTACGGATGACATGAAATCGGGCGCACGTCCGAAGCGC²⁰⁶
                  ::    ::  ::  ::  ::  ::        :  ::  :  :  :  :  :  :  :::  :
HSP               CGAGGCGAAAATCG-AGGACGTCGAGGAAGTCCTCGAAGACGAGGACGCCACGAAAGAGG¹⁶⁷²
                  E   A   K   I   E   D   V   E   E   V   L   E   D   E   D   A   T   K   E


                      *               .       *           .               *
                  L   T   E   I   G   C   R   N   R   R   R   C   T   G   R   G   R   G   C   R
180               TTAACGGAGATCGGGTGTCGAAACCGTCGTCGCTGCACCGGTCGCGGCCGAGGATGCCG-²⁶⁶
                  :::  :   :::          :::  :::  :   ::  :  ::  :   ::  :  :  :  :  :
HSP               ACTACGAGG--CGGTCACCGAGACCCTGAGCGAAGAACTGCAGGAGATCG-GCAAGCAGA¹⁷²⁹
                  D   Y   E       A   V   T   E   T   L   S   E   E   L   Q   E   I       G   K   Q


                          .       .       .       .                       *       .   .
                  P   A   A   S   I   H   R   R   L   R   G   W   C       R   P   V   S   A   C
180               GCCAGCAGCGTCCATTCACCGCCGATTGCGCGGATGGTGTC---GTCCAGTGTCAGCATG³²³
                  :  :::  ::  ::  :  :::  :  :::::  :  ::::  :   :  :       :::  :  ::
HSP               TGTATCAG-GATCAGGCCCAGCAG---GCGCAGGCGGTGCCGCGGGCGCTGGTCCGGGTG¹⁷⁸⁵
                  M   Y   Q       D   Q   A   Q   Q       A   Q   A   V   P   R   A   L   V   R   V


                      *       .   *       .       *           *       .   *       *   *   .       *
                  R   Y   A   A   A   D   R   L   N   H   R   K   A   V   Q   A   V   T   E
180               CCGATATGCAGCTGCTGATCGGCTCAATCACCGCAAGGCGGTGCAGGCCGTGACCGAGC³⁸²
                  ::      ::  :   :  :  :::::      ::::      ::::    ::::::::  :  ::::::
HSP               GCG─CGGCCGGCCCCGGGCGCGCTGCCGGACCGGGCGGCGCAGCAGGCCGCGGCCGAGC¹⁸⁴²
                  A   R   P   A   P   G   A   L   P   D   R   A   A   Q   Q   A   A   A   E


                      *   *   *   .           .   .
                  Q   G   A   Q   A   F   A   H   D
180               AGGGTGCGCAGGCGTTCGCTCATGAC⁴⁰⁸
                  :::::::  :::  ::  ::    :  :
HSP               AGGGTGCTGAGGAGTACGTCGACGCAGACTTTGAAGACGTCGAGGAAAGCGACGAAGACG¹⁹⁰³
                  Q   G   A   E   E   Y   V   D   A   D   F   E   D   V   E   E   S   D   E   D   E*


protein similarity (79aa from 136aa) 58%
```

Fig. 15

at residue 40 (A beta 40) was not detected until approx age 30, a time when degenerating neurites around A beta immunoreactive plaques were first observed. Apo E was detectable in a small subset of A beta 42 immunoreactive plaques beginning at age 12 and rose steadily with age; it clearly followed the deposition of A beta.

Nothing has been reported so far at the stage of brain development and our finding of upregulated apo E3 transcript is shedding new lights and challenging more allele specific studies on apo E3 in brain of adult and fetal DS.

A sequence homologous to the transcription factor (TF) sox-4 was downregulated (Table 2, Fig. 14). It is a Sry-like HMG (high mobility group) box protein and a transcriptional activator in lymphocytes (van de Wetering, 1993). It was shown to be expressed in bone marrow, gonads and thymus where it facilitates thymocyte differentiation (Schilham, 1997). Human SOX4 is encoded on chromosome 6p distal to the MHC region (Farr, 1993). During murine embryogenesis Sox-4 is expressed at several sites including the heart where targeted gene disruption was leading to malformation of the heart (impaired development of endocardial ridges, a finding observed in DS patients) but in adult mice expression is restricted to immature B and T cells (Schilham, 1996). The expression of Sox-4 and other sox family members was reported by Uwanogho and coworkers (1995): Three chicken Sox genes have been identified that show an interactive pattern of expression in the fetal embryonic nervous system and the switch of expression from immature to mature neurons may indicate a role in the developmental programme of the CNS. Recently, Kuhlbrodt and coworkers (Kuhlbrodt, 1998) found in adult glial cells of the oligodendrocyte lineage abundant Sox4 and Sox11 and showed the interaction with the POU- system of TFs. TFs are essential for development and wiring of the brain both in development and in adult brain where it contributes to the modulation of plasticity and deranged TFs have already been reported in DS brain (Labudova, 1998). Sox TFs occur in the animal kingdom as early as in oogenesis (Hiraoka, 1997). Kanda and coworkers (1998) demonstrated the presence of the sox family in rainbow trout oocytes where it acts as a transcriptional regulator.

And as the interspecies homology is extraordinarily high in the sox family, one may suggest that SOX4 may have a role in early development of DS phenotype.

A second TF was shown to be downregulated in DS brain: Tax helper protein 1 (syn: TRE-2S-binding protein) (Table 2), is expressed in thymus, lung, spleen and the brain (Benvenisty, 1992) and the tax-responsive element has been shown to be present in human brain (Nyunoya, 1994).

Fu and coworkers described the transactivation of proenkephalin gene by tax1 protein in glial cells (Fu, 1997). The tax1 induced cFos protein levels and the concurrently increased association of c-Fos/c-Jun transcription factors at the AP-1 site imply a strong functional significance in the activation of proenkephalin gene expression in tax1 expressing glial cells. The downregulation of tax1 in brain of patients with DS may therefore interfere with neuropeptide (proenkephalin) functions in the CNS as mentioned above.

Mavria and coworkers found that tax modulated the expression of MHC class I gene expression in rat oligodendroglial cells (Mavria, 1998). Dysregulation of MHC class I molecules by cytokines and viral infections has been associated with a number of neurological disorders and the downregulation of tax1 in DS brain along with known multiple immune defects in DS (Epstein, 1995) may indicate a link between the immune system and the CNS pathology. Cowan and coworkers (1997) observed that tax induced tumor necrosis factor alpha (TNF alpha) in human neuronal cells, a finding that may link the TNFalpha-pathway of apoptosis to the well-documented apoptosis taking place unequivocally in DS brain.

The finding of downregulation of hsp70 in our DS fetal brain samples (Table 2, Fig. 15) fits perfectly increased apoptosis in brain of patients with DS: Wei and coworkers showed that in tumour cells Hsp70 antisense treatment inhibited hsp70 expression which in turn inhibited cell proliferation and induced apoptosis in tumour cells (Wei, 1995).

The role of hsp70 in apoptosis is, however, not unequivocal, as Mesner and coworkers, reporting a time table of events during programmed cell death induced by trophic factor withdrawal from neuronal PC12 cells, clearly showed that a large series of immediate — early genes were expressed but hsp70 remained unchanged (Mesner et al., 1995).

Also Mailhos (1994) reported that hsp70 and hsp90 protect neuronal cells from thermal stress but not from programmed cell death a finding confirmed by Wyatt and coworkers (1996).

Mori and coworkers performed morphological analysis of germ-cell apoptosis during postnatal testis development in normal and hsp70-2 knockout mice and revealed that in pachytene spermatocytes an additional increased level of apoptosis on day 17 was found (Mori, 1997).

He and Fox (1997) generated results in HL-60 cells suggesting that inducible hsp70(hsp72) but not hsp73 is involved in the development or prevention of apoptosis. Mosser and coworkers (1997) proposed hsp70 — mediated protection against stress-induced apoptosis in vitro by inhibition of the stress-activated protein kinase/c-jun N-terminal kinase signaling and apoptotic protease effector steps.

Hohfeld recently stressed the link between apoptosis and the cellular chaperone machinery (Hohfeld, 1998). The regulation of the chaperone activity of the heat shock cognate Hsc70 protein in the mammalian cell involves a cooperation with chaperone cofactors such as Hsp40, the hsp70-interacting protein Hip, and the hsc70/hsp90-organizing protein Hop and Bag-1, initially identified as an anti-apoptotic molecule and binding partner of the cell death inhibitor bcl-2. BAG-1 appears to fulfill ist cellular function through modulation of hsc70's chaperone activity. Studies from Kwak and coworkers demonstrated that hsp70 plays a crucial role in the differentiation of myeloid cells, participating in cell cycle control and phenotypic changes, with protecting effects on apoptosis induced by different pathways (Kwak et al., 1998).

No systematic studies have been performed on the many apoptotic pathways in human brain, both, physiological or in DS or AD in context with hsp70 and this is the first observation of decreased mRNA steady state levels of hsp70 in DS brain. HSP70 is encoded on human chromosome 21 (altough not confirmed by others) and one would rather expect increased levels, on the other hand, the gene dosage hypothesis of overexpressed genes of chromosome 21 determining the DS phenotype has already been challenged (Greber-Platzer, 1999).

Apart from a tentative role of downregulated hsp70 for the apoptotic mechanism(s) the major biological meaning of downregulated hsp70 may be the following: Dix and coworkers (1996) performed targeted disruption of the hsp70-2 which resulted in failed meiosis, a finding of fundamental importance for DS genetics and pathology in general.

In conclusion, we provided evidence for the dysregulated expression of several functional elements, candidates helping to understand the complex pathogenesis and molecular biology of Down Syndrome brain. We are reporting an emerging pattern of differential expression in DS brain.

Our data support and explain pathological features and phenotype of DS as e.g. metabolic, hormonal, transport derangement; deterioration of brain structure and function could be reflected by altered expression of transcription factors and channels. The expressional difference, set at the mRNAs-level of 1:3, makes allelic differences for the explication of results highly improbable. We are aware of the pitfalls of gene hunting methods as differential display and subtractive hybridization, but they are still the most effective and potent way to the identification of differentially expressed patterns and new genes in the brain (Bertioli, 1995; Sompayrac, 1995; Bauer, 1994).

We are currently complementing our transcriptional data by protein and activity level studies but are readily providing sequences from our subtractive library to the scientifique community on request.

Acknowledgements

We are highly indebted to the Red Bull Company, Salzburg, Austria, for generous financial support of the study.

References

Arai Y, Mizuguchi M, Ikeda K, Takashima S (1995) Developmental changes of apolipoprotein E immunoreactivity in Down syndrome brains. Brain Res Dev Brain Res 87: 228–232

Bauer D, Warthoe P, Rhode I, Struss M (1994) PCR methods and applications. Cold Spring Harbor Laboratory Press, Cold Spring Harbor, NY, pp 97–108

Benvenisty N, Ornitz DM, Bennett GL, Sahagan BG, Kuo A, Cardiff RD, Leder P (1992) Brain tumours and lymphomas in transgenic mice that carry HTLV-I LTR/c- myc and Ig/tax genes. Oncogene 7: 2399–2405

Bertioli DJ, Schlichter UH, Adams MJ, Burrows PR, Steinbiss HH, Antoniw JF (1995) An analysis of differential display shows a strong bias towards high copy number mRNAs. Nucl Acids Res 23: 4520–4523

Brodsky G, Barnes T, Bleskan J, Becker L, Cox M, Patterson D (1997) The human GARS-AIRS-GART gene encodes two proteins which are differentially expressed during human brain development and temporally overexpressed in cerebellum of individuals with Down syndrome. Hum Mol Genet 6: 2043–2050

Casciola-Rosen LA, Miller DK, Anhalt GJ, Rosen A (1994) Specific cleavage of the 70-kDa protein component of the U1 small nuclear ribonucleoprotein is a characteristic biochemical feature of apoptotic cell death. J Biol Chem 269: 30757–30760

Casciola-Rosen L, Nicholson DW, Chong T, Rowan KR, Thornberry NA, Miller DK, Rosen A (1996) Apopain/CPP32 cleaves proteins that are essential for cellular repair: a fundamental principle of apoptotic death [see comments]. J Exp Med 183: 1957–1964

Casiano CA, Martin SJ, Green DR, Tan EM (1996) Selective cleavage of nuclear autoantigens during CD95 (Fas/APO-1)- mediated T cell apoptosis. J Exp Med 184: 765–770

Choi KS, Lim IK, Brady JN, Kim SJ (1998) ICE-like protease (caspase) is involved in transforming growth factor beta1-mediated apoptosis in FaO rat hepatoma cell line. Hepatology 27: 415–421

Cowan EP, Alexander RK, Daniel S, Kashanchi F, Brady JN (1997) Induction of tumor necrosis factor alpha in human neuronal cells by extracellular human T-cell lymphotropic virus type 1 Tax. J Virol 71: 6982–6989

Dammann T, Wohlleben W (1992) A metalloprotease gene from Streptomyces coelicolor "Muller" and its transcriptional activator, a member of the LysR family. Mol Microbiol 6: 2267–2278

de la Monte SM, Xu YY, Hutchins GM, Wands JR (1996) Developmental patterns of neuronal thread protein gene expression in Down syndrome. J Neurol Sci 135: 118–125

Dix DJ, Allen JW, Collins BW, Mori C, Nakamura N, Poorman-Allen P, Goulding EH, Eddy EM (1996) Targeted gene disruption of Hsp70-2 results in failed meiosis, germ cell apoptosis, and male infertility. Proc Natl Acad Sci USA 93: 3264–3268

Drucker DJ (1998) Glucagon-like peptides. Diabetes 47: 159–169

Epstein CJ (1992) Down syndrome (Trisomy 21). In: Scriver CR, Beaudet AL, Sly WS, Valle D (eds) The metabolic and molecular basis of inherited disease. McGraw Hill, New York

Epstein CJ (1995) Epilogue: toward the twenty-first century with Down syndrome–a personal view of how far we have come and of how far we can reasonably expect to go. Prog Clin Biol Res 393: 241–246

Fang-Kircher S (1999) Increased steady state mRNA Levels of DNA-repair genes XRCC1, ERCC2 and ERCC3 in brain of patients with Down Syndrom. Life Sci 64: 1689–1699

Farr CJ, Easty DJ, Ragoussis J, Collignon J, Lovell-Badge R, Goodfellow PN (1993) Characterization and mapping of the human SOX4 gene. Mamm Genome 4: 577–584

Fu W, Shah SR, Jiang H, Hilt DC, Dave HP, Joshi JB (1997) Transactivation of proenkephalin gene by HTLV-1 tax1 protein in glial cells: involvement of Fos/Jun complex at an AP-1 element in the proenkephalin gene promoter. J Neurovirol 3: 16–27

Gerbl-Rieger S, Peters J, Kellermann J, Lottspeich F, Baumeister W (1991) Nucleotide and derived amino acid sequences of the major porin of Comamonas acidovorans and comparison of porin primary structures. J Bacteriol 173: 2196–2205

Gerbl-Rieger S, Engelhardt H, Peters J, Kehl M, Lottspeich F, Baumeister W (1992) Topology of the anion-selective porin Omp32 from Comamonas acidovorans. J Struct Biol 108: 14–24

Giaid A, Gibson SJ, Herrero MT, Gentleman S, Legon S, Yanagisawa M, Masaki T, Ibrahim NB, Roberts GW, Rossi ML, et al (1991) Topographical localisation of endothelin mRNA and peptide immunoreactivity in neurones of the human brain. Histochemistry 95: 303–314

Golubnitchaya-Labudova O, Horecka T, Kapalla M, Perecko D, Kutejova E, Lubec G (1998) Thioredoxin from Streptomyccs aureofaciens controls coiling of plasmid DNA. Life Sci 62: 397–412

122   E. Kitzmueller et al.

Goodison KL, Parhad IM, White CLd, Sima AA, Clark AW (1993) Neuronal and glial gene expression in neocortex of Down's syndrome and Alzheimer's disease. J Neuropathol Exp Neurol 52: 192–198

Greber-Platzer S, Schatzmann-Turhani D, Wollenek G, Lubec G (1999) Evidence against the current hypothesis of "gene dosage effects" of trisomy 21: ets-2, encoded on chromosome 21" is not overexpressed in hearts of patients with Down Syndrome. Biochem Biophys Res Commun 254: 395–399

Greidinger EL, Miller DK, Yamin TT, Casciola-Rosen L, Rosen A (1996) Sequential activation of three distinct ICE-like activities in Fas- ligated Jurkat cells. FEBS Lett 390: 299–303

He L, Fox MH (1997) Variation of heat shock protein 70 through the cell cycle in HL-60 cells and its relationship to apoptosis. Exp Cell Res 232: 64–71

Hiraoka Y, Komatsu N, Sakai Y, Ogawa M, Shiozawa M, Aiso S (1997) XLS13A and XLS13B: SRY-related genes of Xenopus laevis. Gene 197: 65–71

Hirota K, Matsui M, Iwata S, Nishiyama A, Mori K, Yodoi J (1997) AP-1 transcriptional activity is regulated by a direct association between thioredoxin and Ref-1. Proc Natl Acad Sci USA 94: 3633–3638

Hohfeld J (1998) Regulation of the heat shock conjugate Hsc70 in the mammalian cell: the characterization of the anti-apoptotic protein BAG-1 provides novel insights. Biol Chem 379: 269–274

Hori K, Katayama M, Sato N, Ishii K, Waga S, Yodoi J (1994) Neuroprotection by glial cells through adult T cell leukemia-derived factor/human thioredoxin (ADF/TRX). Brain Res 652: 304–310

Hyman BT, West HL, Rebeck GW, Lai F, Mann DM (1995) Neuropathological changes in Down's syndrome hippocampal formation. Effect of age and apolipoprotein E genotype. Arch Neurol 52: 373–378

Kanda H, Kojima M, Miyamoto N, Ito M, Takamatsu N, Yamashita S, Shiba T (1998) Rainbow trout Sox24, a novel member of the Sox family, is a transcriptional regulator during oogenesis. Gene 211: 251–257

Khan IA, Luduena RF (1991) Possible regulation of the in vitro assembly of bovine brain tubulin by the bovine thioredoxin system. Biochim Biophys Acta 1076: 289–297

Kuhlbrodt K, Herbarth B, Sock E, Enderich J, Hermans-Borgmeyer I, Wegner M (1998) Cooperative function of POU proteins and SOX proteins in glial cells. J Biol Chem 273: 16050–16057

Kwak HJ, Jun CD, Pae HO, Yoo JC, Park YC, Choi BM, Na YG, Park RK, Chung HT, Chung HY, Park WY, Seo JS (1998) The role of inducible 70-kDa heat shock protein in cell cycle control, differentiation, and apoptotic cell death of the human myeloid leukemic HL-60 cells. Cell Immunol 187: 1–12

Labudova O, Lubec G (1998) cAmp upregulates the transposable element mys-1: a possible link between signaling and mobile Dna. Life Sci 62: 431–437

Labudova O, Fang-Kircher S, Cairns N, Moenkemann H, Yeghiazaryan K, Lubec G (1998) Brain vasopressin levels in Down syndrome and Alzheimer's disease. Brain Res 806: 55–59

Labudova O, Kitzmueller E, Rink H, Cairns N, Lubec G (1999) Increased phosphoglycerate kinase in the brains of patients with Down's syndrome but not with Alzheimer's disease. Clin Sci (Colch) 96: 279–285

Lemere CA, Blusztajn JK, Yamaguchi H, Wisniewski T, Saido TC, Selkoe DJ (1996) Sequence of deposition of heterogeneous amyloid beta-peptides and APO E in Down syndrome: implications for initial events in amyloid plaque formation. Neurobiol Dis 3: 16–32

Lippoldt A, Padilla CA, Gerst H, Andbjer B, Richter E, Holmgren A, Fuxe K (1995) Localization of thioredoxin in the rat brain and functional implications. J Neurosci 15: 6747–6756

Mailhos C, Howard MK, Latchman DS (1994) Heat shock proteins hsp90 and hsp70 protect neuronal cells from thermal stress but not from programmed cell death. J Neurochem 63: 1787–1795

Mann DM, Pickering-Brown SM, Siddons MA, Iwatsubo T, Ihara Y, Asami-Odaka A, Suzuki N (1995) The extent of amyloid deposition in brain in patients with Down's syndrome does not depend upon the apolipoprotein E genotype. Neurosci Lett 196: 105–108

Marks A, D OH, Lei M, Percy ME, Becker LE (1996) Accumulation of S100 beta mRNA and protein in cerebellum during infancy in Down syndrome and control subjects. Brain Res Mol Brain Res 36: 343–348

Marsault R, Vigne P, Breittmayer JP, Frelin C (1990) Astrocytes are target cells for endothelins and sarafotoxin. J Neurochem 54: 2142–2144

Mavria G, Hall KT, Jones RA, Blair GE (1998) Transcriptional regulation of MHC class I gene expression in rat oligodendrocytes. Biochem J 330: 155–161

Mentlein R, von Kolszynski M, Sprang R, Lucius R (1990) Proline-specific proteases in cultivated neuronal and glial cells. Brain Res 527: 159–162

Mesner PW, Epting CL, Hegarty JL, Green SH (1995) A timetable of events during programmed cell death induced by trophic factor withdrawal from neuronal PC12 cells. J Neurosci 15: 7357–7366

Mori C, Nakamura N, Dix DJ, Fujioka M, Nakagawa S, Shiota K, Eddy EM (1997) Morphological analysis of germ cell apoptosis during postnatal testis development in normal and Hsp 70-2 knockout mice. Dev Dyn 208: 125–136

Mosser DD, Caron AW, Bourget L, Denis-Larose C, Massie B (1997) Role of the human heat shock protein hsp70 in protection against stress- induced apoptosis. Mol Cell Biol 17: 5317–5327

Nyunoya H, Morita T, Sato T, Honma S, Tsujimoto A, Shimotohno K (1994) Cloning of a cDNA encoding a DNA-binding protein TAXREB302 that is specific for the tax-responsive enhancer of HTLV-I. Gene 148: 371–373

Oyama F, Cairns NJ, Shimada H, Oyama R, Titani K, Ihara Y (1994) Down's syndrome: up-regulation of beta-amyloid protein precursor and tau mRNAs and their defective coordination. J Neurochem 62: 1062–1066

Pash J, Smithgall T, Bustin M (1991) Chromosomal protein HMG-14 is overexpressed in Down syndrome. Exp Cell Res 193: 232–235

Poteryaev DA, Zakharov IS, Balaban PM, Belyavsky AV (1998) A novel neuropeptide precursor gene is expressed in the terrestrial snail central nervous system by a group of neurons that control mating behavior. J Neurobiol 35: 183–197

Risser D, You ZB, Cairns N, Herrera-Marschitz M, Seidl R, Schneider C, Terenius L, Lubec G (1996) Endogenous opioids in frontal cortex of patients with Down syndrome. Neurosci Lett 203: 111–114

Saitoh M, Nishitoh II, Fujii M, Takeda K, Tobiume K, Sawada Y, Kawabata M, Miyazono K, Ichijo H (1998) Mammalian thioredoxin is a direct inhibitor of apoptosis signal-regulating kinase (ASK) 1. Embo J 17: 2596–2606

Scaife RM, Margolis RL (1997) The role of the PH domain and SH3 binding domains in dynamin function. Cell Signal 9: 395–401

Schilham MW, Oosterwegel MA, Moerer P, Ya J, de Boer PA, van de Wetering M, Verbeek S, Lamers WH, Kruisbeek AM, Cumano A, Clevers H (1996) Defects in cardiac outflow tract formation and pro-B-lymphocyte expansion in mice lacking Sox-4. Nature 380: 711–714

Schilham MW, Moerer P, Cumano A, Clevers HC (1997) Sox-4 facilitates thymocyte differentiation. Eur J Immunol 27: 1292–1295

Seidl R, Greber S, Schuller E, Bernert G, Cairns N, Lubec G (1997) Evidence against increased oxidative DNA-damage in Down syndrome. Neurosci Lett 235: 137–140

Seidl R, Fang-Kircher S, Bidmon B, Cairns N, Lubec G (1999) Apoptosis-associated proteins p53 and APO-1/Fas (CD95) in brains of adult patients with Down syndrome [In Process Citation]. Neurosci Lett 260: 9–12

Sompayrac L, Jane S, Burn TC, Tenen DG, Danna KJ (1995) Overcoming limitations of the mRNA differential display technique. Nucl Acids Res 23: 4738–4739

Sullivan AM, Morton AJ (1996) Endothelins induce Fos expression in neurons and glia in organotypic cultures of rat cerebellum. J Neurochem 67: 1409–1418

Tewari M, Beidler DR, Dixit VM (1995) CrmA-inhibitable cleavage of the 70-kDa protein component of the U1 small nuclear ribonucleoprotein during Fas- and tumor necrosis factor- induced apoptosis. J Biol Chem 270: 18738–18741

Torre E, McNiven MA, Urrutia R (1994) Dynamin 1 antisense oligonucleotide treatment prevents neurite formation in cultured hippocampal neurons. J Biol Chem 269: 32411–32417

Tyrrell J, Cosgrave M, Hawi Z, McPherson J, C OB, McCalvert J, McLaughlin M, Lawlor B, Gill M (1998) A protective effect of apolipoprotein E e2 allele on dementia in Down's syndrome. Biol Psychiatry 43: 397–400

Urrutia R, Henley JR, Cook T, McNiven MA (1997) The dynamins: redundant or distinct functions for an expanding family of related GTPases? Proc Natl Acad Sci USA 94: 377–384

Uwanogho D, Rex M, Cartwright EJ, Pearl G, Healy C, Scotting PJ, Sharpe PT (1995) Embryonic expression of the chicken Sox2, Sox3 and Sox11 genes suggests an interactive role in neuronal development. Mech Dev 49: 23–36

van de Wetering M, Oosterwegel M, van Norren K, Clevers H, Farr CJ, Easty DJ, Ragoussis J, Collignon J, Lovell-Badge R, Goodfellow PN (1993) Sox-4, an Sry-like HMG box protein, is a transcriptional activator in lymphocytes Characterization and mapping of the human SOX4 gene. Embo J 12: 3847–3854

Wei YQ, Zhao X, Kariya Y, Teshigawara K, Uchida A (1995) Inhibition of proliferation and induction of apoptosis by abrogation of heat-shock protein (HSP) 70 expression in tumor cells. Cancer Immunol Immunother 40: 73–78

Wyatt S, Mailhos C, Latchman DS (1996) Trigeminal ganglion neurons are protected by the heat shock proteins hsp70 and hsp90 from thermal stress but not from programmed cell death following nerve growth factor withdrawal. Brain Res Mol Brain Res 39: 52–56

Zhang W, Brooun A, McCandless J, Banda P, Alam M (1996) Signal transduction in the archaeon Halobacterium salinarium is processed through three subfamilies of 13 soluble and membrane-bound transducer proteins. Proc Natl Acad Sci USA 93: 4649–4654

Zukowska-Grojec Z, Karwatowska-Prokopczuk E, Rose W, Rone J, Movafagh S, Ji H, Yeh Y, Chen WT, Kleinman HK, Grouzmann E, Grant DS (1998) Neuropeptide Y: a novel angiogenic factor from the sympathetic nerves and endothelium. Circ Res 83: 187–195

Authors' address: Prof. Dr. G. Lubec, Department of Pediatrics, University of Vienna, Währinger Gürtel 18, A-1090 Vienna, Austria, e-mail: gert.lubec@akh-wien.ac.at

J Neural Transm (1999) [Suppl] 57: 125–136

Gene expression in fetal Down Syndrome brain as revealed by subtractive hybridization

O. Labudova[1], E. Kitzmueller[1], H. Rink[2], N. Cairns[3], and G. Lubec[1]

[1] Department of Pediatrics, University of Vienna, Austria
[2] Department of Radiobiology, University of Bonn, Federal Republic of Germany
[3] Brain Bank, Institute of Psychiatry, University of London, United Kingdom

Summary.   Information on gene expression in brain of patients with Down Syndrome (DS, trisomy 21) is limited and molecular biological research is focussing on mapping and sequencing chromosome 21. The information on gene expression in DS available follows the current concept of a gene dosage effect due to a third copy of chromosome 21 claiming overexpression of genes encoded on this chromosome.

Based upon the availability of fetal brain and recent technology of gene hunting, we decided to use subtractive hybridization to evaluate differences in gene expression between DS and control brains.

Subtractive hybridization was applied on two fetal brains with DS and two age and sex matched controls, 23rd week of gestation, and mRNA steady state levels were evaluated generating a subtractive library. Subtracted sequences were identified by gene bank and assigned by alignments to individual genes.

We found a series of up-and downregulated sequences consisting of chromosomal transcripts, enzymes of intermediary metabolism, hormones, transporters/channels and transcription factors (TFs).

We show that trisomy 21 or aneuploidy leads to the deterioration of gene expression and the derangement of transcripts described describes the involvement of chromosomes other than chromosome 21, explains impairment of transport, carriers, channels, signaling, known metabolic and hormones imbalances. The dys-coordinated expression of transcription factors including homeobox genes, POU-domain TFs, helix-loop-helix-motifs, LIM domain containing TFs, leucine zippers, forkhead genes, maybe of pathophysiological significance for abnormal brain development and wiring found in patients with DS. This is the first description of the concomitant expression of a large series of sequences indicating disruption of the concerted action of genes in that disorder.

Introduction

Molecular biological approaches are focussing on mapping and gene hunting of chromosome 21, but data showing differences between gene expression in normal and Down Syndrome (DS) brain are limited. This maybe due to the fact that a major current concept incriminates gene dosage effects of a third copy of chromosome 21 as responsible for the development of the DS phenotype (Epstein, 1992).

Among the few publications of gene expression in the brain of patients with DS Pash and coworkers showed the overexpression of chromosomal protein HMG-14, which is mapped to chromosome 21 (Pash et al., 1991) and the authors propose a pathogenetic role for this upregulated structural chromatin — related protein.

Goodison and coworkers (1993) described that in frontal cortex of patients with DS chromosome-21 encoded mRNAs, neurofilament light subunit, amyloid precursor transcripts, were increased. The upregulation of beta-amyloid precursor was confirmed by Oyama and coworkers (1994) and the authors showed the disruption of coordinated expression between the amyloid precursor APP-751 and four-repeat tau.

Not all genes present in the critical "Down Syndrome region" of chromosome 21 are, however, overexpressed in the brain: Marks and coworkers reported comparable mRNA and protein levels of S100, a calcium-binding protein encoded on chromosome 21 and accumulating during CNS maturation in mammals, in controls and DS cerebellum (Marks et al., 1996).

De la Monte et al. (1996) demonstrated that early in life neuronal thread protein was overexpressed in cerebral cortex of patients with DS, a tentative candidate for the development of Alzheimer's disease pathology which occurs from the fourth decade in DS brain.

Transient overexpression of a trifunctional protein with glycinamide ribonucleotide synthase, aminoimidazole ribonucleotide synthase and glycinamide ribonucleotide formyltransferase activity, key enzymes for purine de novo biosynthesis, in cerebellum of infants with DS has been published by Brodsky and coworkers (1997).

No gene hunting approach is known to us in peer reviewed accessible literature, testing the differences in gene expression between DS and control brain. However important the known reports on different expression on individual genes are, no pattern or derangement of the concerted action of genes involved in DS pathology has emerged yet. The availability of age and sex matched fetal brain samples of DS and controls on the one and the gene hunting technique of subtractive hybridization on the other hand made us create a subtractive librarywhich enabled us to reveal up-and downregulated sequences which were subsequently assigned to genes by gene bank work.

In this paper we describe expressional differences in terms of mRNA steady state levels of transcription factors, channels, transporters, chromosomal transcripts, which may well serve to explain structural and functional abnormalities in DS brain pathology.

Methods

Subtractive hybridization was selecedf as a gene hunting method and carried out as given previously (Labudova and Lubec, 1998a,b):

Fetal brain samples of two fetuses with DS and two age and sex matched controls, 23rd week of gestation, were obtained from the Brain Bank of the Institute of Psychiatry (Denmark Hill, London, UK). Hippocampus (gyrus parahippocampalis) was taken into liquid nitrogen and ground for the isolation of mRNA. Isolation of mRNA was performed using the Quick Prep Micro mRNA purification kit (Pharmacia Biotech Inc., Uppsala, Sweden). One microgram of mRNA from each preparation was quality — checked by cDNA cloning kit (Gibco, Life Technologies, Eggenstein, Germany, cat. 18248-013) using the incorporation of [$\square$-^{32}P]dATP (Amersham, Buckinghamshire, UK) with subsequent electrophoresis on 1% agarose followed by autoradiography. The Reflection film (Dupont, Germany) was exposed to the gel for a period of two hours at room temperature. Construction of the substractive library: 10 micrograms each of mRNA from brain of DS and control were biotinilated by UV irradiation at 360 nm according to the instructions supplied in the subtractor kit (Invitrogen, Leek, Netherlands). One microgram of mRNA-pools each from the DS brain sample was subject to reverse transcriptase reaction (subtractor kit, Invitrogen) and the cDNA-pools were hybridized with the corresponding biotinilated mRNAs from controls. The substractive hybridization mixture was incubated with streptavidin according to the subtractor kit given above and thus the biotinilated molecules (non-induced biotinilated mRNAs and the hybrid [biotinilated mRNAs/cDNAs]) complexed. The streptavidine complexes were removed by repeated phenol-chloroform extraction and subtracted cDNAs were separated from the aqueous phase by alcohol precipitation (subtractor kit). In order to amplify and clone subtracted cDNAs, they were ligated with Not I linkers followed by Not I-digestion. The Not I-linked cDNAs were ligated to Not I site of sPORT 1 cloning vector (cDNA cloning kit, Gibco). To enable visualization of subtracted cDNAs, the cloned cDNAs were amplified using universal primers I 5'-GTAAAACGACGGCCAGT-3' II 5'-ACAGCTATGACCATG-3' from multiple cloning site of the sPORT-1 vector (cDNA cloning kit, Gibco). Amplified cDNAs were analyzed by electrophoresis on 1% agarose. Cloning of subtracted Not I-linked cDNAs Not I-linked cDNAs ligated with sPORT 1 vector were used for the transformation of highly competent INFalpha F E. coli cells (Invitrogen, Leek, Netherlands) and plated clones were analysed by plasmid isolation kit (Quiagen, Hilden, Germany) and digestion with Eco RI/Hind III. Recombinant clones were sequenced by K. Granderath, MWG — Biotech (Ebersberg, Germany). Homologies were determined by computer assisted comparison of data from the genebank sequence library: fastA@ebi.ac.uk

(GBALL, EMBL, Gen Bank Heidelberg, Germany). Subtractive hybridization was performed cross-wise, i.e. DS sample mRNA subtraction from control and vice versa at the 1:3 level (DSmRNAs: control mRNAs).

Results

Tables 1 shows the sequences overexpressed in fetal DS brain and Table 2 lists the sequences which were downregulated in fetal DS.

Chromosomal transcripts of chromosomes 7, 9, 11, 16, 22 and X were found to be overexpressed, whereas chromsosomal transcripts of chromosomes 12 and 19 were downregulated.

Glucose handling enzymes phosphoglycerate kinase, glyceraldehyde-3-phosphate dehydrogenase werc upregulated whereas phosphoglucose isomerase was downregulated.

Table 1. Genes with higher transcriptional level in DS

No of clone	Name of gene	EMBL accession number	Organism	Length of cDNA (bp)	%homology of cDNA
7	chromosome 7 region 7q3 (from 13,710 nt)	Ac002080	hs	150	71
8	chromosome 7 region 7q3 (from 88,346 nt)	Ac002528	hs	86	79
12	chromosome 9 region 9p34 (from 22,375 nt)	Ac000387	hs	147	71
2	chromosome 11 region 11 17e5 (from 5,958 nt)	U73639	hs	86	66
1	chromosome 16 region 16p13.11 (from 158,360 nt)	U95738	hs	158	76
3	chromosome 16 region 16p13.1 (from 37,640 nt)	U95743	hs	144	72
5	chromosome 16 region 16p13 (from 12,988 nt)	U91318	hs	94	77
6	chromosome 16 region 16p13.11 (from 80,593 nt)	U95741	hs	144	71
4	chromosome 16 region 16p13.11 (from 38,317 nt)	U91319	hs	157	66
9	chromosome 21 Down Syndrome region (from 75,588 nt)	Ac000002	hs	149	70
10	chromosome 21 Down Syndrome region (from 1,395 nt)	U44740	hs	149	70
11	chromosome 21 Down Syndrome region (from 406 nt)	U44743	hs	149	70
14	chromosome 22 region 22q11 (from 32,796 nt)	Ac000091	hs	144	66

(continued)

Table 1. *Continued*

No of clone	Name of gene	EMBL accession number	Organism	Length of cDNA (bp)	%homology of cDNA
13	chromosome X region Xp28 (from 3,427 nt)	U52112	hs	89	80
109	PGK (phosphoglycerate kinase)	V00572	hs	1,258	100
104	gapB (glyceraldehyde 3-phosphate dehydrogenase)	J02642	hs	846	100
89	scleraxis (basic helix-loop--helix-type transcriptional factor)	U58681	hs	339	100
294	neuronal alpha 1A Ca^{++} channel	Af004883	hs	327	100
6–19	GLNP (glutamine transport protein)	P76825	E. coli	285	75
221	H+-ATPase	L31770	Bos taurus	227	73
203	gene in citrate utilization locus	M11559	E. coli	987	72
77	4-hydroxybutyrate dehydrogenase	L36817	Alcaligenes eutrophus	677	65
80	ABC-transporter	Ae000129	E. coli	410	63
99	acetyl-CoA carboxylase, β-sujunit	M68934	E. coli	565	62
119	NADH dehydrogenase (nuoG, subunit)	Ae00317	E. coli	625	57
100	dnrS, glycosyl transferase	L47164	Streptomyces pencetius	676	55
265	TSH (thyroid stimulating hormone receptor)	M34842	rat	155	67

130 O. Labudova et al.

Table 2. Genes with higher transcriptional level in the control

No	Name of gene	EMBL acces. number	Organism	Length of cDNA (bp)	%homology of cDNA
15	chromosome 12 (from 19,680 nt)	Ac002070	hs	444	60
133	chromosome 19, region 19p13.3, (7,160–7,340 nt)	Ac005339	hs	177	60
193	junD	X56681	hs	965	100
189	ALDH1 (aldehyde dehydrogenase 1)	K03000	hs	778	100
274	PGI (phosphoglucose isomerase)	X83462	hs	308	100
267	NF-kappa B	Z47747	hs	295	100
157	2′,3′-cyclic nucleotide 3′ phosphodiesterase	M19650	hs	240	100
268	Human growth hormone-releasing hormone gene	U42224	hs	150	100
302	HOXA 13 (human transcriptional factor)	U82827	hs	147	100
292	hemB gene (substrate-binding protein)	D85613	E. coli	598	95
123	nirE, uro gen-III c-methyltransferase (heme d1 synthesis of cytochrome cd1)	D84475	Pseudomonas aeruginosa	832	66
139	ahc Y gene (S-adenosyl-homocysteine hydrolase)	D90911	Synechocystis sp. PCC6803	622	65
134	amino acid ABC transporter	U75364	Rhodopseudomonas palustris	890	63
188	betaine aldehyde dehydrogenase	Ad000006	Mycobacterium tuberculosis	780	60
264	Ptx1 (homeodomain protein transcriptional factor)	U71206	mouse	215	58
177	hydroxylase (cts4, cts6, cts8 genes)	D38214	Streptomyces aureofaciens	936	57
303	estC = esterase III	S69066	Pseudomonas flu	419	56
293	tap gene (efflux, multidrug resistance protein)	Aj000283	Mycobacterium fortuitum	485	55

Enzymes of intermediary metabolism as acetyl-CoA carboxylase, NADH dehydrogenase (nuoG), dnrs glycosyltransferase, genes for citrate utilization, 4-hydroxybutyrate dehydrogenase were upregulated and the sequences of metabolic factors and enzymes aldehyde dehydrogenase, hem B, uro gen — III c-methyltransferase hem d1 for hem d1 biosynthesis, S-adenosyl-homocysteine hydrolase, betaine aldehyde dehydrogenase, hydroxylases (homologous to cts4, cts6, cts8) esterase III were downregulated.

Endocrine factors as the thyroid stimulating hormone receptor were upregulated or downregulated as human growth hormone — releasing hormone gene.

mRNA steady state levels of transporters were increased in the cases of the glutamine transport protein GLNP, H+-ATPase L 31770 and ABC transporter Ae000129 and of the tap gene corresponding sequence Aj000283 were found to be decreased.

The channel biomolecule: neuronal alpha 1A Ca++ channel was upregulated.

The signaling compound 2′,3′-cyclic nucleotide 3′ phosphodiesterase was downregulated.

The transcription factors scleraxis was upregulated and junD, NF-kappa B, HOXA-13, Ptx1 were downregulated.

Discussion

Subtractive hybridization, i.e. the generation of a subtractive library of DS and control brain, led to the discovery of a series of sequences and many of them could be assigned to corresponding genes by gene bank work. In this paper we have omitted the non-identified sequences but present those with a homology higher than 50%. Some of them could not be assigned to human genes and were matching procaryotic or eucaryotic counterparts only, mainly as no human data were available. The high homology, however, makes the identification highly probable. Moreover, some sequences were never reported in the human or even in the mammalian system and we are the first to describe their presence in human brain. In this paper we are not providing the 46 alignments of the sequences to the gene bank data but they are available on request from the corresponding author.

A significant portion of genes expressed differently at least threefold (see the setting of 1:3 of mRNAs used for subtraction) could be assigned to chromosomal transcripts. As expected, the chromosomal transcripts of chromosome 21, DS region (accession numbers Ac0000002, U44740, U44743), were overexpressed, although not at the 1.5 fold but at the threefold level. The most innovative finding in this context, however, is the upregulation of other chromosomal transcripts as e.g. that of chromosome 7 (region 7q3), chromosome 9 (region 9p34), chromosome 11 (region 11 17e5), chromosome 16 (regions 16p13.11, 16p13.1), chromosome 22 (region 22p11), chomosome X (region Xp28). The chromosomal transcripts of chromosome 12 (region

12p13) and chromosome 19 (region 19p13.3) were downregulated thus most probably demonstrating the (regulatory-modulating) effect of trisomic chromosome 21 aneuploidy, on other chromosomes.

A series of dysregulated enzymes involved in basic metabolic functions complementing findings of altered purine metabolism (see above) was found: brain glucose metabolism is known to be deranged in DS and has been established in vivo by positron emission tomography findings (Pietrini et al., 1997; Schapiro et al., 1992).

We provide evidence for the transcriptional changes of glucose handling enzymes phosphoglycerate kinase, glyceraldehyde — 3-phosphate dehydrogenase, phosphoglucose isomerase and 3-phosphoglycerate dehydrogenase, supporting findings of deteriorated glucose metabolism at the transcriptional level.

Among upregulated sequences assinged to genes involved in intermediary metabolism are a gene from the citrate utilization, NADH dehydrogenase, 4-hydroxybutyrate dehydrogenase, glycosyltransferase, acetyl-CoA-carboxylase, and among downregulated sequences are hemB, urogen III-c-methyltransferase, required for heme biosynthesis, S-adenosyl-homocysteine hydrolase, betaine aldehyde dehydrogenase, hydroxylase cts4, esterase III and aldehyde dehydrogenase, a major aldehyde detoxification enzyme which was found to be reduced at the protein level in five regions of adult DS brain (herein).

Hormonal dysfunction is a major finding in patients with DS and literature on thyroid dysfunction is abundant. Here we report the dysregulation of the TSH receptor in fetal brain, a finding, already published in quantitative terms in adult DS brain (Labudova et al., 1999). The intriguing finding of downregulated human growth hormone — releasing hormone-gene in fetal brain with DS — the sequence showed 100% homology with the human sequence — fits the clinical situation of growth hormonal status in children with DS (Castells et al., 1996; Pueschel, 1993).

Transport of biomolecules is a vital function in the brain and the basis of physiological equilibria. We observed upregulation of the ABC- (ATP-binding-cassette-) transporter Ae000129 of the ABC transporter superfamily, the glutamine transport protein GLNP and the H+-ATPase L31770; and indeed ATPases were located to chromosome 21 (Chen et al., 1995) and deficits of the cation-carriers (Naylor et al., 1993) were reported in DS. The tag gene, an efflux, multidrug resistance protein from the same superfamily was downregulated: Multidrug resistance proteins are responsible for antibiotic resistance in bacteria, for failure of chemotherapy in cancer treatment and for externalization of compounds by cytopempsis in general. In DS brain impaired transport may well be explaining derangement of neurotransmitter (axonal?) transport, compartmentalization and handling, thus contributing to the well-documented neurotransmitter-abnormalities.

In excitable tissues ion fluxes are mediated through channels and this forms the basis for neurophysiological function. In DS brain we found the upregulation of Ca++ channel B and this first observation of a modified channel expression in Down Syndrome should challenge quantitative testing of our and other ion channels described to be encoded on chromosome 21 (Gosset et al., 1997; Ohira et al., 1997; Malo et al., 1995), which in turn should help to understand the intrinsic electrophysiolial and EEG-changes in DS brain (Kakigi and Kuroda, 1992).

In all cells second-messenger related molecules work in tight cooperation with channels thus constituting the signaling network. In our subtractive library 2′,3′-cyclic nucleotide 3′-phosphodiesterase was downregulated. This is of particular interest as phosphodiesterases are controlling cAMP levels in the cell and cAMP in turn could be activating phosphokinases but even more important, cyclic nucleotide — gated ion channels:

A number of cyclic nucleotide activated ion channels have been reported involved in a variety of cell functions including ion permeation, channel gating and modulation (Frinn et al., 1996).

The development of the central nervous system is performed and controlled by the about 5,000 transcription factors of the mammalian brain (Stewart and Davies, 1997). These transcription factors (TFs) include homeobox genes (Stein et al., 1996), TFs containing the POU-domain (Sharp and Morgan, 1996), helix-loop-helix motifs (Kageyama et al., 1995), the LIM domain (Sanchez-Garcia et al., 1994), leucine zippers, a large number of zinc-finger — containing genes (Struhl, 1989), forkhead genes (Kaufmann and Knöchel, 1996) and others. We found that a helix-loop-helix type TF, scleraxis, was overexpressed in fetal DS brain. This TF was never reported in humans and never reported in human brain and is considered a TF regulating gene expression in mesenchymal cell lineages. We are therefore not able to assign any tentative role in normal or DS neurogenesis.

The downregulation of junD in fetal DS was described previously by our group in several regions of adult DS brain also at the protein level and decreased junD may well be responsible for impaired wiring of DS brain (Labudova et al., 1998).

Functional and pathophysiological relevance maybe also assigned to the decrease of mRNA steady state levels for the TF NF-kappa-B in fetal DS brain and the 100% identity of our sequence with the designed NF-kappa-B from gene bank seems convincing. Moreover, NF-kappa-B is a redox controlled TF in the brain (Kaltschmidt et al., 1993) and its dysregulation would also be compatible with oxidative stress, thought to play a role in DS pathogenesis and neuronal (apoptotic) death (Busciglio and Yankner, 1995).

HOXA-13 is a TF which was never described in human brain and is a coordinator of the development of the body axis (Sordino et al., 1996). This gene family is strongly conserved among vertebrates and its downregulation in brain of DS fetus may well be involved in abnormal brain modelling in DS.

Finally, the homeobox TF Ptx1 was downregulated in fetal DS brain. Ptx1 is a member of the rapidly expanding family of bicoid — related vertebrate homeobox genes, which seem to play a role in the development of anterior structures including the face and brain (Crawford et al., 1997; Lanctot et al., 1997) and therefore the downregulation may fit dysmorphic development of the DS fetus.

134 O. Labudova et al.

In conclusion, we provided evidence for the dysregulated expression of several functional elements, candidates helping to understand the complex pathogenesis and molecular biology of Down Syndrome brain. We are reporting an emerging pattern of differential expression in DS brain.

Our data support and explain pathological features and phenotype of DS as e.g. metabolic, hormonal, transport derangement; deterioration of brain structure and function could be reflected by altered expression of transcription factors and channels. The expressional difference, set at the mRNAs-level of 1:3, makes allelic differences for the explication of results highly improbable. We are aware of the pitfalls of gene hunting methods as differential display and subtractive hybridization, but they are still the most effective and potent way to the identification of differentially expressed patterns and new genes in the brain (Bauer et al., 1994; Bertioli et al., 1995; Sompayrac et al., 1995).

We are currently complementing our transcriptional data by protein and activity level studies but are readily providing sequences from our substractive library to the acientifique community on request.

Acknowledgements

We are highly indebted to the Red Bull Company, Salzburg, Austria, for generous financial support of the study.

References

Bauer D, Warthoe P, Rohde L, Struss M (1994) In: PCR methods and applications. Cold Spring Harbor Laboratory Press, Cold Spring Harbor, NY, pp S97–S108

Bertioli DJ, Schlichter UHA, Adams MJ, Burrow PR, Steinbiss HH, Antoniew JF (1995) An analysis of differential display shows a strong bias towards high copy number messenger RNAs. Nucl Acids Res 23: 4520–4523

Brodsky G, Barnes T, Bleskan J, Becker L, Cox M, Patterson D (1997) The human GARS-AIRS-GART gene encodes two proteins which are differentially expressed during human brain development and temporally overexpressed in cerebellum of individuals with Down Syndrome. Hum Mol Genet 6: 2043–2050

Busciglio J, Yankner BA (1995) Apoptosis and increased generation of reactive oxygen species in Down Syndrome. Nature 378: 776–779

Castells S, Beaulieu I, Torrado C, Wisniewski KE, Zarny S, Gelato MC (1996) Hypothalamic versus pituitary dysfunction in Down's Syndrome as cause of growth retardation. J Intellect Disabil Res 40: 509–517

Chen H, Morris MA, Rossier C, Blouin JL, Antonorakis SE (1995) Cloning of the cDNA for the human ATP synthase OSCP subunit (ATP50) by exon trapping and mapping to chromosome 21q22.1–q22.2. Genomics 28: 470–476

Crawford MJ, Lanctot C, Tremblay JJ, Jenkins N, Gilbert D, Copeland N, Beatty B, Drouin J (1997) Humana and murine PTX1/Ptx1 gene maps to the region for Treacher-Collins syndrome. Mamm Genome 8: 841–850

De la Monte SM, Xu YY, Hutchins GM, Wands JR (1996) Developmental patterns of neuronal thread protein gene expression in Down Syndrome. J Neurol Sci 135: 118–125

Epstein CJ (1992) Down Syndrome (Trisomy 21). In: Scriver CR, Beaudet AL, Sly WS, Valle D (eds) The metabolic and molecular basis of inherited disease. McGraw Hill, New York, pp 749–794

Finn JT, Grunwald ME, Yau KW (1996) Cyclic nucleotide-gated ion chann els: an extended fa mily with diverse functions. Ann Rev Physiol 58: 395–426

Goodison KL, Parhad IM, White CL, Sima AA, Clark AW (1993) Neuronal and glial gene expression in neocortex of Down's syndrome and Alzheimer's disease. J Neuropathol Exp Neurol 52: 192–198

Gosset P, Ghezala GA, Korn B, Yaspo ML, Poutska A, Lehrach H, Sinet PM, Creau N (1997) A new inward rectifier potassium channel gene localized on chromosome 21 in the Down Syndrome chromosomal region 1 (DCR1). Genomics 44: 237–241

Kageyama R, Sasai Y, Akazewa C, Ishibashi M, Takebayashi K, Shimizu C, Tomita K, Nakanishi S (1995) Regulation of mammalian neural development by helix — loop — helix transcription factors. Crit Rev Neurobiol 9: 177–188

Kakigi R, Kuroda Y (1992) Brain-stem auditory evoked potentials in adults with Down Syndrome. Electroencephalogr Clin Neurophysiol 84: 293–295

Kaltschmidt B, Baeuerle PA, Kaltschmidt C (1993) Potential involvement of the transcription factor NF-kappa-B in neurological disorders. Mol Aspects Med 14: 171–190

Kaufmann E, Knöchel W (1996) Five years on the wings of forkhead. Mech Dev 57: 3–20

Labudova O, Lubec G (1998) cAMP upregulates the transposable element mys-1: a possible link between signaling and mobile DNA. Life Sci 62: 431–437

Labudova O, Cairns N, Yeghiazaryan K, Lubec G (1998) The upregulation of vasopressin in brain of patients with Down Syndrome and Alzheimer's disease. Brain Res 806: 55–59

Labudova O, Kitzmüller E, Köck Th, Cairns N, Lubec G (1998) Overexpression of the thyroid stimulating hormone receptor in brain of patients with Down Syndrome. Life Sci 64: 1037–1044

Labudova O, Krapfenbauer K, Cairns N, Lubec G (1998) Decreased junD in brain of patients with Down Syndrome. Neurosci Lett 252: 159–162

Lanctot C, Lamolet B, Drouin J (1997) The bicoid-related homeoprotein Ptx1 defines the most anterior domain for the embryo and differentiates posterior from anterior lateral mesoderm. Development 124: 2807–2817

Malo MS, Srivastava K, Ingram VM (1995) Gene assignment by polymerase chain reaction: localization of the human potassium channel IsK gene to the Down's syndrome region of chromosome 21q11.1–q22.2. Gene 159: 273–275

Marks A, O'Hanlon D, Lei M, Percy ME, Becker LE (1996) Accujmulation of S100 beta mRNA and protein in cerebellum during infancy in Down Syndrome and control subjects. Brain Res-Mol Brain Res 36: 343–348

Naylor GJ, Semple M, Irvine EA (1993) Erythrocyte membrane cation carrier in Down Syndrome. Clin Genet 43: 9–10

Ohira M, Seki N, Nagase T, Suzuki E, Nomura N, Ohara O, Hattori M, Sakaki Y, Eki T, Murakami Y, Saito T, Ichikawa H, Ohki M (1997) Gene ideentification in 1.6-Mb region of the Down syndrome region on chromosome 21. Genome Res 7: 47–58

Oyama F, Cairns NJ, Shimada H, Oyama R, Titani K, Ihara Y (1994) Down's syndrome: upregulation of beta-amyloid protein precursor and tau mRNAs and their defective cooredination. J Neurochem 62: 1062–1066

Pash J, Smithgall T, Bustin M (1991) Chromosomal protein HMG-14 is overexpressed in Down Syndrome. Exp Cell Res 193: 232–235

Pietrini P, Dani A, Furey ML, Alexander GE, Freo U, Grady CL, Mentis MJ, Mangot D, Simon EW, Horwitz B, Haxby JV, Schapiro MB (1997) Low glucose metabolism during brain stimulation in older Down's syndrome subjects at risk for Alzheimer's disease prior to dementia. Am J Psychiatry 154: 1063–1069

Pueschel SM (1993) Growth hormone response after administration of L-dopa, clonidine, and growth hormone releasing hormone in children with Down Syndrome. Res Dev Disabil 14: 291–298

Sanchez-Garcia I, Rabbitts TH (1994) The Lim-domain: a new structural motif found in zinc-finger like proteins. Trends Genet 10: 315–320

Schapiro MB, Grady CL, Haxby JV (1992) Nature of mental retardation and dementia in Down's syndrome: study with PET, CT and neuropsychology. Neurobiol Aging 13: 723–734

Sharp ZD, Morgan WW (1996) Brain POU-er. Bioassays 18: 347–350

Sompayrac L, Jane S, Burn TC, Tenen DG, Danna KJ (1995) Overcoming limitations of the messenger RNA differential display technique. Nucl Acids Res 23: 4738–4739

Sordino P, Douboule D, Kondo T (1996) Zebrafish Hoxa and Evx-genes: cloning, developmental expression and implications for the functional evolution of posterior Hox genes. Mech Dev 59: 165–175

Stein S, Fritsch R, Lemaire L, Kessel M (1996) Checklist: vertebrate homeobox genes. Mech Dev 55: 91–108

Stewart GJ, Davies RW (1997) In: Davies RW, Morris BJ (eds) Molecular biology of the neuron. Bios Scientifique publishers, Oxford UK, pp 1–18

Struhl K (1989) Helix-turn-helix, zinc finger, and leucine-zipper motifs for eukaryotic transcriptional regulatory proteins. Trends Biochem Sci 14: 137–140

Authors' address: Prof. Dr. G. Lubec, Department of Pediatrics, University of Vienna, Währinger Gürtel 18, A-1090 Vienna, Austria, e-mail: gert.lubec@akh-wien.ac.at

J Neural Transm (1999) [Suppl] 57: 137–159

Molecular misreading of genes in Down syndrome as a model for the Alzheimer type of neurodegeneration

F. W. van Leeuwen and **E. M. Hol**

Netherlands Institute for Brain Research, Amsterdam, The Netherlands

Summary. The occurrence of +1 frameshifted proteins, such as amyloid precursor protein (APP^{+1}) and ubiquitin-B (UBB^{+1}) in Down syndrome (DS) has been linked to the onset of Alzheimer's disease (AD). In DS and AD patients, but also in elderly non-demented persons, these co-called +1 proteins accumulate in the neuropathological hallmarks (neurofibrillary tangles, dystrophic neurites of the neuritic plaques and neuropil threads) and may have deleterious effects on neuronal function. Frameshifts are caused by dinucleotide deletions in GAGAG motifs in messenger RNA and are now thought to be the result of unfaithful transcription of normal DNA by a novel process termed "molecular misreading". In the present review some of the critical events in molecular misreading are discussed, the emphasis being on DS.

Introduction

Trisomy 21 is the most common genetic cause of mental retardation (1/700 live births) and is associated with the numerous complex phenotypes observed in Down syndrome (DS). As a result of deregulation of gene expression (Hernandez and Fischer, 1996; Antonarakis, 1998) most DS patients already start to develop the neuropathological hallmarks of Alzheimer's disease (AD) around the age of 40. The numerous plaques and neurofibrillary tangles in DS have been linked to the disproportional overexpression of the β amyloid precursor protein (βAPP) gene located at 21q21.2 of chromosome 21 (Neve et al., 1989; Rumble et al., 1989). Therefore it has been generally accepted that DS is instrumental to studying the Alzheimer type of neurodegeneration, the latter being the most common cause of cognitive decline in the aged population, the incidence of which is expected to double over the next thirty years. During the past decade most attention in AD research has been focused on the forms with an autosomal dominant inheritance pattern. Mutations in the genes for βAPP on chromosome 21 and presenilin 1 and 2 on chromosome 14 and 1, respectively, were identified, and they have been shown to be causal in the development of AD. Nevertheless, all these autosomal dominant forms account for no more than 5% of all AD cases. ApoE-E4 alleles on chromosome 19 and more recently a suggested deletion at a splice site of the α-2 macroglobulin gene on chromosome 12 are important risk factors for the development of AD (Blacker et al., 1998; Van Broeckhoven, 1998). These factors are not causative for AD, but they determine the age of onset of AD, and thus can be regarded as timing factors. However, for the majority of the cases (about 95%) no clear genetic effect has been reported yet (Fig. 1). For these non-autosomal dominant forms of AD, including the sporadic cases as the most frequent form, a different mechanism must exist that initiates neurodegeneration. Recently we found evidence for the existence of such a mechanism that, in contrast to what has been reported in most studies so far, is not due to faulty DNA, but is involved in generating errors in mRNA during transcription. We call this novel process "molecular misreading", which is defined as the inaccurate conversion of genomic information into mRNA and the subsequent translation into aberrant proteins (Van Leeuwen et al., 1998a).

DS brain autopsies have been instrumental in this regard as molecular misreading is positively correlated with overexpression of genes (Van Leeuwen et al., 1998a). Our recent discovery of dinucleotide deletions in βAPP and Ubiquitin-B (UBB) mRNAs in AD and DS patients was in fact the first example of molecular misreading in a human disease. The aberrant mRNAs (or so-called "nonsense transcripts") are translated in the +1 reading frame as "+1 proteins", i.e. proteins with a wild-type N-terminus and an aberrant, frameshifted and often truncated C-terminus (Fig. 2).

Many factors may contribute to the increased frequency of AD during aging and its acceleration in DS, e.g. activational state of neurons, oxidative damage, DNA damage and synapse loss (Evans et al., 1995; Mann, 1997; Swaab, 1998). The high degree of molecular misreading in DS is most probably another major factor and it has been suggested that it could help explain the early development of neuropathology in DS (Vogel, 1998).

How molecular misreading was discovered

Some years ago, curious to understand the presence of vasopressin (VP) precursor products that theoretically could not exist (Van Leeuwen et al., 1989; Finch and Goodman, 1997), we discovered a novel type of transcript variability in homozygous Brattleboro rats (di/di) (Evans et al., 1994). These rats suffer from hypothalamic diabetes insipidus as a result of a single-base germ-line mutation in the VP gene that encodes an aberrant VP precursor (Valtin, 1982). The mutant VP precursor is not admitted to the secretory pathway since it is arrested in the membranes of the endoplasmatic reticulum (Schmale et al., 1984) (Fig. 3). We surprisingly found that in a small number of solitary magnocellular neurons in the supraoptic and paraventricular nucleus an additional mutation (ΔGA) in their VP transcripts results in a restoration of the wild-type reading frame (Figs. 4, 5). This VP precursor protein, which is still partially mutated but nevertheless able to enter the secretory pathway, undergoes axonal transport, during which enzymatic

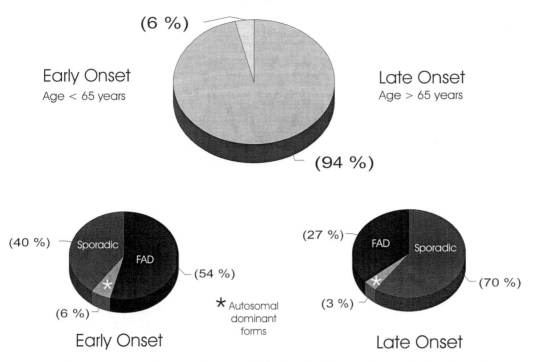

Fig. 1. Different types of dementia and Alzheimer's disease. A population based cross-section study of all dementias has shown that AD is the most frequently occurring neurodegenerative disorder (72%), followed by vascular dementia (16%), Parkinson's disease (6%) and other dementias (5%; e.g. Pick's disease) (Ott et al., 1995). Since not all studies supply data on the relative proportions of AD, here the most recent insights are presented (data composed in collaboration with Prof. C. van Broeckhoven and Dr. M. Cruts, both University of Antwerp, Belgium, and Dr. C. M. van Duijn, Erasmus University, Rotterdam, The Netherlands). An additional reason to present these data is that the terminology used regarding AD is sometimes inconsistent, and thus confusing. For example, familial types and autosomal forms of AD are often intermingled although they represent different groups. It is clear from the diagram that the sporadic and non-autosomal forms of familial AD account for 95% of all AD cases. It has been reported that in these latter cases molecular misreading of correct genomic information may play an important role (Van Leeuwen et al., 1998a,b; Vogel, 1998). Early (<65 years) and late (>65 years) onset (EOAD and LOAD) forms of AD are distinguished. Familial means that AD was observed in relatives of the first degree. This study is based upon Van Duijn et al. (1994), Ott et al. (1995), Van Broeckhoven (1995), Cruts et al. (1998). Data between brackets represent a subpercentage. In familial EOAD the majority (54%) is not yet linked to a chromosome, whereas 6% is inherited in an autosomal dominant way and linked to chromosome 1 (PS2, <1%), chromosome 14 (PS1, 33%), 21 (APP, 5%), whereas 62% of the autosomal dominant forms is still not linked. PS1 and PS2 mutations can also be of the sporadic type (Cruts et al., 1998). In familial LOAD, the majority (90%) is not yet linked to a chromosome, whereas 10% is inherited in an autosomal dominant way. A subset has been linked to chromosome 12 (Pericak-Vance et al., 1997). *Risk factors*: 65% of all EOAD and 25% of all LOAD cases display ApoE4 polymorphism (one or two E4 alleles), which is the best-known risk factor. ApoE4 data in early onset AD are based upon a study by Van Duijn et al. (1994) (n = 175 patients). Several other risk factors for late onset AD have been reported (for details, see Van Broeckhoven, 1998) and most recently alpha 2 macroglobulin and the lipoprotein receptor-like protein LRP1; Blacker et al., 1998). *Total picture (EOAD and LOAD)*: Familial AD (FAD) represents about 40% of all AD cases. Most of these FAD cases (i.e. 35% of all AD patients) do not inherit AD as an autosomal dominant trait. Thus, the non-familial or sporadic form of AD comprises approximately 60% of all AD cases and together with the non-autosomal dominant forms of FAD it accounts for 95% of the cases

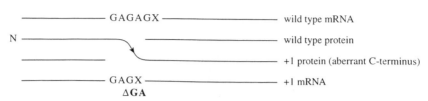

Fig. 2. A schematic representation of a frameshift mutation. A +1 protein is formed as a result of a dinucleotide deletion (for example ΔGA) and consists of a wild-type N-terminus (blue). Downstream of the dinucleotide deletion, the transcript is translated in the +1 reading frame and gives rise to the aberrant (red) C-terminus of the +1 protein (taken from Van Leeuwen et al., 1998b, with permission)

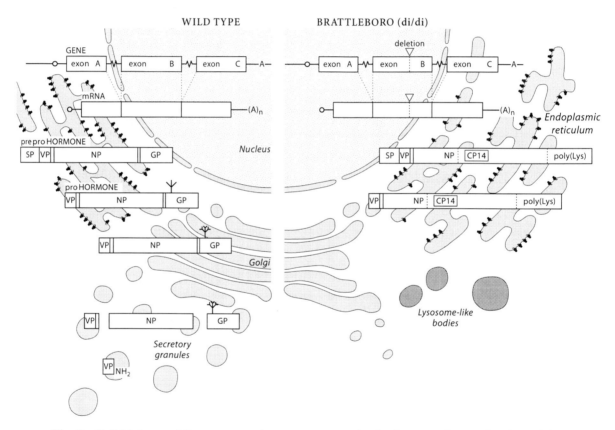

Fig. 3. Cell biology of the vasopressin precursor synthesis. Structural organization of the vasopressin (VP) genes from wild-type (left panel) and homozygous Brattleboro (di/di) (right panel) rats: their transcription, translation into a precursor protein that is posttranslationally processed (e.g. amidation) in magnocellular neurons. *ER* endoplasmic reticulum; *SP* signal peptide; *VP* vasopressin; *NP* neurophysin; *GP* C-terminal glycopeptide of propressophysin; *CP14* stretch of 14 amino acids predicted from the frame-shifted DNA sequence and used to generate mutant-specific antibodies. In the homozygous Brattleboro rat (di/di) the mutant precursor transport from the ER to the Golgi apparatus is disturbed, most probably by incorrect folding of the prohormone (see Schmale et al., 1989). However, a co-expressed peptide such as dynorphin is packaged in secretory granules that are much smaller (diameter 100 nm) than usual (150 nm) (picture adapted from Ivell et al., 1990)

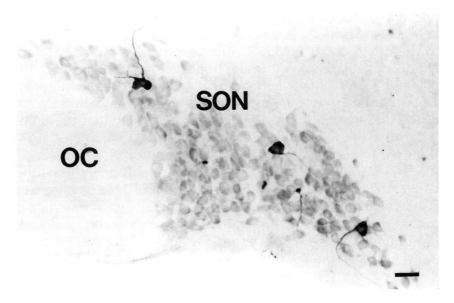

Fig. 4. Solitary neurons in Brattleboro rats. Vibratome section of the hypothalamus of a homozygous Brattleboro rat showing three solitary neurons expressing glycopeptide immunoreactivity in the perikaryon and its neurites. *OC* optic chiasm, *SON* supraoptic nucleus, bar = 25 μm

VP PRECURSOR AND MUTANT FORMS

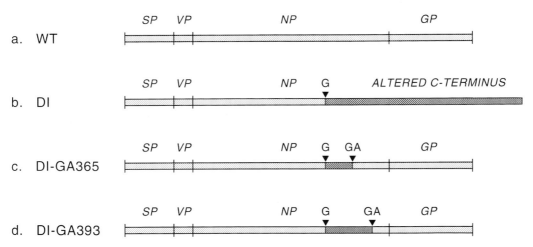

Fig. 5. VP precursors and mutant forms. Different forms of the vasopressin (VP) precursor in the wild-type (WT) and homozygous diabetes insipidus Brattleboro (di) rat. The VP gene encodes a precursor consisting of a signal peptide (SP), VP, Neurophysin (NP), and Glycopeptide (GP). a): wt rat VP precursor. b): rat VP precursor. By a G-deletion in the VP gene, the C-terminal part of NP and GP is altered into a new amino acid sequence. c,d): +1 di-VP precursors (di-GA 365 + 393) expressed in reverted solitary magnocellular neurons. These precursors are encoded by VP transcripts with a GA-deletion at two different sites downstream the G-deletion and which result in a restored C-terminal NP and GP moiety. Compared to the wt VP precursor a stretch of 13 (c) or 22 (d) amino acids within the NP region is still mutated (for details see Evans et al., 1994)

processing into functional VP takes place (Evans et al., 1994; Sonnemans et al., 1996). Two sites of the dinucleotide deletion (ΔGA) were found that occurred preferentially in GAGAG motifs of the VP RNA. Moreover, the mutation rate depended upon transcriptional activity (Evans et al., 1994) and the number of these solitary neurons was age-related (van Leeuwen et al., 1989). It was subsequently demonstrated that a similar mutation also occurs in wild-type rat and human VP gene transcripts, meaning that these mutations are not restricted to the VP cells of Brattleboro rats (Evans et al., 1994, 1996). In wild-type VP transcripts, contrary to the restoration of the reading frame as seen in the Brattleboro rat, the outcome is a frameshift and the synthesis of a VP mutated protein (VP^{+1}). These findings suggest the existence of a novel mechanism which may very well be related to aging and age-related diseases in general, in both rodents and humans. These observations prompted us to initiate a study on the presence of dinucleotide deletions in other neuronal genes associated with the age-related neuropathology of AD and DS.

The frequency of GAGAG-motifs in transcripts

In the VP transcripts of the di/di rat the GAGAG motif is a hotspot for the occurrence of a dinucleotide deletion (Evans et al., 1994, 1995). In the human transcript population the GAGAG motif may also be a hotspot for this type of deletion. The chance that a GAGAG motif occurs in a sequence is one in every 1024 (i.e. $1:4^5$) bases. The human haploid genome consists of 3×10^9 bases, 9×10^7 (3%) of which code for transcripts. This implies that there are 9×10^4 GAGAG motifs within coding sequences. Since the human genome consists of 65,000–80,000 genes with a mean length of 2.2 kb, an average of 2.1 GAGAG motifs per gene can be expected (Strachan and Read, 1996). We selected a panel of genes associated with AD pathology, calculated the expected number of GAGAG motifs (1:1024) and determined the actual number of these motifs (Table 1). Both coding sequences of βAPP and UBB showed an actual higher number of GAGAG motifs compared to the predicted number of GAGAG motifs: βAPP seven motifs instead of the 2.2, and UBB two motifs instead of 0.7. A dinucleotide deletion in or adjacent to the GAGAG motif results in a translation of the mRNA in the +1 reading frame into a +1 protein. The presence of βAPP and UBB frameshift mutations and their corresponding +1 proteins was studied in the brains of DS, AD and Parkinson patients and age-matched non-demented controls (see http://www.knaw.nl/nih: Van Leeuwen et al., 1998a).

β amyloid precursor protein

The βAPP gene is located on chromosome 21 and has 18 exons, and by alternative splicing three different proteins are formed: APP 695 (695 amino acids (aa) Δ exon 7 and 8), APP 751 (751 aa, Δ exon 8) and APP 770 (770 aa); Yoshikai et al., 1990). Especially APP 695 is abundantly expressed in neurons

Table 1. Genes associated with Alzheimer's disease

Gene	Number of base pairs Coding sequence, longest form	GAGAG motifs Expected number	Actual number	Exon from which +1 peptides are derived	+1 Peptide sequence
βAPP	2234	2.2	7	9/10	RGRTSSKELA
Ubiquitin-B	687	0.7	2	2	DHHPGSGAQ
Tau	1056	1.1	—	13	HGRLAPARHAS[a]
Presenilin I	1392	1.4	3	9	SIQKFQV
Presenilin II	1346	1.3	3	3	VEKPGERGGR
Apolipoprotein E4	951	0.9	—	4	GAPRLPPAQAA[a]
MAP2b[b]	5595	5.6	13	18	KTRFQRKGPS
Neurofilament-light	1625	1.6	3	1	PGNRSMPGHE
Neurofilament-medium	2748	2.8	3	3	EAEGEGSPS
Neurofilament-heavy	3063	3.1	2	1	VGAARDSRAA
GFAP[c]	1299	1.3	6	6	EDRGDAGWRGH

[a] Based upon a GAGA motif. [b] Microtubule associated protein 2B. [c] Glial fibrillary acidic protein

and lacks the Kunitz type serine protease domain. Mutations found in the FAD cases in the APP gene are all located in exons 16 and 17, which is the region coding for $A\beta_{42-43}$. Misprocessing of the APP precursor in FAD results in extracellular aggregation of $A\beta_{42-43}$ (Van Broeckhoven, 1998). Theoretically, the coding sequence of βAPP (2234 bases) is expected to contain 2.2 GAGAG-motifs. However, in the sequence 7 GAGAG-motifs are present, indicating that βAPP has a high risk of dinucleotide deletions in and around GAGAG motifs, especially in exons 9 and 10, where three GAGAG-motifs are clustered (see Table 1).

A dinucleotide deletion in either exon 9 or 10 gives rise to the same +1 protein. Antibodies were raised against a part of the C-terminus of this +1 protein: RGRTSSKELA (see Table 1). Theoretically, the APP^{+1} protein is a 38kDa protein, but due to an acidic-rich domain in the N-terminus the molecular weight appears to be 60kDa (Hol, in preparation). However, in our Western blot analyses we reported a 38kDa APP^{+1} protein that somehow (e.g. by degradation of full-length APP^{+1}) may have been generated during tissue processing. In fact, our data were in agreement with an earlier report (Cole et al., 1989) in which a 35kDa APP-like protein was mentioned. This protein was detected in hippocampus homogenates of AD patients by an antibody raised against the N-terminus of wild-type βAPP (βAPP 175–186), which is also present in APP^{+1}. In addition, it has been reported that the βAPP present in dystrophic neurites of neuritic plaques is an N-terminal fragment of the βAPP protein (Cole et al., 1991). Similar data were obtained with an N-terminal APP antibody raised against βAPP 45–62 by Palmert et al. (1988). The βAPP proteins described in these three reports may have been the first indication for the presence of βAPP^{+1} in AD.

Paraffin sections of brains (temporal and frontal cortex and hippocampus) of AD and DS patients and age, sex and post-mortem delay-matched controls were incubated with this βAPP^{+1} antibody (Hol et al., 1998a; Van Leeuwen et al., 1998a; Table 2). The APP^{+1} protein accumulates in the dystrophic neurites of the neuritic plaques, tangles and neuropil threads in brain sections of early (<65 years) and late onset (>65 years) AD and DS patients and one Parkinson patient with initial AD neuropathology (Table 3, Fig. 6). No βAPP^{+1} immunoreactivity was found in paraffin sections of cerebral cortex and hippocampus of non-demented controls and in the nigrostriatal system of ten Parkinson patients (Hol et al., 1998a; Van Leeuwen et al., 1998a; see http://www.knaw.nl/nih). The pooled data of the frontal and temporal cortex and the hippocampus show that βAPP^{+1} accumulates in 71% of the AD patients and in 88% of the demented DS patients studied. It is remarkable that one DS patient (#96015), who had no symptoms of dementia, did not show any sign of neuropathology or APP^{+1} either in all three areas studied. However, in the transenthorhinal cortical region the pre-α-islands (Braak et al., 1996) showed intense APP^{+1} and UBB^{+1} immunoreactivity (Fig. 7).

Using an expression-cloning strategy, we indeed confirmed the proposed deletions (ΔGA) in GAGAG motifs of exons 9 and 10 (Van Leeuwen et al., 1998a; Table 4). A deletion in exon 9 would result in the loss of the growth-promoting (neurotrophic) domain in βAPP (Jin et al., 1994) and due to the

Table 2. Clinicopathological information of post-mortem material used in this study

NBB/ autopsy no.	Age (years)	Sex (m/f)	Dementia duration (years)	GDS	Post-mortem delay (h)	Fixation duration (days)	Brain weight (g)	Cause of death
Non-demented controls								
89003	34	m	—	—	<17	1,124	1,348	empyema of pleura, fibrous pleuritis and fibrous pericarditis, AIDS
81021	43	m	—	—	23	53	1,260	non-Hodgkin lymphoma
94119	51	f	—	—	8	41	1,156	progressive liposarcoma and ileus
94125	51	m	—	—	6	47	1,518	sepsis
88037	58	m	—	—	<41	1,088	1,797	lung carcinoma, massive hemorrhage
90073	65	f	—	—	<41	403	1,234	pulmonary embolism
90079	72	m	—	—	4	126	1,330	myocardial infarction, cardiogenic shock
91026	80	f	—	—	36	65	1,205	cardiogenic shock
91027	82	f	—	—	<55	38	1,100	myocardial infarction, ventricular fibrillation
90080	85	m	—	—	5	126	1,050	cardiac failure, myo-cardial infarction, coronary sclerosis, lung emphysema
81007	90	f	—	—	12	48	1,110	postoperative infections
90083	90	f	—	—	5	143	1,040	metabolic acidosis
Down's syndrome								
93162	54	f**	11	na	<17	614	730	DS, bronchopneumonia
92080	58	f**	8	7	10	140	712	DS, epilepsia, pneumonia, decompensatio cordis
98105	58	m***	10	7	7	48	1,155	DS, pneumonia
89055	59	f**	5	7	5	29	812	DS, cardiac arrest
93161	62	f**	11	na	<17	585	1,100	DS, pneumonia
96015	63	f**	—	—#	<24	87	980	DS, cardiac-respiratory insufficiency
94058	64	m***	3	7	7	47	875	DS, dehydration, pneumonia
93028	67	f**	8	7	11	104	859	DS, pneumonia

NBB Netherlands Brain Bank. **/*** = karyotype 47XX21/47XY21, *na* not available, *GDS* Global deterioration scale, #non-demented

Table 3. βAPP+1 and UBB+1 immunoreactivity in frontal and temporal cortex and hippocampus

NBB autopsy no.	age (years)	sex (m/f)	Frontal cortex				Temporal cortex				Hippocampus			
			neuropathological state*		βAPP+1	UBB+1	neuropathological state*		βAPP+1	UBB+1	neuropathological state*		βAPP+1	UBB+1
			plaques	tangles			plaques	tangles			plaques	tangles		
Non-demented controls														
89003	34	m	–	–	–	–	–	–	–	–	–	–	–	–
81021	43	m	–	–	–	–	–	–	–	–	–	–	–	–
94119	51	f	–	–	–	–	–	–	–	–	–	–	–	–
94125	51	m	–	–	–	–	–	–	–	–	–	–	–	–
88037	58	m	–	–	–	–	–	–	–	–	–	+[a]	–	–
90073	65	f	–	–	–	–	–	–	–	–	–	–	–	–
90079	72	m	+[a]	–	–	–	–	–	–	–	+[b]	+[c]	–	+
91026	80	f	–	–	–	–	–	–	–	–	+[b]	+[c]	–	+
91027	82	f	–	–	–	–	–	–	–	–	–	+[c]	–	+
90080	85	m	+[b]	+[a]	–	–	+[b]	+[b]	–	+	+[b]	+[b]	–	+
81007	90	f	–[b]	+[a]	–	–	+[a]	+[a]	–	–	+[b]	+[b]	–	+
90083	90	f	+[b]	–	–	–	+[b]	–	–	–	–	+[a]	–	+
% pos. staining					0%	0%			0%	8%			0%	50%
Down's syndrome														
93162	54	f	+[c]	+[c]	+	+	+[c]	+[c]	+	+	+[b]	+[c]	+	+
92080	58	f	+[b]	+[c]	+	+	+[b]	+[c]	+	+	+[c]	+[c]	+	+
98105	58	m	+[c]	+[c]	+	+	+[c]	+[c]	+	+	+[c]	+[c]	+	+
89055	59	f	+[b]	+[c]	+	+	+[b]	+[c]	+	+	+[b]	+[c]	+	+
93161	62	f	+[c]	+[c]	+	+	+[c]	+[c]	+	+	+[c]	+[c]	+	+
96015	63	f	–	–	–	–	+[a]	–	–	–	–	–	–	–
94058	64	m	+[b]	+[c]	+	+	+[b]	+[b]	+	+	+[b]	+[c]	+	+
93028	67	f	+[c]	+[c]	+	+	+[c]	+[c]	+	+	+[c]	+[c]	–	+
% pos. staining					88%	88%			88%	88%			75%	88%

NBB Netherlands Brain Bank, * Number of plaques (all types) and tangles as revealed by Congo red and Bodian silver staining: a) few, b) moderate and c) many. ** APP+1 and UBB+1 proteins are present in the enthorinal cortex (Fig. 7)

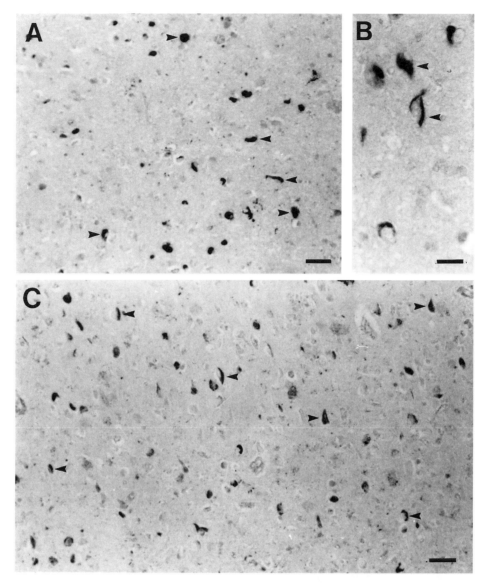

Fig. 6. +1 Proteins in DS cortex. APP^{+1} (**A,B**) and UBB^{+1} (**C**) immunoreactivity in paraffin sections of the temporal cortex of two DS patients (a,b: #94058, BA11, layers 2, 3 and 5; and c: the subiculum of #92080; Table 2, 3). Note the numerous immunoreactive tangle-like structures (arrowheads). Scale bar A and C = 50 μm, B = 20 μm

premature termination codon in principle no Aβ$_{42-43}$ is produced. However, the production of βAPP^{+1} does not exclude the production of Aβ$_{42-43}$. There is an additional transcription initiation site in βAPP downstream of the mutation (Citron et al., 1993) that might be activated due to the dinucleotide deletion, leading to misprocessing of βAPP and the deposition of Aβ$_{42-43}$. Furthermore, wild-type βAPP is still expressed and can account for the Aβ$_{42-43}$ production as well.

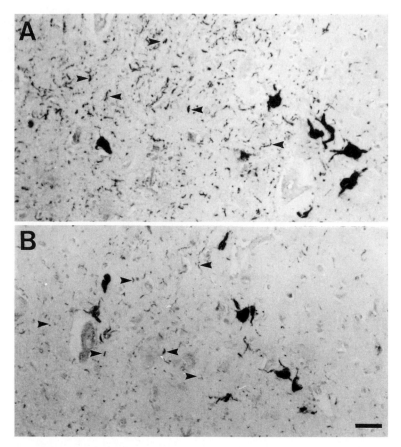

Fig. 7. +1 Proteins in DS enthorhinal cortex. APP^{+1} (**A**) and UBB^{+1} (**B**) immuno-reactivity in two adjacent sections of the transenthorhinal cortex of a DS patient (#96015). The tangle-like cells of a pre-α-island show intense staining. The frontal and temporal cortex and the hippocampus were not immunoreactive (Table 3). Arrowheads point to neuropil threads. Bar = 10 μm

Enhanced gene transcription in Down Syndrome patients and molecular misreading

The occurrence of a dinucleotide deletion in transcripts seems to be promoted by the transcriptional activity of the VP gene, as seen in the di/di rat. These magnocellular neurons normally express high levels of the VP transcript, as VP expression is activated in di/di rats due to osmotic stress (Sherman and Watson, 1988; Szot et al., 1992). Suppression of the metabolic activity (i.e. VP transcription) by substitution with exogenous VP in the magnocellular neurons in the supraoptic and paraventricular nucleus of the di/di rats leads to a decrease in the number of cells with a reverted phenotype (Evans et al., 1994). These experiments indicate that a higher incidence of dinucleotide deletions is observed when expression of certain transcripts is enhanced.

DS patients are of particular interest as almost all of these patients exhibit a similar neuropathology as AD patients (Wisniewski et al., 1985). The AD

Table 4. Overview of the deletions found in β-APP and UBB transcripts of controls and DS patients

Patient nr:	age	disease	total number of clones	number of positive clones	ΔGA
βAPP					
94125	51	control	20,000	0	—
94119	51	control	20,000	0	—
90079	72	control	20,000	0	—
91092	54	Alzheimer	20,000	10	+ (ex9)
88073*	66	Alzheimer	20,000	2	+ (ex10)
90015*	81	Alzheimer	20,000	5	+ (ex9 or ex10)
93048	92	Alzheimer	20,000	5	+ (ex9)
92080*	58	Down syndrome	20,000	2	+ (ex9)
94058*	64	Down syndrome	20,000	12	+ (ex9)

Patient nr:	age	disease	total number of clones	number of positive clones	dinucleotide deletion
Ubiquitin-B					
94125	51	control	20,000	0	—
94119	51	control	3,000	14	Δgt
90079	72	control	20,000	13	Δgt or Δct
89057	40	Alzheimer	5,000	15	Δgt
88073*	66	Alzheimer	1,500	138	Δgt
90015*	81	Alzheimer	2,000	32	Δgt
92080*	58	Down syndrome	800	44	Δgt
89055	59	Down syndrome	1,000	2	Δgt or Δct
94058*	64	Down syndrome	20,000	93	Δgt

*indicates patients with a dinucleotide deletion in both βAPP and UBB transcripts.
[1] Clinicopathological information can be found in Hol et al. (1998a) or www.knaw.nl/index.html (click on Science cover)

neuropathology in DS patients usually becomes apparent 40 years earlier compared to the pathology in sporadic AD. DS patients highly express βAPP, due to an extra copy of chromosome 21. However, the expression level of the βAPP gene, which has increased seven-fold as compared to the βAPP expression in controls, exceeds the expression level expected from the 1.5 fold gene dosage (trisomy 21) alone (Neve et al., 1988; Rumble et al., 1989; Kola, personal communication). Indeed, we detected a transcript mutation in βAPP and the expression of APP[+1] protein in 100% of the DS patients (Hol et al., 1998a; Van Leeuwen et al., 1998a) (see http://www.knaw.nl/nih).

Ubiquitin-B

Ubiquitin is an evolutionary highly conserved 76-residue protein that plays an essential role in a large number of processes including apoptosis and

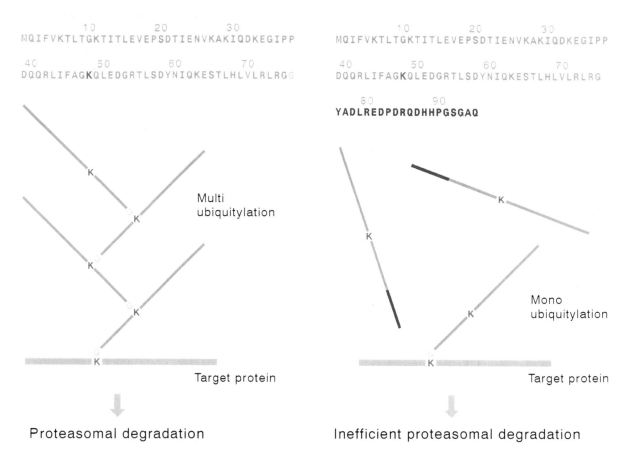

Fig. 8. The ubiquitin-proteosomal pathway to degrade aberrant proteins. Under normal circumstances (left side) a lysine (K) residue of the target protein (thick line, for example, hyperphosphorylated Tau or any other protein that is degraded via the proteosomal pathway) is recognized by the C-terminal glycine (G, yellow) residue of ubiquitin-B (UBB). In turn, a lysine residue of UBB at position 48 is recognized by another UBB molecule. If this process is repeated, multiubiquitylation occurs, the proteasomal pathway is triggered and the protein degraded. The right panel shows that a GT deletion in the UBB transcript results in the loss of the G residue and the formation of an aberrant C-terminus (red), which might block the process of multiubiquitylation. Only the wild-type transcripts are still able to ubiquitylate. In AD, however, the Tau protein is mainly mono-ubiquitylated (Morishima-Kawashima et al., 1993). It is hypothesized that in time (AD lasts for approximately eight years) the proteasomal degradation becomes increasingly inefficient resulting in a piling up of aberrant proteins in the neuron whose function is increasingly disturbed (taken from Van Leeuwen et al., 1998b, with permission)

ATP-dependent proteasomal breakdown of proteins (Varshavsky, 1997; Ciechanover, 1999). Prior to protein degradation, ubiquitin is tagged to several target proteins through an isopeptide bond between a lysin moiety in the target protein and the C-terminal glycine of ubiquitin (Varshavsky, 1997). The C-terminal glycine of one ubiquitin molecule can bind to the ε amino group of a lysine in an adjacent ubiquitin moiety leading to multi-ubiquitylation (Fig. 8, left panel). The presence of a multi-ubiquitin chain on an aberrant protein, a process designated as multi-ubiquitylation, triggers the proteasomal breakdown of this particular protein. Ubiquitin is a component of neurofibrillary tangles in AD (Mori et al., 1987; Mayer et al., 1991, 1996) and DS patients (Mann et al., 1989), and it is conjugated to tau protein in paired helical filaments. Most ubiquitin in paired helical filaments appears to represent mono-ubiquitylated tau (Morishima-Kawashima et al., 1993), indicating that multi-ubiquitylation does not occur and consequently the proteasomal breakdown of these mono-ubiquitylated proteins is inefficient.

Ubiquitin is a multi-gene family (i.e. Ubiquitin-A, -B and -C; cf. Hol et al., 1998a), with some genes containing multiple repeats of ubiquitin coding sequences (Wiborg et al., 1985). UBB and -C are expressed in the human brain (Adams et al., 1992). UBB consists of three repeats (Baker et al., 1987) and comprises two GAGAG motifs one in first repeat and one in the second at amino acid (aa) position 75. A dinucleotide deletion in either of the repeats will lead to a similar aberrant protein of 95 aa instead of 76 aa (Fig. 8, right panel). This UBB^{+1} protein lacks the essential C-terminal glycine and is unable to multi-ubiquitylate target proteins. Consequently, the presence of UBB^{+1} would result in an inefficient proteasomal breakdown and thus could account for the lack of poly-ubiquitylation of tau protein in paired helical filaments, as reported by Morishima-Kawashima et al. (1993). An antibody was raised against the C-terminus of the UBB^{+1} sequence (RQDHHPGSGAQ; Van Leeuwen et al., 1998a; Table 1). In early and late onset sporadic AD cases we found UBB^{+1} in neuritic plaques, neuropil threads and tangles (Fig. 6a). Brain section of DS patients revealed a similar result (Van Leeuwen et al., 1998a). In addition, in paraffin sections of hippocampi (CA1 and subiculum) of elderly control patients (>72 years) and of one Parkinson patient with initial AD neuropathology UBB^{+1} immunoreactivity was found. In young non-demented controls (<72 years), no UBB^{+1} immunoreactivity was found (Van Leeuwen et al., 1998a). However, recent studies on vibratome sections showed that UBB^{+1} im-munoreactivity was already present at an earlier age (>51 years; #94119) in the hippocampus and temporal cortex of young non-demented controls (for a detailed overview of the neuropathological status of the patients studied and the presence of UBB^{+1} immunostaining in frontal cortex, temporal cortex and hippocampus, see http://www.knaw.nl/nih or Hol et al., 1998a). Taking all these data together, UBB^{+1} was expressed in 100% of the AD and demented DS patients studied. UBB^{+1} is probably an early marker for neuro-degeneration as immunoreactivity is also present in non-demented elderly controls in brain structures that are known to be an early target for AD neuropathology, such as the entorhinal cortex, CA1 and subiculum (Braak et al., 1996). The predicted dinucleotide deletion was indeed found, however, contrary to VP and APP, not within the GAGAG motif but as a GT deletion directly adjacent to it (Hol et al., 1998a; Van Leeuwen et al., 1998a; Table 4).

Genomic mutation or transcript modification

In the di/di rat an age-dependent increase in the number of VP cells with a revertant phenotype was found, which we originally interpreted as somatic mutations in the genes occurring at an exceptionally high frequency (Evans et al., 1995; Finch and Goodman, 1997). However, we were unable to show a mutation at the genomic level, whereas in transcripts the mutation (ΔGA) could be readily determined (Evans, Spence, Burbach and Van Leeuwen, unpublished work). The same results were obtained in transcripts and genomic DNA of βAPP and UBB using even more sensitive approaches and huge amounts of genomic DNA (Van Leeuwen et al., 1998a). Using the most sensitive approach, we were able to amplify as a positive control ten copies of mutant plasmid DNA out of a background of 500 ng genomic DNA corresponding to about 160,000 copies of DNA or 80,000 cells. Using this very sensitive method, a total of 5 µg DNA for each control, AD and DS patients was investigated (i.e. a total of 1.6×10^6 copies of βAPP and UBB), but no mutations were found (Van Leeuwen et al., 1998a). Consequently, it is very unlikely that these dinucleotide deletions take place in the genome. We therefore favor the opinion that transcript mutation occurs, the more so since different mutant proteins coexist in the same neurons (Van Leeuwen et al., 1998a,b).

The mechanism of RNA mutations and RNA quality control

RNA mutations may take place either co- or post-transcriptionally. Substitutional RNA editing has been described in the nervous system (Newman, 1994). Knowledge of the mechanisms by which RNA polymerases produce errors in repeats, such as GAGAG or CUCU (e.g., slippage or stuttering; Jacques and Kolakovsky, 1991; Kolakovsky and Hausmann, 1998) in the brain tissue of mammals is lacking. From our experiments it appears that the generation of dinucleotide deletions by molecular misreading of correct DNA is promoted by their enhanced transcription, as illustrated both in VP cells of homozygous Brattleboro rats (Szot et al., 1992; Evans et al., 1994) and in βAPP transcripts of the cerebral cortex of DS patients (Van Leeuwen et al., 1998a). The finding that two different +1 proteins, arising from two different transcripts, coexist in neurons of AD patients points to a general controlling mechanism (the "common denominator"; Van Leeuwen et al., 1998a) that shows a failure to detect mutated mRNA in AD patients (Van Leeuwen et al., 1998a). It is well-known that such a proofreading mechanism for DNA acts during DNA replication resulting in an error rate of $1:10^{13}$ (Radman and Wagner, 1988; Kunkel, 1995). In the case of RNA, mRNA surveillance or nonsense-mediated mRNA decay (NMD) was described as a mechanism that monitors for premature stop codons in RNA (Pulak and Anderson, 1993; Ruiz-Echevarria et al., 1996; Culbertson, 1999; Hentze and Kulozik, 1999). This system increases the fidelity of gene expression by eliminating incompletely translated RNAs and could act as the common denominator controlling the mRNA quality. Indeed, a human homologue of a yeast gene with potential "mRNA surveillance" activity (i.e. human up frameshift protein, HUPF1) has recently been cloned (Applequist et al., 1997; Perlick et al., 1996). A declining accuracy of such an mRNA surveillance system later in life could therefore be an important aging factor, in addition to those already known (e.g., oxidative stress and DNA damage; cf. Finch, 1990). Recently, a similar mechanism has been suggested to play a role in sporadic amyotrophic lateral sclerosis, during which, as a result of RNA processing errors (e.g. exon skipping), aberrant RNA molecules of the astrocyte glutamate transporter (EEAT2) occur, whose protein may exert a dominant negative effect after translation (Lin et al., 1998). It will be of interest to determine in which way molecular misreading of the APP and Ubiquitin B genes in DS is related to mRNA surveillance.

Other neurodegenerative and age-related diseases

Several genes thought to be involved in other age-related neurodegenerative diseases indeed contain vulnerable sequences such as GAGAG motifs (Table 1). Candidates are diseases with inclusion bodies such as frontal lobe dementia or Pick's disease, progressive supranuclear palsy, argyrophilic grain disease, amyotrophic lateral sclerosis, diffuse Lewy body disease, multisystem atrophy and Huntington, all of which contain ubiquitin in their characteristic inclusions, and possibly UBB[+1] as well. The resulting differential pathology may be determined by various other factors (such as other +1 proteins, cell type and risk factors; Mann et al., 1997). If in addition to Parkinson's disease (Van Leeuwen et al., 1998) in all these diseases no UBB[+1] or APP[+1] proteins are present, this implies a specificity of these +1 proteins for AD. However, it cannot be excluded that molecular misreading of other genes occurs in these diseases. If molecular misreading occurs in neuronal genes, it might be possible that a similar process occurs outside the nervous system, viz. in proliferating tissues. Recently we have indeed obtained evidence that molecular misreading occurs in rat VP transgenes expressed ectopically in the male reproductive tract of the mouse (Van Leeuwen et al., in prep.) and in UBB transcripts in human liver disease (McPhaul et al., 1999). In these cases, molecular misreading (ΔGA, ΔGU, ΔCU) may take place in, or adjacent to, stretches of GAGAG and CUCU repeats as reported before (van Leeuwen et al., 1998a). As the 65,000–80,000 genes expressed by the human genome (Strachan and Read, 1996) with an estimated mean length of 2.2 kb contain an average of two GAGAG motifs per gene, numerous transcripts may be mutated in a similar manner as VP, APP and UBB mRNA (van Leeuwen et al., 1998b). A large array of gene transcripts, for instance those involved in, or associated with, pathologies, such as cancer, cardiovascular diseases and diabetes mellitus, may undergo a similar modification and are therefore interesting to study.

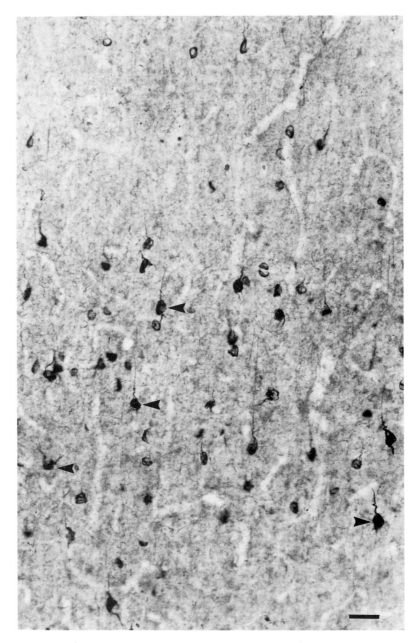

Fig. 9. Ubiquitin-B^{+1} immunoreactivity in a 50-μm-thick vibratome section of the temporal cortex of a 92-year-old Alzheimer patient. Intense immunoreactivity was present in the neurofibrillary tangles (arrowheads) in layers 2, 3 and 5. Scale bar = 25 μm

Concluding remarks and future studies

The novel mechanism of molecular misreading has already raised a large number of key questions, three of which will be briefly addressed below. Our major challenge in the coming years will be to show that molecular misreading of genomic information is an early event in AD and DS. Transgenic mice

MOLECULAR MISREADING

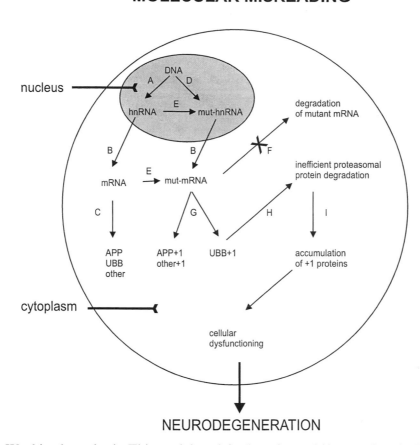

NEURODEGENERATION

Fig. 10. Working hypothesis. This model explains how frameshift mutations in RNA and the subsequent +1 proteins could accumulate in the neurons of DS and AD patients. In the cell nucleus the genomic information is intact and is transcribed into heteronuclear RNA (hnRNA) (A). This hnRNA is spliced and transported to the cytoplasm (B), where it is translated into wild-type βAPP, UBB and other proteins (B). In the event that the coding sequence contains GAGAG motifs, the sequence is susceptible for dinucleotide deletions (D,E). Molecular misreading can be caused by stuttering or slippage of the RNA poylmerase during transcription (D) or by editing of the mRNA or hnRNA after transcription (E). In DS and AD patients the mRNA decay pathway, termed mRNA surveillance, may be impaired (F) and therefore the +1 mRNA is translated into +1 protein (G). In particular, UBB^{+1} may have a prominent role as it will probably directly interfere with the ubiquitin/proteasome system and lead to an inefficient protein breakdown through the proteasomal pathway (H) (Fig. 8). Together, these changes lead to a gradual accumulation of aberrant proteins in neurons, cellular dysfunction and ultimately neurodegeneration

overexpressing +1 proteins, such as APP^{+1} and UBB^{+1}, will be instrumental to find out whether they are causal in the onset of AD neuropathology. Moreover, if the transgenic overexpression of these +1 proteins does not result in neuropathology, it is plausible that the accumulation of these mutated proteins in DS and AD cortical neurons will have deleterious effects on neuronal function.

A second key question is the mechanism of molecular misreading in and around simple sequences such as GAGAG or CUCU (Van Leeuwen et al., 1998a). All we know at present is that molecular misreading is correlated with enhanced transcription of genes (Evans et al., 1994; Van Leeuwen et al., 1998b). Trisomy 16 mice, whose chromosome 16 shows a high degree of synteny with the human chromosome 21 and a seven-fold overexpression of the βAPP gene (Kola and Hertzog, 1997, 1998; Sago et al., 1998), will be useful to confirm these data. Presently we are studying the mechanism of molecular misreading in neuroblastoma and peripheral neuroepithelioma cell lines that have been observed to display molecular misreading (Hol et al., 1998b) and that can be modulated in their transcriptional activity and be transfected with gene reporter constructs as well.

A third important question is whether molecular misreading (e.g. of the UBB gene) is a process restricted to Alzheimer's pathology or more general processes occurring in other neurodegenerative disorders and other age-related diseases.

We anticipate that molecular misreading of genes will result in a loss of cellular function. The presence of various +1 proteins in one neuron points to a lack of mRNA quality control by a common denominator. The mRNA surveillance system is a likely candidate for this role. A possible coincidence of decreased activity of mRNA surveillance and an increased degree of molecular misreading in elderly people (Fig. 10) might explain why age is the greatest risk factor for developing neural and non-neural pathologies.

Acknowledgements

The authors are grateful to Dr. D. Fischer for critically reading the manuscript and to H. Stoffels, G. van der Meulen and O. Pach for their technical and secretarial assistance.

References

Adams MD, Dubnick M, Kerlavage AR, Moreno R, Kelley JM, Utterback TR, Nagle JW, Fields C, Venter JC (1992) Sequence identification of 2,375 human brain genes. Nature 355: 632–634

Antonarakis SE (1998) 10 Years of genomics, chromosome 21, and Down syndrome. Genomics 51: 1–16

Applequist SE, Selg M, Raman C, Jäck H-M (1997) Cloning and characterization of HUPF1, a human homolog of the Saccharomyces cerevisiae nonsense mRNA-reducing UPF1 protein. Nucl Acids Res 25: 814–821

Baker RT, Board PG (1987) The human ubiquitin gene family: structure of a gene and pseudogenes from the Ub B subfamily. Nucl Acids Res 15: 443–463

Blacker D, Wilcox MA, Laird NM, Rodes L, Horvath SM, Go RCP, Perry R, Watson Jr B, Bassett SS, McInnis MG, Albert MS, Hyman BT, Tanzi RE (1998) Alpha-2 macroglobulin is genetically associated with Alzheimer disease. Nature Genet 19: 357–360

Braak H, Braak E, Yilmazer D, de Vos RAI, Jansen EHN, Bohl J (1996) Pattern of brain destruction in Parkinson's disease and Alzheimer's disease. J Neural Transm 103: 455–490

Ciechanover A (1998) The ubiquitin-proteasome pathway: on protein death and cell life. EMBO J 17: 7151–7160

Citron M, Haass C, Selkoe DJ (1993) Production of amyloid-β-peptide by cultured cells: no evidence for internal initiation of translation at Met$_{596}$. Neurobiol Aging 14: 571–573

Cole GM, Masliah EM, Huyn TV, DeTeresa R, Terry RD, Okuda C, Saitoh T (1989) An antiserum against amyloid β-protein precursor detects a unique peptide in Alzheimer brain. Neurosci Lett 100: 340–346

Cole GM, Masliah E, Shelton ER, Chan HW, Terry RD, Saitoh T (1991) Accumulation of amyloid precursor fragment in Alzheimer plaques. Neurobiol Aging 12: 85–91

Cruts M, Van Duijn CM, Backhovens H, Van den Broeck M, Wehnert A, Serneels S, Sherrington R, Hutton M, Hardy J, St George-Hyslop PH, Hofman A, Van Broeckhoven C (1998) Estimation of the genetic contribution of presenilin-1 and -2 mutations in a population-based study of presenile Alzheimer disease. Hum Mol Genet 7: 43–51

Culbertson, MR (1999) RNA surveillance, unforseen consequences for gene expression, inherited genetic disorders and cancer. TIG 15: 74–79

Evans DAP, Van der Kleij AAM, Sonnemans MAF, Burbach JPH, Van Leeuwen FW (1994) Frameshift mutations at two hotspots in vasopressin transcripts in post-mitotic neurons. Proc Natl Acad Sci USA 91: 6059–6063

Evans DAP, Burbach JPH, Van Leeuwen FW (1995) Somatic mutations in the brain: relationship to aging? Mutat Res 338: 173–182

Evans DAP, Burbach JPH, Swaab DF, Van Leeuwen FW (1996) Mutant vassopressin precursors in the human hypothalamus: evidence for neuronal somatic mutations in man. Neuroscience 71: 1025–1030

Finch CE (1990) Longevity, senescence and the genome. University of Chicago Press, Chicago

Finch CE, Goodman MF (1997) Relevance of 'adapative' mutations arising in non-dividing cells of microorganisms to age-related changes in mutant phenotypes of neurons. TINS 20: 501–507

Hentze MW, Kulozik AE (1999) A perfect message: RNA surveillance and nonsense-mediated decay. Cell 96: 307–310

Hernandez D, Fisher EMC (1996) Down syndrome genetics: unraveling a multifactorial disorder. Hum Mol Genet 5: 1411–1416

Hol EM, Neubauer A, De Kleijn DPV, Sluijs JA, Ramdjielal RDJ, Sonnemans MAF, Van Leeuwen FW (1998a) Dinucleotide deletions in neuronal transcripts: a novel type of mutation in non-familial Alzheimer's disease and Down syndrome patients. Progr Brain Res 117: 379–394

Hol EM, Sluijs JA, Sonnemans MAF, Versteeg R, Van Leeuwen FW (1998b) Frameshift mutations in β-amyloid precursor protein and ubiquitin-B RNA in human neuronal cell lines. Abstr Soc Neurosci Meeting 24: 722

Ivell R, Burbach JPH, Van Leeuwen FW (1990) The molecular biology of the Brattleboro rat. Front Neuroendocrinol 11: 313–338

Jacques JP, Kolakofsky D (1991) Pseudo-templated transcription in prokaryotic and eukaryotic organisms. Genes Dev 5: 707–713

Jin L-W, Ninomiya H, Roch J-M, Schubert D, Masliah E, Otero DAC, Saitoh T (1994) Peptides containing the RERMS sequence of amyloid β/A4 protein precursor bind cell surface and promote neurite extension. J Neurosci 14: 5461–5470

Kola I, Hertzog PJ (1997) Animal models in the study of the biological function of genes on human chromosome 21 and their role in the pathophysiology of Down syndrome. Hum Mol Genet 6: 1713–1727

Kola I, Hertzog PJ (1998) Down syndrome and mouse models. Curr Opin Genet Dev 8: 316–321

Kolakofsky D, Hausmann S (1998) Co-transcriptional paramyxovirus mRNA editing; a contradiction in terms? In: Grosjean H, Benne R (eds) Modification and editing of RNA. ASM Press, Washington DC, pp 413–420

Kunkel TA (1995) The intricacies of eukaryotic spell-checking. Curr Biol 5: 1091–1094

Lin C-LG, Bristol LA, Jin L, Dykes-Hoberg M, Crawford T, Clawson L, Rothestein JD (1998) Aberrant RNA processing in a neurodegenerative disease: the cause for absent EAAT2, a glutamate transporter, in amyotrophic lateral sclerosis. Neuron 20: 589–602

Mann DMA (1997) Molecular biology's impact on our understanding of aging. BMJ 315: 1078–1081

Mann DMA, Prinja D, Davies CA, Iahara Y, Delacourte A, Défossez A, Mayer RJ, Landon M (1989) Immunocytochemical profile of neurofibrillary tangles in Down's syndrome patients of different ages. J Neurol Sci 92: 247–260

Mayer RJ, Arnold J, Lásló L, Landon M, Lowe J (1991) Ubiquitin in health and disease. Biochim Biophys Acta 1089: 141–157

Mayer RJ, Tipler C, Arnold J, Lásló L, Al-Khedhairy A, Lowe J, Landon M (1996) Endosome-lysosomes, ubiquitin and neurodegeneration. Adv Exp Med Biol 389: 261–269

McPhaul L, Wang J, Yuan SW, French SW, Van Leeuwen FW (1999) Frameshift mutants of ubiquitin-B in Mallory body formation in human liver. FASEB J 13: A736

Mori H, Kondo J, Ihara Y (1987) Ubiquitin is a component of paired helical filaments in Alzheimer's disease. Science 235: 1641–1644

Morishima-Kawashima M, Hasegawa M, Takio K, Suzuki M, Titani K, Ihara Y (1993) Ubiquitin is conjugated with amino-terminally processed tau in paired helical filaments. Neuron 10: 1151–1160

Neve RL, Finch EA, Dawes LR (1988) Expression of the Alzheimer amyloid precursor gene transcript in the human brain. Neuron 1: 669–677

Newman AJ (1994) Pre-mRNA splicing. Curr Opin Gen Dev 4: 298–304

Ott A, Bretler MMB, Van Harskamp F, Claus JJ, Van der Cammen TJM, Grobbee DE, Hofman A (1995) Prevalence of Alzheimer's disease and vascular dementia: association with education. The Rotterdam study. Br Med J 310: 970–973

Palmert MR, Podlisny MB, Witker DS, Oltersdorf T, Younkin LH, Selker DJ, Younkin SG (1998) Antisera to an amino-terminal peptide detect the amyloid precursor of Alzheimer's disease and recognize senile plaques. Biochem Biophys Res Commun 136: 432–437

Pericak-Vance MA, Bass MP, Ya,aoka LH, Gaskell PC, Scott WK, Terwedow HA, Menold MM, Connealy PM, Small GW, Vance JM, Satinders AM, Roses AD, Haines JL (1997) Complete genomic screen in late-onset familial Alzheimer disease. Guidance for a new locus on chromosome 12. JAMA 278: 1237–1241

Perlick HA, Medghalchi SM, Spencer FA, Kendzior Jr RJ, Dietz HC (1996) Mammalian orthologues of a yeast regulator of nonsense transcript stability. Proc Natl Acad Sci USA 93: 10928–10932

Pulak R, Anderson P (1993) mRNA surveillance by the Caenor habditis elegans Smg genes. Genes Dev 7: 1885–1887

Radman M, Wagner R (1988) The high fidelity of DNA duplication. Sci Am August: 24–30

Ruiz-Echevarría MJ, Czaplinski K, Peltz SW (1996) Making sense of nonsense in yeast. Trends Biochem Sci 21: 433–438

Rumble B, Retallack R, Hilbich C, Simms G, Multhaup G, Martins R, Hockey A, Montgomery P, Beyreuther K, Masters CL (1989) Amyloid A4 protein and its precursor in Down's syndrome and Alzheimer's disease. N Engl J Med 320: 1446–1452

Sago H, Carlson EJ, Smith DJ, Kilbridge J, Rubin EM, Mobley WC, Epstein CJ, Huang T-T (1998) Ts1Cje, a partial trisomy 16 mouse model for Down syndrome, exhibits learning and behavioral abnormalities. Proc Natl Acad Sci USA 95: 6256–6261

Schmale H, Borowiak B, Holt-Grez H, Richter D (1989) Impact of altered protein structures on the intracellular traffic of a mutated vasopressin precursor from Brattleboro rats. Eur J Biochem 182: 621–627

Sherman TG, Watson SJ (1988) Differential expression of vasopressin alleles in the Brattleboro heterozygote. J Neurosci 8: 3797–3811

Sonnemans MAF, Evans DAP, Burbach JPH, Van Leeuwen FW (1996) Immunocytochemical evidence for the presence of vasopressin in intermediate sized neurosecretory granules of solitary neurohypophyseal terminals in the homozygous Brattleboro rat. Neuroscience 72: 225–231

Strachan T, Read AP (1996) Human molecular genetics. Bios Scientific Publishers Ltd, Oxford

Swaab DF, Lucassen PJ, Salehi A, Scherder EJA, Van Someren EJW, Verwer RWH (1998) Reduced neuronal activity and reactivation in Alzheimer's disease. Progr Brain Res 117: 343–377

Szot P, Dorsa DM (1992) Cytoplasmic and nuclear vasopressin RNA in hypothalamic and extrahypothalamic neurons of the Brattleboro rat: an in situ hybridization study. Mol Cell Neurosci 3: 224–236

Valtin H (1982) The discovery of the Brattleboro rat, recommended nomenclature, and the question of proper controls. Ann NY Acad Sci 394: 1–9

Van Broeckhoven CL (1995) Molecular genetics of Alzheimer disease: identification of genes and gene mutations. Eur Neurol 35: 8–19

Van Broeckhoven C (1998) Alzheimer's disease: identification of genes and genetic risk factor. In: Van Leeuwen FW et al (eds) Neuronal degeneration and regeneration: from basic mechanisms to prospects for therapy. Progr Brain Res 117: 315–326

Van Duijn CM, De Knijff P, Cruts M, Wehnert A, Havekes LM, Hofman A, Van Broeckhoven C (1994) Apolipoprotein E4 allele in a population-based study of early onset Alzheimer's disease. Nature Genet 7: 74–78

Van Leeuwen FW, Van der Beek EM, Seger M, Burbach JPH, Ivell R (1989) Age-related development of a heterozygous phenotype in solitary neurons of the homozygous Brattleboro rat. Proc Natl Acad Sci USA 86: 6417–6420

Van Leeuwen FW, De Kleijn DPV, Van den Hurk HH, Neubauer A, Sonnemans MAF, Sluijs JA, Köycü S, Ramdjielal RDJ, Salehi A, Martens GJM, Grosveld FG, Burbach JPH, Hol EM (1998a) Frameshift mutants of β amyloid precursor protein and ubiquitin-B in Alzheimer's and Down patients. Science 279: 242–247

Van Leeuwen FW, Burbach JPH, Hol EM (1998b) Mutations in RNA: a first example of molecular misreading in Alzheimer's disease. TINS 21: 331–335

Varshavsky A (1997) The ubiquitin system. Trends Biochem Sci 22: 383–387

Vogel G (1998) Possible new cause of Alzheimer's disease found. Science 279: 174

Wiborg O, Pederson MS, Wind A, Berglund LE, Marcker KA, Vuust J (1985) The human ubiquitin multigene family: some genes contain multiple directly repeated ubiquitin coding sequences. EMBO J 4: 755–759

Wisniewski KE, Wisniewski HM, Gy W (1985) Occurrence of neuropathological changes and dementia of Alzheimer's disease in Down's syndrom. Ann Neurol 17: 278–282

Yoshikai S-I, Sasaki H, Doh-ura K, Furuya H, Sakaki Y (1990) Genomic organization of the human amyloid beta-protein precursor gene. Gene 87: 257–263

Authors' address: Dr. F. W. van Leeuwen, Netherlands Institute for Brain Research, Meibergdreef 33, 1105 AZ Amsterdam, The Netherlands, email: f.van.leeuwen@nih.knaw.nl

J Neural Transm (1999) [Suppl] 57: 161–177

Expression of the dihydropyrimidinase related protein 2 (DRP-2) in Down Syndrome and Alzheimer's disease brain is downregulated at the mRNA and dysregulated at the protein level

G. Lubec[1], M. Nonaka[2], K. Krapfenbauer[1], M. Gratzer[1], N. Cairns[3], and M. Fountoulakis[4]

[1] Department of Pediatrics, University of Vienna, Austria
[2] Department of Biochemistry, Nagoya City University, Nagoya, Japan
[3] Brain Bank, Institute of Psychiatry, University of London, United Kingdom
[4] F. Hoffmann — La Roche, Basel, Switzerland

Summary. Deteriorated migration, axonal pathfinding and wiring of the brain is a main neuropathological feature of Down Syndrome (DS). Information on the underlying mechanisms is still limited, although basic functions of a series of growth factors, cell adhesion molecules, guidance factors and chemoattractants for brain histogenesis have been reported.

We used proteomics to detect differences in protein expression between control, DS and Alzheimer's disease brains: In five individual brain regions of 9 individuals of each group we performed two dimensional electrophoresis with MALDI — identification of proteins and determined mRNA levels of DRP-2.

Significantly decreased mRNA levels of DRP-2 in four brain regions of patients with DS but not with AD as compared to controls were detected. 2D electrophoresis revealed variable expression of DRP-2 proteins, which showed a high heterogeneity per se. Dysregulation of DRP-2 was found in brains of patients with DS and AD presenting with an inconsistent pattern, which in turn may reflect the inconsistent neuropathological findings in patients with DS and AD.

The decrease of mRNA DRP-2 steady state levels in DS along with deteriorated protein expression of this repulsive guidance molecule of the semaphorin/collapsin family, may help to explain deranged migration and histogenesis of DS brain and wiring of AD brain.

Introduction

Although there is considerable disagreement about the frequency and nature of anatomic abnormalities of the brain in Down Syndrome (DS), the present consensus indicates that no consistent or specific neuropathologic finding exists (Epstein, 1995). Brain weight is in the low normal range and the size of the cerebellum and brain stem may be reduced (Crome, 1966); frontal-occipital length is shortened, probably secondary to reduced frontal lobe growth, there is narrowing of temporal gyri in about a third of cases (Schmidt-Sidor, 1990) and the anterior commissure in adults with DS is reduced in cross-sectional area (Sylvester, 1986). Nerve cell heterotopias in the white layers of the cerebellum are indicating disturbance of embryonic cell migration (Rehder, 1981). Abnormalities in morphology and number of dendritic spines including atrophy of the dendritic tree of visual cortex continuing into adulthood and becoming even more pronounced when Alzheimer disease develops (Marin-Padilla, 1976; Becker, 1986, 1991; Takashima, 1989) have been described.

Other defects in brain histogenesis include description of poverty of granular cells throughout the cortex, decreased neuronal densities in layers II and IV of the occipital cortex (Wisniewski, 1986), diminution in the number of hypothalamic neurons (Wisniewski, 1991) and there is accumulating evidence for abnormalities of neuronal differentiation and abnormal migration in fetal and infant brain (Epstein, 1995). A consistent picture of the "wiring" of the brain in DS has not emerged yet.

Current knowledge of developmental brain research is providing new concepts and potent tools for studying the migration and wiring in the brain, highlighting the use of growth factors, cell adhesion molecules, repulsive guidance factors and chemoattractants. In this report we describe the downregulation of DRP-2 (dihydropyrimidinase related protein — 2) in brain of patients with DS, which may well be involved in the deteriorated migration / axonal pathfinding and wiring (Hamajima, 1996).

Methods

Northern and dot blot determination and quantification of DRP-2 transcript

Northern and dot blot analyses were carried out on postmortem brain samples from adult patients with DS, AD and controls.

The brain regions temporal, frontal, occipital, parietal cortex and cerebellum of patients with DS (n = 9; 3 females, 6 males; 56.1 ± 7.1 years old), AD (n = 9; 6 females, 3 males; 72.3 ± 7.6 years old) and controls (n = 9; 5 females, 4 males; 72.6 ± 9.6 years old) were mainly characterized in a previous publication (7) and used for the studies at the transcriptional level. Briefly, post mortem brain samples were obtained from the MRC London Brain Bank for Neurodegenerative Diseases, Institute of Psychiatry). In all DS brains there was evidence of abundant beta A plaques and neurofibrillary tangles. The AD patients fulfilled the National Institute of neurological and Communicative Disorders and Stroke and Alzheimer Disease and Related Disorders Association (NINCDS/ADRDA) criteria for probable AD (Tierney, 1988). The histological diagnosis of AD was established and was consistent with the CERAD criteria (Mirra, 1991) for a "definite" diagnosis of AD. The controls were brains from individuals with no history of neurological or psychiatric illness. The major cause of death was bronchopneumonia in DS and AD patients and heart disease in controls. Post mortem interval of brain dissection in AD, DS and controls was 34.1 ± 13.7, 30.6 ± 17.5 and 34.8 ± 15.0 hrs. Tissue samples were stored at −70°C and the freezing chain was never interrupted.

Biopsies from frontal, temporal, parietal, occipital cortex and cerebellum were obtained at autopsy and taken immediately into liquid nitrogen.

Northern and slot blots were performed according to the principle given previously (Hardmeier, 1997): Frozen samples were ground and mRNA extraction was performed using the TotalRNA Purification kit (Pharmacia). Subsequently, RNA was applied onto 1.4% agarose gel after denaturation with glyoxal and dimethylsulfoxide according to the method by McMaster and Carmichael (1977) and electrophoresed at 3–4 V/cm for 2.5 hrs in circulating 0.01 M phosphate buffer pH 7.0. RNA was then transferred to a positively charged nylon membrane (Hybond N+, Dupont, NEF 986) by capillary blotting and fixed with 0.05 N sodium hydroxide for 5 min at room temperature and finally equilibrated at pH 7.0 with 3 washes in 2XSSC.

The probe for human beta-actin (ATCC 9800) was bought and the cDNA for DRP-2 was obtained from Prof.Dr.M.Nonaka, Nagoya City University, Medical School, Mizuho — ku, Nagoya, Japan (Hamajima, 1996).

The probe was denatured prior to labelling by boiling for 5 min and subsequent cooling on ice, and labeled with fluorescein-12-dUTP using the Renaissance Random Primer Fluorescein-12 d-UTP labeling kit (Dupont, NEL 203).

After fixation of bound RNA, the nylon membrane was incubated in pre-hybridization solution (0.25 M phosphate buffer pH 7.2, containing 5% SDS w/v 1 mM EDTA and 0.5% blocking reagent [from Dupont NEL 203]) for 12 hrs at 65°C in a hybridization oven. The blots were hybridized overnight at 65°C with the labelled probes each (50 ng/ml of pre-hybridization buffer).

After hybridization, nonspecifically bound material was removed by post-hybridization washes with 0.5X and 0.1X pre-hybridization buffer 2 × 10 min each at 65°C. The 0.5X and 0.1X pre-hybridization buffer was brought up to 65°C prior to use and the second wash was performed at room temperature.

Hybridized blots were blocked with 0.5% blocking reagent in 0.1 M Tris HCl pH 7.2 and 0.15 M NaCl for 1 hr at room temperature. Membranes were then incubated with antifluorescein HRP antibody (Dupont NEL 203) at a 1:1,000 dilution in the solution given above for 1 hr under constant shaking.

Membranes were washed 4X for 5 min each in the solution given above.

The Nucleic Acid Chemiluminescence Reagent (Dupont NEL 201) was added to the membranes and incubated for 1 min. Excess detection reagent was removed by the use of filter paper, the membrane was placed in Sarawrap paper and exposed to autoradiography Reflection films (Dupont NEF 496) for 15 min at room temperature.

Slot blots were performed according to the method of White and Bancroft (1982). This procedure consisted of placing 2 ug of total RNA dissolved in 10 ul double distilled water mixed with 500 ul of 100% formamide, 162 ul of 37% formaldehyde, 100 ul of 10X MOPS(4-morpholinepropanesulfonic acid) buffer. This mixture was incubated for 10 min at 65°C and cooled down subsequently on ice. Samples were placed onto the membrane by the Manifolds filtration equipment (slot blot apparatus Bio slot TM, BIORAD) and the hybridization was performed as described above.

Denistometry of films was performed using the Hirschmann elscript 400 densitometer and calibration performed as given in the instrument's software program (Germany).

2D electrophoresis for the evaluation of DRP-2 proteins

Sample preparation

Brain tissue was suspended in 0.5 ml of sample buffer consisting of 40 mM Tris, 5 M urea (Merck, Darmstadt, Germany), 2 M thiourea (Sigma, St.Louis, MO, USA). 4% CHAPS (Sigma), 10 mM PMSF and 1 microg/ml of each pepstatin A, chymostatin, leupeptin and antipain. The suspension was sonicated for approximately 30 s and centrifuged at

10,000 × g for 10 min and the supernatant was centrifuged further at 150,000 × g for 45 min to sediment undissolved material. The protein content in the supernatant was determined by the Coomassie blue method (Bradford, 1976). The average protein concentration was 8 mg/ml.

Two dimensional (2D) electrophoresis

The 2D electrophoresis was performed essentially as reported (Langen, 1997). Samples of approx. 1.5 mg were applied on immobilized pH 3–10 nonlinear gradient strips (IPG, immobilized pH-gradient strips, Pharmacia Biotechnology, Uppsala, Sweden), at both, the basic and acidic ends of the strips. The proteins were focused at 300 V for 1 hr, after which the voltage was gradually increased to 3,500 V within 6 hrs. Focusing was continued at 3,500 V for 12 hrs and at 5,000 V for 48 hrs. The proteins were then separated on 9–16% linear gradient polyacrylamide geis (chemicals from Serva, Heidelberg, Germany and Bio-Rad, Hercules, CA, USA). After protein fixation with 40% methanol containing 5% phosphoric acid for 12 hrs, the gels were stained with colloidal Coomassie blue (Novex, San Diego, CA, USA) for 48 hrs. The molecular mass was determined by running standard protein markers at the right side of selected gels. The size markers (Gibco, Basel, Switzerland) covered the range 10–200 kDA, pl values were used as given by the supplier of the IPG strips. The gels were destained with water and scanned in a Molecular Dynamics Personal densitometer. The images were processed using Adobe Photoshop and Power-Point software. Protein spots were quantified using the imageMaster 2D Elite software (Pharmacia Biotechnology).

Matrix assisted laser desorption ionization mass spectroscopy

MALDI-MS analysis was performed as described (Fountoulakis, 1997) with minor modifications. Briefly, spots were excised, destained with 50% acetonitrile in 0.1 M ammonium bicarbonate and dried in a speedvac evaporator. The dried gel pieces were reswollen with 3 microliters of 3 mM Tris-HCl, pH 8.8, containing 50 ng trypsin (Promega, Madison, WI, USA). After 15 min, 3 microliters oif water were added and left at room temperature for 12 hrs. Two microliters of 30% acetonitrile, containing 0.1% trifluoroacetic acid were added, the content was vortexed, centrifuged for 3 min and sonicated for 5 min. One microliter was applied onto the dried matrix spot. The matrix solution consisted of 15 mg nitrocellulose (Bio-Rad) and 20 mg alphacyano 4 hydroxy cinnamic acid (Sigma) dissolved in 1 ml acetone: isopropanol 1 : 1 (v/v). 0.5 microliters of the matrix solution were applied on the sample target. Specimen were analysed in a time-of-flight Voyager Elite mass spectrometer (PerSeptive Biosystems, Cambridge, MA, USA) equipped with a reflectron. An accelerating voltage of 20 kV was used. Calibration was internal to the samples. The peptide masses were matched with the theoretical peptide masses of all proteins from all species of the SWISS-PROT database. For protein search, monoisotopic masses were used and a mass tolerance of 0.0075% was allowed.

Results

DRP-2 mRNA steady state levels

Northern blot analysis showed a single band at 4.9 kb (Hamajima, 1996) reflecting the specificity of the probe.

Table 1. Results of dot blot analysis of mRNA DRP-2 brain levels as normalized versus the housekeeping gene mRNA beta-actin (arbitrary units from densitometry)

	Frontal	Parietal	Temporal	Occipital	Cerebellum
Control	0.57 ± 0.25	0.67 ± 0.11	0.37 ± 0.13	0.66 ± 0.38	0.97 ± 0.17
Alzheimer's disease	0.59 ± 0.29	0.59 ± 0.37	0.49 ± 0.25	0.52 ± 0.39	0.87 ± 0.28
Down Syndrome	0.26 ± 0.11*	0.49 ± 0.31	0.12 ± 0.10*	0.41 ± 11*	0.27 ± 0.21*

An asterisks (*) reveals significant difference from controls

Densitometry of ot blots revealed significantly decreased mRNA levels for DRP-2 transcipts in brain regions frontal, occipital, temporal cortex and cerebellum of patients with DS but no significant changes were observed at the DRP-2 transcriptional level in brains of patients with AD (Table 1).

Quantification of DRP-2 protein

The protein extracts from five brain regions, frontal, parietal, temporal and occipital cortex and cerebellum of 9 patients with DS, 9 with AD and 9 controls were separated by 2D-gel electrophoresis. The 2D-gels with proteins from the corresponding brain regions of controls, DS and AD patients were compared with each other. Variable expression levels for several proteins were observed, which could partially represent allelic differences.

Figure 1 shows an example of analysis of a brain protein from a control (Fig. 1A–E) and a patient with DS (Fig. 1F–J). The protein shows a high heterogeneity and is in most cases represented by five spots divided into two locations. A shorter and more acidic protein form is represented by two adjacent spots with approximate pI and Mr values of 5.8 and 55kDa (designated here as the 55-kDa-form). A larger and more neutral form is represented by three adjacent, usually strong spots with approximate pI and Mr values of 6.9 and 65kDa (designated as the 65-kDa-form). The protein from all five spots was unambiguously identified by MALDI-MS as dihydropyrimidinase-related protein 2 (DRP-2, SWISS-PROT accession number Q16555). Figure 2A and 2B show the mass spectra of spots a and b, respectively (Fig. 1A) representing the 55- and 65kDa forms of DRP-2.

The proteins from the five spots were subjected to N-terminal sequencing analysis. However, no sequence was obtained, indicating that the N-termini are blocked. Ion spray mass spectrometry analysis of the proteins did not result into a clear definition of the difference between the two forms. We do not know presently the reasons and biological significance of the multiple heterogeneity which are further investigated.

In the four cortical regions, the spots representing the 65-kDA-form are stronger in comparison with those representing the 55-kDa-form. In the

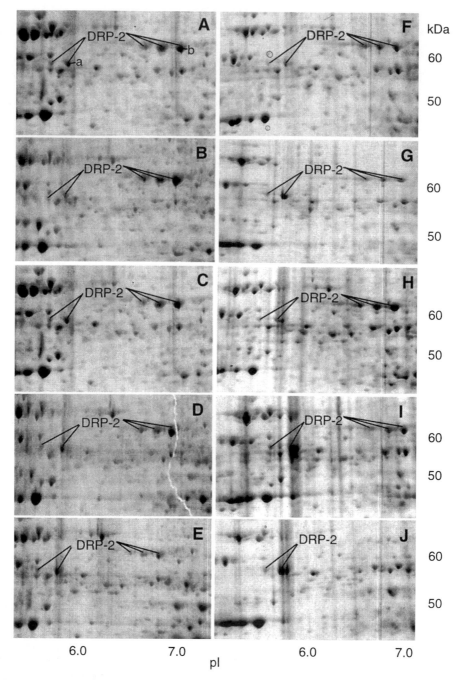

Fig. 1. Two-dimensional gel analysis of human brain samples from a control (**A–E**) and a patient with Down syndrome (**F–J**). The protein samples were prepared as stated under Materials and Methods and separated on pH 3–10 nonlinear IPG strips, followed by 9–16% SDS-gels. The gels were stained with Coomassie blue. Corresponding parts of the gels are shown. Samples were taken from the frontal (**A,F**), parietal (**B,G**), temporal (**C,H**) and occipital (**D,I**) lobes of the cortex, and the cerebellum (**E,J**). The spots representing dihydropyrimidinase-related protein 2 (DRP-2, forms 55 and 65 kDa) are indicated. **A** the spots a and b are indicated, the mass spectra of which are shown in Fig. 2

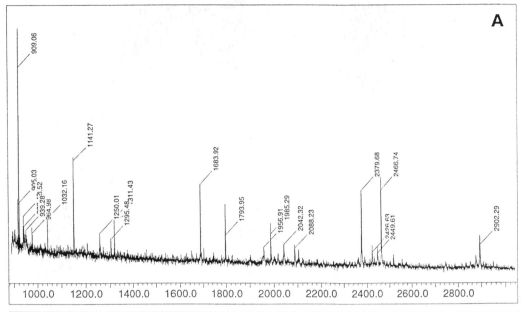

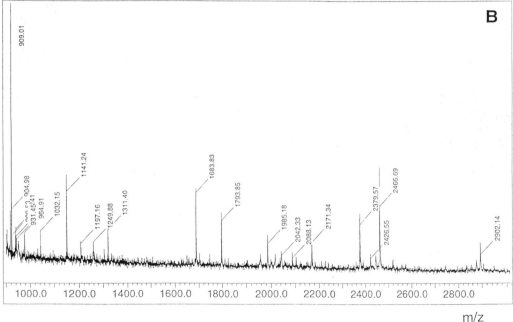

Fig. 2. Mass spectrometry analysis of the 55- (**A**) and 65-kDa (**B**) forms of DRP-2. The spots a and b, representing the 55- and 65-kDa forms of the protein (Fig. 1A) were excised from the gel and digested with trypsin and the peptides generated were analyzed by MALDI-MS as stated under Materials and Methods. The two mass spectra are practically identical. Search in the SWISS-PROT database resulted in the unambiguous identification of the protein as dihydropyrimidinase-related protein 2 (DRP-2) with 11 matching peptides out of 20 total peptides. Masses of matching peptides: 909.06, 1032.16, 1141.27, 1295.48, 1311.43 (oxidized methionine), 1683.92, 1793.95, 1985.29, 2088.23, 2426.63, 2902.29 (masses are not corrected, mass tolerance window 0.065%)

168   G. Lubec et al.

Table 2. Ratio of the 55- to the 65-kDa forms of DRP-2

	Frontal	Parietal	Temporal	Occipital	Cerebellum
Control	0.50	0.29	0.29	0.34	1.59
Alzheimer's disease	0.47	0.67	0.57	0.63	2.00
Down Syndrome	0.73	0.48	0.65	0.68	3.65

The proteins from the various brain regions of the members of the three groups were separated by 2-D gels. The gels were stained with Coomassie blue and the spots were quantified using the ImageMaster 2D software and the mean values of the spot intensities were determined. The sum of the intensities of the three spots representing the DRP-2 65-kDa form and the sum of the intensities of the two spots representing the 55-kDa form were considered

normal brain, the ratio of the average volume of the 55 — to the 65-kDa-form was approximatelz 0.5 in frontal, 0.3 in parietal, temporal and occipital cortex and 1.6 in the cerebellum (Table 2). In the AD and DS brain, the ratio of the two forms increased in comparison with the control brain (Table 2).

Differences concerning the DRP-2 expression levels were observed in the five brain regions. Gel cmparison of samples from the various brain regions of the same donor showed that in most cases the intensity of the three spots representing the 65-kDa-form was stronger in the samples from the frontal and temporal cortex and weaker in all the samples of the cerebellum (compare Fig. 1E with 1A–D). Levels of the 55-kDa-form were usually comparable and in the cerebellum higher in comparison with the four cortical lobes.

The heterogeneity concerning DRP-2 forms 55 and 65 kDa was more evident in the DS and AD groups. Figure 3A and 3B show examples of DRP-2 in samples from the frontal cortex of two different DS patients. In the sample of Fig. 3A only one of the three spots usually representing the 65-kDa-form is visible, whereas the 55-IDa-form is represented by strong spots. In the sample of Fig. 3B, essentially no difference between DS and control (Fig. 1A) can be seen. Figure 3C shows the 2-D-gel analysis of a DS brain sample from the cerebellum. The three spots corresponding to the 65-kDa form are not clearly visible, whereas two strong spots representing the 55-kDa-form are present.

Similar cases of heterogeneity were observed with the AD brain. Figure 4A and 4B show the 2-D gel analyses of samples from the frontal region of cortex of two patients with AD. In Fig. 4A, the 65-kDa-form is represented by stronger spots, whereas the 55-kDa form is represented by weak spots. In Fig. 4B, the spot distribution essentially resembles that of the control samples, however, the spot intensity was weaker. In the samples from parietal and temporal brain regions (Fig. 4C and 4D), the spots corresponding to DRP-2 are either partially missing or very weak.

The spots corresponding to DRP-2 were quantified in selected gels using specific software. For the 55-kDa-form the results were inconsistent (Fig. 5). Only in frontal cortex the DRP-2 decreased in the DS and AD groups

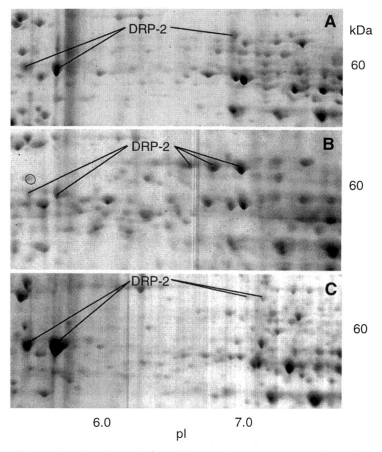

Fig. 3. Two-dimensional gel analysis of human brain proteins from Down syndrome patients. The samples, from the frontal lobe of cortex (**A,B**) and the cerebellum (**C**), were analyzed as stated under legend to Fig. 1. Only parts of the gels are shown. The spots corresponding to dihydropyrimidinase-related protein 2 (DRP-2, forms 55 and 65 kDa) are indicated

(Fig. 5A). The levels of the 65-kDa-form were decreased in most of the patients with DS and in particular with AD in the four cortex regions (Fig. 6A–D). In the cerebellum, the levels for the 65-kDa-form were reduced in the majority of patients with DS, whereas the reduction was less clear in the AD patients (Fig. 6E). In the 2-D gel images shown in Fig. 1,2 and 3, more differences in the expression level of other proteins were observed. Some of them maybe disease-related which have not been evaluated yet.

Discussion

Pathfinding and guidance of axons and cell migration are fundamental questions not only in developmental neurobiology but also in brain plasticity.

Major advances for the understanding of pathfinding and migration were provided just in this decade coming from Drosophila and Caenorhabditis

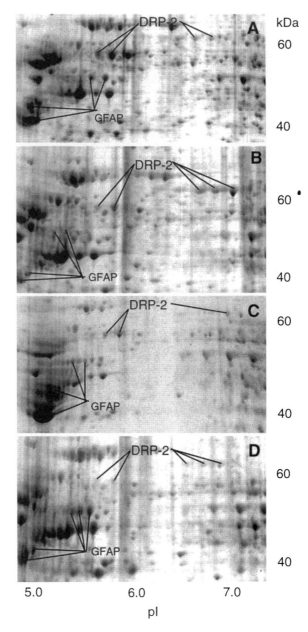

Fig. 4. Two-dimensional gel analysis of human brain proteins from Alzheimer's disease patients. The samples, from the frontal (**A,B**), parietal (**C**) and temporal (**D**) lobe of cortex, were analyzed as stated under legend to Fig. 1. Only parts of the gels are shown. The spots corresponding to dihydropyrimidinase-related protein 2 (DRP-2, forms 55 and 65 kDa) are indicated. The spots representing glial fibrillary acidic protein (GFAP), a marker of neurodegeneration, are indicated as well

elegans neurogenetic studies revealing candidate genes for guidance molecules and regulatory pathways. Mutations in the commissureless and roundabout genes were affecting growth cones guidance at the midline (Seeger, 1993), several genes have been identified that are necessary for axons

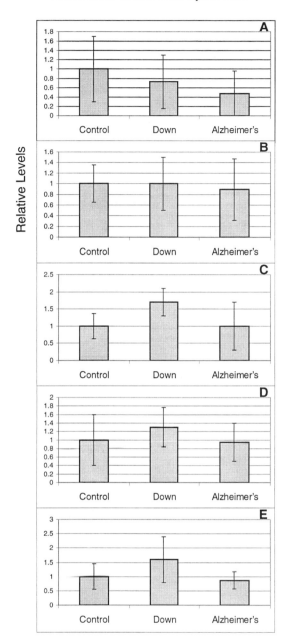

Fig. 5. Quantification of the DRP-2 55 kDa form levels in the brain regions of patients with DS and AD. The proteins from the frontal (**A**), parietal (**B**), temporal (**C**), occipital (**D**) cortex lobes and the cerebellum (**E**) of the patients and the control group were separated on 2-D gels and visualized following stain with colloidal Coomassie blue. In gel images, including total brain proteins, the intensities of the spots representing DRP-2 55 kDa form were quantified, compared to the total proteins present. The quantification was performed using the ImageMaster 2D software. In each donor, the percentage of the volume of the two spots representing DRP-2 55 kDa form was determined. The DRP-2 55 kDa form mean values in the DS and AD groups were normalized to the mean value of the controls and the relative values are indicated. The bars indicate the standard deviation of DRP-2 55 kDa form levels in the three groups

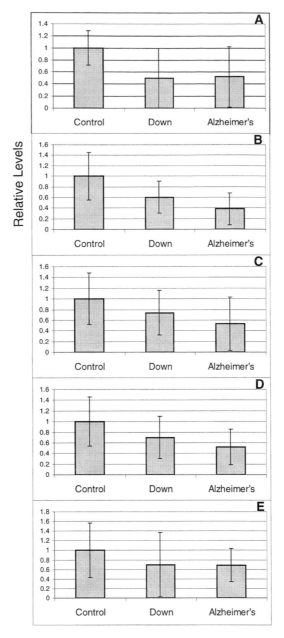

Fig. 6. Quantification of the DRP-2 65 kDa form levels in the brain regions of patients with DS and AD. The proteins from the frontal (**A**), parietal (**B**), temporal (**C**), occipital (**D**) cortex lobes and the cerebellum (**E**) of the patients and the control group were separated and quantified as stated under legend to Fig. 5. The bars indicate the standard deviation of DRP-2 65 kDa form levels in the three groups

and motoneurons to navigate past particular choice points in the pathfinding process and for the ability of motoneurons to recognize their appropriate muscle targets (Van Vactor, 1993). Guidance of retinal growth cones by sequential cues as they navigate from the eye to the tectum have been reported (Kuwada, 1995). The detection of chemoattractants, called netrins, were detected by their function (Kennedy, 1994; Serafini, 1994), and other guidance molecules as e.g. UNC-6 and laminin were reported in addition (Jokusch, 1997). Various cell adhesion molecules (CAMs) have been implicated as axon guidance molecules, usually on the basis of suggestive expression patterns or cell culture studies. Fasciclin II, e.g., is related to the N-CAM neural cell adhesion molecule family mediating homophilic cell-adhesion in culture (Grenningloh, 1991) and its ectopic expression leads to subsequent errors in pathfinding.

In contrast to chemoattractants repulsive guidance molecules have been described as well: Two candidate proteins of molecular masses of 33 kDa and 25 kDa which show repulsive activities have been purified (Stahl, 1990; Drescher, 1995). In several species a variety of experimental approaches have indicated that members of the semaphorin/collapsin family play a repulsive role in axon guidance, target recognition or both. Fasciclin IV (i.e. semaphorin I) and the growth cone collapse-inducing chick protein collapsin were the first members of this family. Based on sequence homologies, many more related genes were identified in several species including humans, encoding both, secreted and transmembrane proteins. This protein family is characterized by the presence of a conserved semaphorin domain of 500 amino acids and 16 conserved cysteine residues (Kolodkin, 1993). Whether the semaphorins act locally, being anchored on a cell surface, or also exert more distant effects on pathfinding is not yet clear. Hamajima and coworkers (1996) detected a novel gene family defined by human dihydropyrimidinase and three related proteins with differential tissue distribution and function. One member of this gene family is the dihydropyrimidinase related protein 2 (DRP-2). This protein is highly homologous to a series of proteins with known effect on pathfinding and migration: DRP-2 is 99% homologous (amino acid identities given) to rat TOAD-64, 98% to mouse unc-33 and 98% homologous to chicken CRMP-62.

Performing protein brain mapping and gene hunting using 2-D electrophoresis with MALDI-identification of proteins, we observed 5 spots that could be clearly assigned to DRP-2 proteins by in-gel digestion of the spot with subsequent MALDI-identification. The differences between 55- and 65 kDa forms may be assigned to different splicings as shown on the northern blots, where a major band at at 4.9 kb and up to 3 minor transcripts were detected, or to different posttranslational modifications as truncation or phosphorylation and indeed, the homolog unc-33 is a phosphoprotein. Gaetano and coworkers (1997) described phosphorylation states of unc-33 with a molecular mass from 58-70 kDa in neural cells.

Different phosphorylation may also explain the heterogeneity of spots in control and brain samples with DS and AD. As shown in table 1 the ratio of the 55-kDa-forms to the 65-kDa forms was increased in brains of AD and DS probably suggesting decreased phosphorylation of DRP-2 protein in these disorders and a first clue for phosphorylation deficit was published recently showing that cyclin dependent kinase is significantly decreased in brain of patients with DS (Bernert, 1996).

Although at the mRNA steady state level significantly lower DRP-2 levels were observed in four brain regions of patients with DS, no consistent pattern was found at the protein level. This reflects the histological (and clinical) heterogeneity of the DS brain and very large groups of DS and AD brains is required to perform a fair statistical analysis of DRP-2 findings. Moreover, although we used five individual brain regions, the investigation of individual areas may be needed when clear cut results are expected.

Dysregulated DRP-2 protein and downregulated DRP-2-mRNA steady state levels in DS, may well explain brain pathology of adult DS and AD: The 99% of DRP-2-homology to TOAD-64 strongly suggests a similar function and although TOAD-64 itself disappears in the adult mammalian nervous system (Wang, 1996; Gaetano, 1997) DRP-2 could play a TOAD-64 — like role in the machinery underlying axonal outgrowth and pathfinding, in the signal transduction processes permitting axons to choose correct routes and targets (Minturn, 1995). The 98% of DRP-2-homology to unc-33 may point to unc-33 or Ulip family (unc-33-like phosphoprotein) functions of DRP-2 proteins (Byk, 1998). Mutations in the unc-33 gene of the nematode Caenorhabditis elegans lead to severely uncoordinated movement, abnormalities in the guidance and outgrowth of the axons of many neurons (Li, 1992). The highly homologous DRP-2 maybe therefore involved in the control of neuronal differentiation, neuritic outgrowth and plasticity, functions impaired in both AD and DS.

The high DRP-2-homology of 98% to CRMP-62, the collapsin response mediator protein of relative molecular mass of 62 kDa, implies a function for DRP-2 protein(s) as a repulsive guidance cue: collapsin, a member of the semaphorin family, contributes to axonal pathfinding during neural development by inhibiting growth cone extension, preventing mature axons from regenerating (Luo, 1993) and CRMP-62 is required for collap-sin-induced inward currents in Xenopus laevis oocytes. Introduction of anti-CMRP-62 antibodies into dorsal root ganglion neurons block collapsin-induced growth cone collapse. CRMP-62 seems to be an intracellular component of a signaling cascade initiated by an unidentified transmembrane collapsin-binding protein (Goshima, 1995). By this mechanism dysregulated DRP-2 may well be modulating brain plasticity and wiring in adult DS and AD brain. Recently a novel action of collapsin, the promotion of axoplasmic vesicular transport systems, was reported and this in turn, may contribute to deteriorated transport and neurotransmission known to occur in DS and AD (Goshima, 1997).

We conclude that deranged DRP-2 mRNA and protein levels in brain of patients with DS and at the protein level in AD, however inconsistent they are, may contribute to abnormal plasticity and wiring in DS and AD brain. This preliminary study indicates that in futural studies large study cohorts are mandatory to cope with the inherent problems of multiple heterogeneities at the anatomical, histological and biochemical level. Subgroups with well-defined migration-, differentiation-, dendritic and neuronal density-deficits must be introduced and individual areas rather than brain regions must be examined. At the biochemical (protein-) level, the 55- and 65-kDa-forms must be further identified and brain extracts should be dephosphorylated to quantify DRP-2 protein (s) in the non-phosphorylated state.

Due to the sensitivity and specificity of the 2-D gel electrophoretic method used (5 spots identified for a single protein), this paper reflects the inherent problems of both, the biochemical-analytical and the anatomical-biological approach in neurochemical research, thus also representing a heuristic-hermeneutic consideration.

Acknowledgements

We (G. L.) are highly indebted to the Red Bull Company, Salzburg, Austria, for generous support of the study.

References

Becker LE, Armstrong DL, Chan F (1986) Dendritic atrophy in children with Down's syndrome. Ann Neurol 20: 520–532

Becker L, Mito T, Takashima S, Onodera K (1991) Growth and development of the brain in Down Syndrome. In: Epstein CJ (ed) The morphogenesis of Down Syndrome. Wiley-Liss, New York, p 133

Bernert G, Nemethova M, Cairns N, Lubec G (1996) Decreased cyclin dependent kinase in brain of patients with Down Syndrome. Neurosci Lett 216: 68–70

Bradford M (1976) A rapid and sensitive method for the quantitation of microgram quantities of protein utilizing the principle of protein-dye binding. Anal Biochem 72: 248–254

Byk T, Ozon S, Sobel A (1998) The ulip family phosphoproteins — common and specific properties. Eur J Biochem 254: 14–24

Crome L, Cowie V, Slater E (1966) A statistical note of cerebellar and brain stem weight in mongolism. J Ment Defic Res 10: 69–77

Drescher U, Kremoser C, Handwerker C, Loeschinger J, Noda M, Bonhoeffer F (1995) In vitro guidance of retinal ganglion cell axons by RAGS, a 25 kDa tectal protein related to ligands for Eph — receptor tyrosine kinases. Cell 82: 359–370

Epstein CJ (1995) Down Syndrome (Trisomy 21). In: Scriver CR, Beaudet AL, Sly WS, Valle D (eds) The metabolic and molecular basis of inherited disease. McGraw Hill, New York, pp 749–794

Fountoulakis M, Langen H (1997) Identification of proteins by matrix-assisted laser desorption ionization-mass spectroscopy following in-gel digestion in low — salt, nonvolatile buffer and simplified peptide recovery. Anal Biochem 250: 153–156

Gaetano C, Matsuo T, Thiele CJ (1997) Identification and characterisation of a retinoic acid regulated human homologue of the unc-33-like phosphoprotein gene (hUlip) from neuroblastoma cells. J Biol Chem 272: 12195–12201

Goshima Y, Nakamura F, Strittmatter P, Strittmatter SM (1995) Collapsin-induced growth cone collapse mediated by an intracellular protein related to UNC-33. Nature 376: 509–514

Goshima Y, Kawakami T, Hori H, Sugiyama Y, Takasawa S, Hashimoto Y, Kagoshima-Maezono M, Takenaka T, Misu Y, Strittmatter SM (1997) A novel action of collapsin: collapsin-1 increases antero- and retrograde axoplasmic transport independently of growth cone collapse. J Neurobiol 33: 316–328

Grenningloh G, Rehm EJ, Goodman CS (1991) Genetic analysis of growth cone guidance in Drosophila: fasciclin II functions as a neuronal recognition molecule. Cell 67: 45–57

176   G. Lubec et al.

Hamajima N, Matsuda K, Sakata S, Tamaki N, Sasaki M, Nonaka M (1996) A novel gene family defined by human dihydropyrimidinase and three related proteins with differential tissue distribution. Gene 180: 157–163

Jokusch H, Nave K-A, Grenningloh G, Schmitt-John T (1997) Molecular genetics of nervous and neruromuscular systems. In: Davies RW, Morris BJ (eds) Molecular biology of the neuron. BIOS Scientifique Publishers, Oxford, pp 21–66

Kolodkin AL, Matthes DJ, Goodman CS (1993) The semaphorin genes encode a family of transmembranre and secreted growth cone guidance molecules. Cell 75: 1389–1399

Kuwada J (1995) Development of the zebrafish nervous system: genetic analysis and manipulation. Curr Opin Neurobiol 5: 50–54

Langen H, Roeder D, Juranville J-F, Fountoulakis M (1997) Effect of the protein application mode and the acrylamide concentration on the resolution of protein spots separated by two-dimensional gel electrophoresis. Electrophoresis 18: 2085–2090

Li W, Herman RK, Shaw JE (1992) Analysis of the Caenorhabditis elegans axonal guidance and outgrowth gene unc-33. Genetics 132: 675–689

Luo Y, Raible D, Raper JA (1993) Collapsin: a protein in brain that induces the collapse and paralysis of neuronal growth cones. Cell 75: 217–227

McMaster GK, Carmichael GG (1977) Analysis by single and double stranded nucleic acids on polytacrylamide and agarose gels by using glyoxal and acridine orange. Proc Natl Acad Sci (USA) 74: 4835–4841

Marin-Padilla M (1976) Pyramidal cell abnormalities in the motor cortex of a child with Down's Syndrome. J Comp Neurol 167: 63–75

Minturn JE, Fryer HJ, Geschwind DH, Hockfield S (1995) TOAD-64, a gene expressed early in neuronal differentiation in the rat, is related to unc-33, a C.elegans gene involved in axon outgrowth. J Neurosci 15: 6757–6766

Mirra SS, Heyman A, McKeel D, Sumi S, Crain BJ (1991) The consortium to establish a registry for Alzheimer disease (CERAD). II. Standardisation of the neuro-pathological assessment of Alzheimer's disease. Neurology 41: 479–486

Rehder H (1981) Pathology of trisomy 21. In: Burgio GR, Fraccaro M, Tiepolo L, Wolf U (eds) Trisomy 21. An International Symposium. Springer, Berlin Heidelberg New York Tokyo, p 57

Schmidt-Sidor B, Wisniewski KE, Shepard TH, Sersen A (1990) Brain growth in Down Syndrome subjects 15–22 weeks of gestational age and birth to 60 months. Clin Neuropathol 9: 181–195

Seeger M, Tear G, Ferres-Marco D, Goodman CS (1993) Mutations affecting growth cone guidance in Drosophila: genes necessary for guidance towards or away from the midline. Neuron 10: 409–426

Seidl R, Greber S, Schuller E, Bernert G, Cairns N, Lubec G (1997) Evidence against increased oxidative DNA-damage in Down Syndrome. Neurosci Lett 235: 137–140

Stahl B, Mueller B, von Boxberg Y, Cox EC, Bonhoeffer F (1990) Biochemical characterization of a putative axonal guidance molecule of the chick visual system. Neuron 5: 733–743

Sylvester PE (1986) The anterior commissure in Down's syndrome. J Ment Defic Res 30: 19–25

Takashima S, Ieshima A, Nakamura H, Becker LE (1989) Dendrites, dementia and the Down Syndrome. Brain Dev 2: 131–143

Tierney MC, Fisher RH, Lewis AJ, Torzitto ML, Snow WG, Reid DW, Nieuwstraaten P, Van Rooijen LAA, Derks HJGM, Van Wijk R, Bischop A (1988) The NINCDA-ADRDA work group criteria for the clinical diagnosis of probable Alzheimer's disease. Neurology 38: 359–364

Van Vactor D, Sink H, Fambrough D, Tsoo R, Goodman CS (1993) Genes that control neuromuscular specificity in Drosophila. Cell 73: 1137–1153

Wang LH, Strittmatter SM (1996) A family of rat CRMP genes is differently expressed in the nervous system. J Neurosci 16: 6197–6207

White BA, Bancroft FC (1982) Cytoplasmic dot hybridization. Simple analysis of mRNA levels in multiple small cell or tissue samples. J Biol Chem 257: 8569–8575

Wisniewski KE, Bobinski M (1991) Hypothalamic abnormalities in Down syndrome. In: Epstein CJ (ed) The morphogenesis of Down Syndrome. Wiley — Liss, New York, p 153

Wisniewski KE, Laure-Kamionowska M, Connell F, Wen GY (1986) Neuronal density and synaptogenesis in the postnatal stage of brain maturation in Down syndrome. In: Epstein CJ (ed) The neurobiology of Down Syndrome. Raven, New York, p 29

Authors' address: Prof. Dr. G. Lubec, Department of Pediatrics, University of Vienna, Währinger Gürtel 18-20, A-1090 Vienna, Austria, e-mail: gert.lubec@akh-wien.ac.at

J Neural Transm (1999) [Suppl] 57: 179–195

Gene expression relevant to Down Syndrome: problems and approaches

F. Tassone[1], **R. Lucas**[1], **D. Slavov**[1], **V. Kavsan**[2], **L. Crnic**[3], and **K. Gardiner**[1]

[1] Eleanor Roosevelt Institute, Denver, CO, USA
[2] Institute of Molecular Biology & Genetics, Ukrainian Academy of Sciences, Kiev, Ukraine
[3] Department of Pediatrics, University of Colorado Health Sciences Center, Denver, CO, USA

Summary. The long arm of human chromosome 21 likely contains several hundred genes. To determine which of these are responsible for specific aspects of the Down Syndrome phenotype, protein functional analysis coupled to phenotypic analysis of transgenic mice will be required. Because such experiments are both time consuming and expensive, prioritizing 21q genes for further studies would be advantageous. Here, we discuss expression analysis, specifically the use of Northern analysis, cDNA array screening and RNA tissue in situ hybridization to assess place and time of expression of forty-two genes. For a subset of these, over expression in normal versus trisomy cell lines and mouse tissues is discussed. Lastly, several examples of alternative processing and their potential for generation of brain specific proteins are described. Together, these experiments give information on time, place and level of expression of a number of 21q genes and suggest some interesting candidates worth further investigation for relevance to Down Syndrome. These data also illustrate the complexities and ambiguities inherent in interpretation and use of expression information.

Introduction

Much research on Down Syndrome has as its long term goal the development of therapies that will prevent or ameliorate specific features of the phenotype. This of course requires the identification of responsible genes and ascertainment of gene-phenotype correlations. Before discussing approaches, it is worth considering the scope of the problem this goal presents. The long arm of human chromosome 21 contains 40 Mb of DNA, sufficient to harbor 500–1,000 genes. While earlier work suggested that a subregion of 21q may contain the critical genes (Rahmani et al., 1989; Delabar et al., 1993; Sinet et al., 1994), and therefore should be the focus of gene discovery efforts, it now appears that essentially no segment of 21q can be excluded from containing relevant genes (Korenberg et al., 1994). This is particularly true for aspects of mental retardation. Thus, at present all 500–1,000 genes need to be considered as candidate genes.

Currently, approximately 75 genes with complete coding sequences have been reported in the literature (summarized in Antonarakis, 1998). For many of these, some information on biological association is available. For example, E4TF1-60, AML1, SIM2, ERG and ETS2 are transcription factors; MNB is a kinase; RED1 is an RNA editase; SOD1, GART and CBS function in biosynthetic/biochemical pathways. Even this information, however, can be quite limited; little, if anything, is known regarding the target genes of the transcription factors; few, if any, of the normal substrates of the kinases and editase are known; and even for enzymatic steps in well characterized biochemical pathways, information may be incomplete and misleading: analysis of mutant forms of SOD1 that lead to some cases of Familial ALS suggests that not all functions, and not all critical functions, of this well known protein are known (Patterson et al., 1994). Thus, determining that a gene is relevant to the DS phenotype, and then how it is relevant, are not trivial issues. While relevance can initially be based on biochemical speculation, it must eventually be demonstrated experimentally. Unequivocal demonstration probably depends upon the development of directly relevant phenotypic features in a transgenic mouse overexpressing the gene or genes of interest.

This brings us back to the scope of the problem of gene/phenotype correlations: constructing and analyzing transgenic mice for hundreds of genes is impractical. While only 75 genes are currently in hand, and analyzing even these is a challenge, hundreds more will become available as the complete genomic sequence is analyzed and annotated. Some type of triage system would be useful, some experimental approach that would allow prioritization of genes for more detailed study. Logically, this approach would be based on map, expression and functional information. For example, candidate genes would be expected to fulfill the following criteria:

i)　mapping to chromosome 21
ii)　protein characteristics suggesting a relevant function
iii)　expression in relevant tissues (e.g. specific brain regions)
iv)　expression at a relevant developmental time point
v)　overexpression in the trisomic state
vi)　known target or interacting proteins suggesting relevant functions

While these criteria are logical, practically only (i) and (ii) have been investigated for most genes. Here, we will discuss approaches and results that address (iii)–(v), and then return to (i) to discuss the impact that genomic sequence information is having on the field of DS and gene expression.

Materials and methods

cDNAs

Known genes included S100b, CBS, KCNEI, RED1, TRPD, AMLI, APP, GART, INST, CD18, PFKL, IFNAR, ATP50, SOD1, NNP1, OSC, HRMT, NCAM2, SIM2, SMT3a,

S100β	143G8	121G1	CBS	14016	32C1	SIM2	KCNE1
OSC	RED1	SMT3A	33H3	TRPD	RED1	TRPD	NCAM2
AML1	46A12	44A4	45G3	54G11	ATP50	APP	KCNE1
109C9	GART	SLC5A3	57G11	HO2E2	ITSN	2C6	---
CD18	2C1	EO9	29D4	PFKL	EO9	HRMT	IFNARB
131B1	SOD1	121G1	---	NNP1	114	PWP2	---

Fig. 1. Map of arrayed genes and anonymous cDNAs. Gene names, accession numbers, map locations and brief descriptions can be found at http://www.expasy.chuge.chr21.txt. Sources of anonymous cDNAs are described in the text

SLC5A3 and PWP2 (see http://expasy.hcuge.ch/cgi-bin/lists?humchr21.txt for a summary of gene descriptions). The following anonymous cDNAs were isolated from an arrayed placenta cDNA library by the method of reciprocal cosmid/cDNA library screening (Lee et al., 1995) using the human chromosome 21 specific cosmid library (LL21NCO2-Q) (Soeda et al., 1995): 143G8, 121G1, 146A6 and 131B1 (accession #s: H55740-H55747). cDNA 109C9 (accession #s: H55738/739), obtained by the same method, showed significant homology to Diamine Oxidase, DAO. E09 and H02 were obtained by screening a fetal brain cDNA library with cosmid fragments. cDNAs 114, 2C1 and 2C6 are previously described fragments obtained by cDNA selection (Cheng et al., 1994; Gardiner et al., 1995; Tassone et al., 1995). The remainder of the cDNAs were obtained from the Soares infant brain cDNA library; they were reported by Genethon to map to chromosome 21. Regional localization information for all anonymous cDNAs was obtained using the chromosome 21 hybrid cell mapping panel (Gardiner et al., 1995).

For generation of cDNA arrays, cDNA inserts were prepared by PCR amplification with appropriate vector primers, quantitated by gel electrophoresis and spotted onto duplicate nylon membranes (HybondN+). The order of genes and cDNAs within the arrays is given in Fig. 1.

RNA preparation

Total RNA was prepared from lymphoblastoid cell lines, human adult brain material (autopsy material) and mouse brains (Ts65Dn and normal litter mates) by the standard guanidinium/phenol method (Chomzynski and Sacchi, 1987). PolyA+ RNA was prepared from total RNA by purification through two oligo dT columns (Ambion). Adult whole brain polyA+ RNA was purchased from Clontech. For Northerns, approximately 2 μg of polyA+ RNA was loaded per lane of a 1% agarose, tris-borate gel, electrophoresed and transferred to a nylon membrane (HybondN+). Additional human fetal and adult tissue Northerns were purchased from Clontech.

Probe preparation

For hybridization to the cDNA arrays, approximately 5 μg of polyA+ RNA from human adult whole brain, hippocampus, prefrontal cortex and occipital lobe was reverse transcribed and labelled by random hexamer priming (Feinberg and Vogelstein, 1984). Arrays were hybridized in duplicate with each probe. For hybridization to Northerns, inserts from cDNA clones were gel purified and labelled by random hexamer priming. Both arrays and Northerns were hybridized overnight in 50% formamide buffer at 42 °C and washed under standard conditions (Sambrook et al., 1989). Arrays were exposed to x-ray film for 1–4 days. Northerns were exposed overnight and signals quantitated on a phosphoimager (Molecular Dynamics).

Tissue in situ hybridization

For each gene, approximately 500 bp fragments were cloned into pBluescript. Sense and antisense probes were prepared with T3/T7 polymerases and 35S. Mouse brain slices were prepared and hybridizations carried out essentially as described by Breese et al. (1994) and Fremeau et al. (1989).

Results

Differential novel gene expression in tissues

To increase the number of candidate cDNAs for DS phenotypic features, a number of novel cDNAs mapping to chromosome 21 were isolated. Following the approach of Lee et al. (1995), five novel cDNAs were obtained by reciprocal screening of a placenta cDNA library and the chromosome 21 specific cosmid library. An additional eight cDNAs, reported by Genethon to map to chromosome 21, were obtained from the Soares infant brain arrayed cDNA library. Each cDNA was regionally localized within 21q, and transcript sizes and tissue specificity of expression were determined by Northern analysis. These data are shown in Table 1. Five cDNAs were found to be ubiquitously expressed, with transcripts detected in all fetal and/or adult tissues examined. Of particular relevance to DS, six genes showed expression in brain, fetal and/

Table 1. Northern analysis of anonymous cDNAs. cDNAs were obtained from a) infant brain and b) placenta. Location: cDNAs were assigned band locations based on mapping with a panel of chromosome 21 hybrid cell lines. Tissue: $A+$ hybridization seen in all adult tissues (heart, brain, placenta, pancreas, skeletal muscle, kidney, lung and liver); $A-$ hybridization to no adult tissue; $F+/F-$ similarly for fetal tissues (brain, liver, lung and kidney). Where expression was not ubiquitous, specific tissues are listed

CDNA	Location	Tissue	Transcript Size (kb)
a) infant brain			
29D4	21q11.2	$A+$; $F+$	2.2
57G1	Distal 21q21	A: muscle, liver	4.5, 6
44A4	21q22.1	F: liver, kidney	5.5
33H3	Proximal 21q22.3	$A+$	0.6
45G3	mid 21q22.3	$A-$; F: liver, kidney	6.5
46A12	Mid 21q22.3	$A+$	4.5
54G11	Distal 21q22.3	A: muscle, liver	6.0
32C1	Distal 21q22.3	$A-$: $F+$	4.5, 1.8
b) placenta			
140A6	21q22.1	A: placenta	2.3
131B1	21q22.1	A: placenta, muscle, kidney, pancreas	2.5
109C9	21q22.2	$A+$	1.4
143G8	Mid 21q22.3	placenta	1.0
121G4	Distal 21q22.3	$A+$	7.6

or adult. One appeared not to be expressed in adult tissues, but only in fetal tissues and therefore could be relevant to development. Expression of the other genes was limited to specific tissues; that some of these were derived from an infant brain library suggests either a lower level of expression in adult and fetal brain, one below the sensitivity of these Northerns, or expression that is brain specific in level to the age of infant brain.

In expression studies, Northern analysis has the unique advantage of providing information, not only on tissue of expression, but also on transcript size. This is of practical use in determining when a complete cDNA, and thus complete coding and protein sequence, has been obtained. Northern data also can indicate alternative processing, which may have functional implications in producing differing protein forms. When observed in brain, this latter is both interesting and important to investigate (see below). The disadvantage of Northerns is their expense in production, their limited reuse (especially a problem when large numbers of genes need to be examined) and the potential difficulty in obtaining the required tissues in sufficient quality and quantity.

Differential gene expression in brain regions

Figure 2 shows the results of hybridization to an array of 42 known genes and anonymous cDNA of polyA+ RNA from human adult whole brain, hippocampus, prefrontal cortex and occipital lobe (see Fig. 1 for the order of genes and cDNAs in the array). The signals from some genes are uniformly high among all brain regions. These include S100b, 143G8, CBS and PFKL. Others, indicated in each panel by arrows, show differential expression levels among at least some regions. For example, several cDNAs show relatively lower expression in the individual brain regions compared with whole brain. These include the anonymous cDNAs 44A4, H02E2, and 114, in addition to the genes ITSN, PWP2 and IFNAR. Of these, 44A4 was not expressed by Northern analysis in fetal brain, but had not been tested in adult brain; lack of expression in development could be a legitimate reason for abandoning further analysis of this gene. In contrast, ATP50, NNP1, 45G3 and 46A12 all show relatively higher expression in specific regions than in whole brain. Interestingly, 45G3 had been negative in adult brain by Northern analysis; conceivably while expression is high in hippocampus, prefrontal cortex and occipital lobe, it is not high enough overall for detection in whole brain material. Higher expression specific only to the occipital lobe is seen with OCS and NCAM2.

These experiments examine only genes expressed at relatively high levels and detect only relative differences in expression levels, not absolute expression. For example, genes that are negative in these experiments in all or some regions may be expressed at levels still detectable by Northerns and easily detectable by RT-PCR. The advantage of arrays lies not in their sensitivity, but in the speed with which a large number of genes can be screened. In a triage process, one designed to prioritize genes for further study, those that

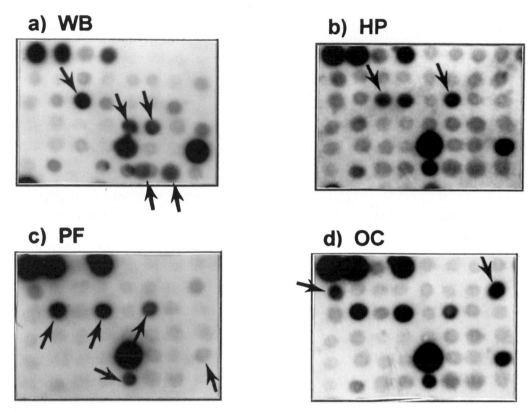

Fig. 2. Screening of cDNA arrays with human adult brain RNAs. Five µg of polyA + RNA from each source was reverse transcribed, labelled and hybridized to duplicate cDNA array filters. The order of genes in the arrays is given in Fig. 1. Only one filter of each pair is shown. RNA was isolated from: **a** whole brain, **b** hippocampus, **c** prefrontal cortex, and **d** occipital lobe. Arrows indicate genes showing significantly different hybridization strength in that region

are expressed in specific subregions may be of particular interest. For example, both hippocampus and prefrontal cortex have been implicated in development of cognitive deficits relevant to DS (Crnic and Pennington, in press; Pennington and Bennetto, 1997; Nadel, 1994). Thus, genes showing high levels of expression in these regions may be of particular interest.

A further example of regional differential expression is shown in Fig. 3. The cDNA for PWP2 was hybridized to adult mouse brain sagittal sections. While undetectable in the array hybridization, PWP2 is clearly expressed in hippocampus, as well as in cerebellum (no doubt contributing to the whole brain positive results in the arrays) and the olfactory bulb. The enlargements show PWP2 signal specific to Purkinje cells and the pyramidal layers CA1 and CA3 of the hippocampus. It is also to be noted that a number of genes show a pattern very similar to that of PWP2 in in situ hybridization, varying, at this level of resolution, only in the intensity of signals (and thus presumably, level of expression). Examples include RED1 (moderate) (Villard et al., 1997), 114 (high) and APP (very high) (data not shown). While more laborious than

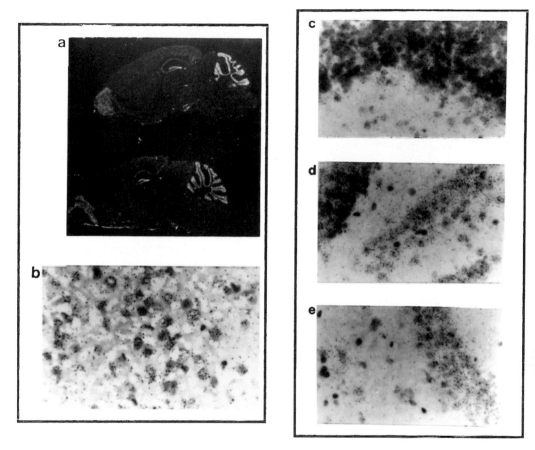

Fig. 3. RNA mouse tissue in situ hybridization with PWP2. **a** adult brain, dark field, showing signals in the cerebellum, hippocampus and olfactory bulb. Emulsions from: **b** a region in the cortex above the hippocampus, **c** Purkinje cells, **d** CA1 pyramidal layer, and **e** CA3 pyramidal layer

array screening, in situ analysis would allow the inclusion in further analysis of more genes expressed in, for example, hippocampus.

Differential gene expression during development

cDNA arrays were also screened with RNA obtained from mouse whole brain of post natal day 7 and day 21. These results are shown in Fig. 4. Clearly expressed at higher levels earlier in post natal development are PWP2 and the anonymous cDNA 114. Also at higher levels earlier, although less dramatically so, are APP and NCAM2, and the anonymous cDNA, 46A12. This allows a further differentiation of gene expression. Both PWP2 and 114 are expressed in hippocampus based on tissue in situ hybridization; both are also more highly expressed during earlier brain development. Because some brain development occurs after birth (both in mouse and in human) and because some of what distinguishes the DS brain from the normal brain (anatomically,

a) 7 Day   b) 21 Day

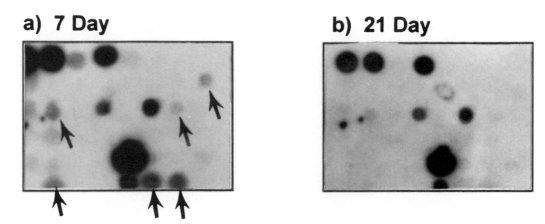

Fig. 4. Screening of cDNA arrays with mouse brain RNAs. Arrays and hybridization were as in Fig. 2. PolyA + RNA was isolated from mouse brain at **a** post natal day 7 and **b** post natal day 21

histologically, etc) appears after birth (Raz et al., 1995; Becker et al., 1991; Wisniewski et al., 1990), these genes become of relatively greater interest for further study. These results also prompt further examination of NCAM2 and 46A12 by tissue in situ for further regional expression discrimination.

Differential gene expression in trisomy

It may be anticipated that the extra copy of genes on chromosome 21, as occurs in DS, will result in their over expression by 50%. This has indeed been observed for SOD1 and GART (where RNA, protein and activity levels have been examined) (Groner et al., 1990; Brodsky et al., 1997), but it has not been examined for most other chromosome 21 genes. To address this question (and with the initial intent of simply verifying 50% expression increases), polyA+ RNA was isolated from normal and DS derived lymphoblastoid cell lines, and quantitative Northern analysis carried out for selected chromosome 21 genes. 50% increases are within, although likely very close to, the limit of quantifiable detection with use of a phosphoimager. For each gene, expression levels were normalized to both actin and HPRT, and normalized levels were then compared between normal and DS RNAs. For each gene, to verify reproducibility, the experiment was repeated using different Northerns containing RNA prepared at different times. Examples of results with STCH, MNB, GART, ETS2 and CD18 are shown in Fig. 5. While both GART and CD18 showed the expected 50% increase, other genes were surprising. Both STCH and MNB were reproducibly over expressed by 3–5 fold, not 0.5 fold. In contrast, ETS2 repeatedly showed no increase in the DS cell line. Table 2 shows the quantitation of relative expression for these and a number of other chromosome 21 genes. Again, most showed a 0.5 fold increase, however, APP like ETS2 showed no increase, and E4TF1-60, like MNB and STCH, showed a significantly greater than 2-fold increase.

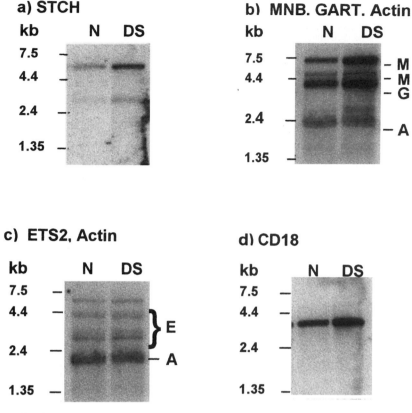

Fig. 5. Quantitative Northern analysis of lymphoblastoid cell lines. Each filter contained 2 μg of polyA + RNA derived from a normal (N) and a Down Syndrome (DS) lymphoblastoid cell line. Size markers are in kb. Probes were: **a** STCH, **b** MNB (two transcripts, M), GART (G) and actin (A), **c** ETS2 (three transcripts, bracketed, E) and actin (A), and **d** CD18. Note the strong increase of STCH signal in the DS cell line and the lack of ETS2 signal change between the N and DS cell lines. Results of quantitation are given in Table 2

Cell lines are generally considered to be, if anything, less transcriptionally active than the parent tissues. In addition, it is anticipated that there will be gene specific, tissue specific differences in expression level and regulation. While the lymphoblastoid data are intriguing and suggestive, they must be extended to tissues. Preliminary experiments have been carried out using tissues obtained from different ages of the Down Syndrome mouse model Ts65Dn (Davisson et al., 1993). This mouse is trisomic for a large region of mouse chromosome 16 that is syntenic with human chromosome 21; it carries three copies of genes from at least APP through MX and displays a number of phenotypic features of relevance to DS (Reeves et al., 1995).

Figure 6 and Table 3 show results of quantitative Northern analysis in staged mouse brains of Ts65Dn and normal litter mate controls. Both APP and MNB, unlike in human lymphoblasts, show 0.5 fold increases in expression levels. Examination of other tissues is more variable. For example, in tissues from 20 week mice, APP shows a more than 2 fold increase in liver and

Table 2. Quantitative Northerns, lymphoblastoid cell lines. Signals were quantitated using the Image Quant system (Molecular Dynamics); gene signals were normalized to actin and HPRT signals. The ratio of the normalized gene signal from the DS cell line to that from the normal cell line is given. Errors are calculated based on results from 2–4 different experiments

Gene	Trisomy: normal expression
STCH	5.4 ± 1.3
APP	1.0 ± 0.1
E4TF1-60	3.2 ± 0.3
SOD1	1.7 ± 0.3
GART	2.0 ± 0.3
MNB	3.0 ± 0.5
ERG	1.5 ± 0.3
ETS2	1.1 ± 0.1
CD18	1.9 ± 0.4

Table 3. Quantitative Northerns, mouse brains. Ratio of expression in Ts65Dn brain to normal littermate control brain is given. Calculations were as in Table 2

Gene	30 day	5–6 weeks	6–7 weeks	13 weeks	20 weeks
APP	1.4	1.3	1.4	1.3	1.3
MNB	1.5	1.5	1.5	1.3	1.3

testes; MNB, more than two fold in testes, and E4TF1-60, more than 2 fold in lung and spleen (data not shown). In other tissues, each of these genes shows only a 0.5 fold increase. Interestingly, SOD1 shows a uniform 0.5 fold increase in all tissues. These experiments need to be repeated with these and other ages, and with an increased number of genes. A difficulty for these experiments is the number of Ts65DN mice that are required for sufficient sensitivity in this Northern analysis.

Differential alternative processing

Alternative processing of a pre-mRNA, whereby differing sets of exons are included in the mature message, is not uncommon. It is a versatile method for generating proteins of precisely varying sequence, adding or subtracting specific domains or blocks of amino acids. As with other facets of differential expression, alternative processing of chromosome 21 genes is of particular interest and potential importance when observed in brain.

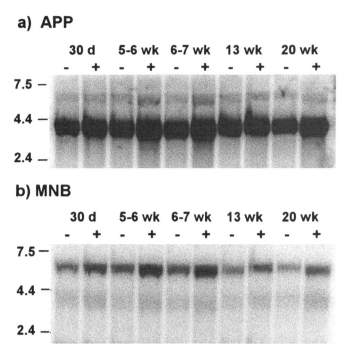

a) APP

b) MNB

Fig. 6. Quantitative Northern analysis of staged mouse brains. PolyA + RNA was isolated from normal (−) and Ts65Dn (+) mice of the indicated ages. Hybridization was with **a** APP and **b** MNB. Results of quantitation are given in Table 3

Figure 7 shows a schematic of the gene content of a ~200 kb segment of distal 21q22.1. This segment has been completely sequenced (included in accession #s: AP000046-48), and verified to contain the complete GART and SON genes, and the 5′ end of the ITSN gene. All three of these genes were known to map within this region (Cheng et al., 1993; Guipponi et al., 1998). Exon prediction programs, dbEST Blast searches, and RT-PCR experiments have identified two novel genes, B17 and B31, between SON and ITSN (Gardiner et al., submitted).

B17 is a compact gene containing 9 exons and spanning only 10 kb. RT-PCR from the putative 5′ end to the 3′ UTR produces a single product in Hela; however, in brain, in addition to this product, two smaller products are obtained (data not shown). The upper schematic in Fig. 7 illustrates the exon content of these three transcripts. Form 1 contains the longest open reading frame (orf) (1323 bp); both forms 2 and 3 are truncated by the alternative splicing downstream of exon 3, resulting in orfs of only 414 and 420 nucleotides. The B17 putative protein has no recognized protein homologies, but it can be speculated that the truncated forms may serve to regulate activity of the longest protein form through deletion of specific functional domains.

B31 shows greater complexity in alternative processing, although data are not yet complete. EST sequences and RT-PCR products confirm 14 exons. As shown in the lower schematic of Fig. 7, however, alternative processing

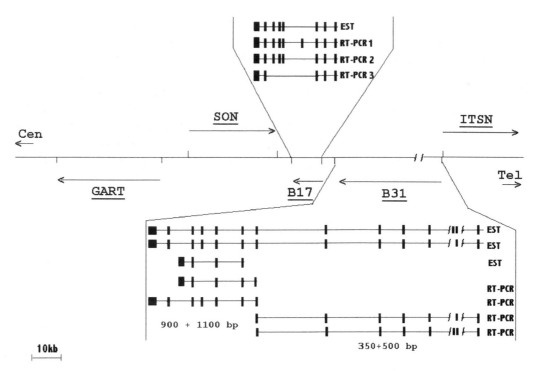

Fig. 7. Structures and alternative processing of novel genes in the GART-SON-ITSN region. The central line drawing indicates the locations and directions of transcription with respect to the centromere and telomere of the characterized genes, GART, SON and ITSN. Two novel genes, B17 and B31, were identified by exon prediction and EST database matches. B17 exon structure is shown above the central line. EST data predicted a single transcript comprised of 8 exons; RT-PCR identified three forms (1,2 and 3) comprised of 9, 8 and 5 exons. Only form 2 is seen in Hela; forms 1–3 are seen in brain. Form 2 has the longest open reading frame. B31 exon structure is shown below the central line. EST data predict two structures comprised of 13 and 14 exons, with alternative exons in the 5′ region, and a third transcript with an alternative 3′ end. RT-PCR confirms the 5′ end products by detecting fragments of 350 and 500 pb. RT-PCR also confirms the EST 3′ data and adds to alternative processing information in the 3′ region of the gene, detecting fragments of 900 and 1100 bp with one 3′UTR, and a a third form with the other 3′UTR

within the 5′ seven exons results in products of 350 and 500 bp and processing within the 3′ seven exons, in products of 900 and 1100 bp. There is also a 3′ truncated EST and RT-PCR product. Together, these may produce as many as 6 different transcripts and proteins.

Adjacent to B31 is Intersectin (ITSN), a gene producing two transcripts (Guipponi et al., 1998). Both encode proteins that contain multiple SH3 (src homology 3) domains and are suggested to be involved in protein interactions and protein trafficking. The longer transcript is brain specific and contains unique guanine nucleotide exchange factor domains. Based on sequence analysis, the 5′ ends of B31 and INST are located within two CpG islands that are less than 400 bp apart; the two genes thus conceivably share a single bi-directional promoter (Gardiner et al., submitted).

The observation of the close proximity of three genes with complex alternative processing in brain and potentially overlapping regulatory sequences begs for further investigation, by defining the brain region specificity and the developmental specificity of the various protein forms. Useful also would be the phenotype of a mouse engineered for simultaneous trisomy of all three genes, in the absence of trisomy of other chromosome 21 genes. As the sequence of chromosome 21 accumulates, more such examples of alternative processing, regulatory sequences and gene organization will become available. While increasing the complexity of the required analyses, it also increases the possibility of determining the genes of greatest relevance to DS.

Discussion

In an effort to assess potential relevance to DS, a number of chromosome 21 genes have been examined for place, time and level of expression. While not unanticipated, results clearly demonstrate that this is not a trivial issue and is not amenable to a straightforward solution applicable in all cases.

Tissue specificity of expression was examined by Northern analysis, showing that 6 of 13 anonymous cDNAs were expressed in fetal and/or adult brain. cDNA array screening further showed that 17 of 42 genes were expressed in adult whole brain or specific subregions. Based on place of expression, at these levels of sensitivity, any of these genes could reasonably be considered as candidates for the mental retardation or cognitive deficits of DS. Highest priority for further analysis could be given to those genes showing specific or higher levels of expression in hippocampus or prefrontal cortex. Inclusion as a candidate gene is, however, much easier than exclusion. Low levels of expression (certainly less than one transcript in 10^{-5}) would not be detected by these methods. Even high expression that is restricted in cell type or within a brain region may go unnoticed. This was evidenced by PWP2 and 114, both of which were positive in hippocampus and prefrontal cortex, but only by tissue in situ hybridization to brain slices and not to arrays. Thus, exclusion from further consideration of 2/3 of the genes analyzed based on Northern or cDNA array results is premature.

Use of mouse brain provided additional support for candidate genes by allowing investigation of time of expression. Examination of only two times of post natal brain showed 6 of 42 genes with relatively higher expression levels at day 7 vs. day 21. Again while this cannot eliminate genes, it can help to prioritize them.

A third strong criterion for inclusion in relevance to DS is over expression in the trisomy state (i.e. at least a 50% increase in expression level relative to the disomy state). Examination of a number of genes in lymphoblastoid cell lines suggests that some genes under at least some growth conditions are regulated such that copy number does not directly define expression level. If a gene is not over expressed (in the relevant tissue), it is reasonable to assume it does not contribute to the DS phenotype. Interestingly, one gene showing no increase in expression in these cell lines was ETS2, one of the few genes that has showed a marked DS relevant phenotype, that of skull and skeletal abnormalities, when over expressed in transgenic mice (Sumarsono et al., 1996). It is also noteworthy, however, that this phenotype is not at all so clear in the Ts65Dn mouse (Reeves et al., 1995), which is also trisomic for ETS2, but for ETS2 in its normal genomic context. In contrast to ETS2, a number of genes actually showed a greater than 0.5 fold increase in expression. This could be due to regulation that involves interaction with one or more other genes that map to chromosome 21 and a cascade effect. In this respect, MNB is of interest, since it is so far the only gene with a direct association with mental retardation from transgenic mouse studies (Smith and Rubin, 1997).

The lymphoblastoid data could, however, be related to lymphocytes in particular or to cell culture and immortalized cell lines. Indeed, when APP (showing no increase in expression in lymphoblastoid cell lines) and MNB (showing >2 fold increase) were examined in mouse brain, results were proportional to copy number and showed 0.5 fold increase, at least in brain. Preliminary data with other tissues and other genes suggests that there is a gene specific pattern of tissue specific regulatory effects. This could be of particular interest in cases such as ITSN and the novel gene B31 which, because of the close proximity of their putative 5′ ends, may share a common bi-directional promoter, and thus could share regulatory sequences. Regardless of details, these data suggest that it is important to verify that a candidate gene is indeed over expressed in trisomy in the tissues of interest at the relevant time of development.

Lastly, accumulating data show that genes vary considerably in the extent, and subtlety, of alternative processing. Detection of brain specific transcript forms could be critical to determining a protein's role in brain development and DS. In some cases, alternative transcripts may be detected by Northern analysis, as it was for ITSN. In other cases, where small or few exons may be included or excluded, they may be refractory to detection unless RT-PCR is employed. Alternative processing may mean that multiple protein functions must be ascertained.

In summary, while there are both sensitive and large scale techniques available for assessing gene expression, their thorough application is neither quick nor trivial nor assured of providing definitive information on the strength of a gene's candidacy for relevance to DS. They can certainly provide data to allow prioritization of genes for further study, but always with the caveat that they may be wrong or misleading. The best evidence for relevance remains the demonstration of a phenotype arising from over expression in a mouse model, followed by correction of this phenotype by therapy that reverts expression (or protein activity) to normal levels. Such experiments remain costly and time consuming. It can, however, be anticipated that assistance will come from a number of avenues. Progress in analysis of human genomic sequence, and equally importantly, in the comparative analysis of mouse genomic sequence, will vastly increase our understanding of gene regulation and our ability to recognize and interpret regulatory sequences. Techniques for expression analysis will also likely improve in both sensitivity and rapidity. As more genes are identified and analyzed, our knowledge of biochemistry and developmental processes will expand. Lastly, advances in the design and interpretation of behavioral tests in both humans and mouse models can be expected and these in turn will help to refine gene/phenotype correlations in transgenic animals. As a contiguous gene syndrome spanning 40 Mb, Down Syndrome is a complex disorder that will require close interaction of a number of disciplines and approaches for its eventual resolution.

Acknowledgements

This work was supported by a grant from the Boettcher Foundation and the National Institutes of Health, HD17749, to KG. This is contribution #1726 from the John C. Mitchell Laboratory for the Study of Human Birth Defects and Genetic Disorders at the Eleanor Roosevelt Institute.

References

Antonarakis SE (1998) 10 years of genomics, chromosome 21, and Down syndrome. Genomics 51: 1–16

Becker L, Mito T, Takashima S, Onodera (1991) Growth and development of the brain in Down syndrome. In: Epstein CJ (ed) The morphogenesis of Down Syndrome. Wiley-Liss, New York, pp 133–152 (Prog Clin Biol Res 373)

Breese CR, D'Costa A, Sonntag WE (1994) Effect of in utero ethanol exposure on the postnatal ontogeny of insulin-like growth factor-1, and type-1 and type-2 insulin-like growth factor receptors in the rat brain. Neuroscience 63: 579–589

Brodsky G, Barnes T, Bleskan J, Becker L, Cox M, Patterson D (1997) The human GARS-AIRS-GART gene encodes two proteins which are differentially expressed during human brain development and temporally overexpressed in cerebellum of individuals with Down syndrome. Hum Mol Genet 6: 2043–2050

Cheng J-F, Boyartchuk V, Zhu Y (1994) Isolation and mapping of human chromosome 21 cDNA: progress in constructing a chromosome 21 expression map. Genomics 23: 75–84

Cheng S, Lutfalla G, Uze G, Chumakov IM, Gardiner K (1993) GART, SON, IFNAR, and CRF2-4 genes cluster on human chromosome 21 and mouse chromosome 16. Mam Gen 4: 338–342

Chomzynski B, Sacchi N (1987) Rapid RNA isolation. Anal Biochem 162: 156–159

Crnic LS, Pennington BF (1999) Contributions of mouse models to understanding Down Syndrome. Prog Infancy Res (in press)

Davisson MT, Schmidt C, Reeves RH, Irving NG, Akeson EC, Harris BS, Bronson RT (1993) Segmental trisomy as a mouse model for Down syndrome. Prog Clin Biol Res 348: 117–133

Delabar JM, Theophile D, Rahmani Z, Chettouh Z, Blouin JL, Prieur M, Noel B, Sinet PM (1993) Molecular mapping of twenty-four features of Down syndrome on chromosome 21. Eur J Hum Genet 1: 114–124

Feinberg A, Vogelstein B (1984) A technique for radiolabelling DNA to high specific activity. Anal Biochem 137: 266–268

Fremeau RT Jr, Autelitano DJ, Blum M, Wilcox J, Roberts JL (1989) Intervening sequence-specific in situ hybridization: detection of the pro-opiomelanocortin gene primary transcript in individual neurons. Brain Res Mol Brain Res 6: 197–201

Gardiner K, Ichikawa H, Ohki M, Patterson D, Chen J-F (1995) Localization of cDNAs to a region poorly represented in the CEPH chromosome 21 YAC contig: candidate genes for genetic diseases mapped to 21q22.3 Genomics 30: 376–379

Groner Y, Elroy-Stein O, Avraham KB, Yarom R, Schickler M, Knobler H, Rotman G (1990) Down syndrome clinical symptoms are manifested in transfected cells and transgenic mice overexpressing the human Cu/Zn-superoxide dismutase gene. J Physiol (Paris) 84: 53–77

Guipponi M, Scott HS, Chen H, Schebesta A, Rossier C, Antonarakis SE (1998) Two isoforms of a human intersectin (ITSN) protein are produced by brain-specific alternative splicing in a stop codon. Genomics 53: 369–376

Korenberg JR, Chen X-N, Schipper R, Sun Z, Gonsky R, Gerwehr S, Carpenter N, Daumer C, Dignan P, Disteche C, Graham JM Jr, Hugdins L, McGillivray B, Miyazaki K, Ogasawara N, Park JP, Pagon R, Pueschel S, Sack G, Say B, Schuffenhauer S, Soukup S, Yamanaka T (1994) Down syndrome phenotypes: the consequences of chromosomal imbalance. Proc Natl Acad Sci USA 91: 4997–5001

Lee CC, Yazdani A, Wehnert M, Zhao ZY, Lindsay EA, Bailey J, Coolbauch MI, Couch L, Xiong M, Chinault AC, Baldini A, Caskey CT (1995) Isolation of chromosome-specific genes by reciprocal probing of arrayed cDNA and cosmid libraries. Hum Mol Gen 4: 1373–1380

Nadel L (1994) Neural and cognitive development in Down syndrome. In: Nadel L, Rosenthal D (eds) Down Syndrome: living and learning in the community. Wiley-Liss, New York, pp 312–321

Patterson D, Warner HR, Fox LM, Rahmani Z (1994) Superoxide dismutase, oxygen radical metabolism, and amyotrophic lateral sclerosis. Mol Genet Med 4: 79–118

Pennington BF, Bennetto L (1998) Towards a neuropsychology of mental retardation. In: Burack JA, Hodapp RM, Zigler EF (eds) Handbook of mental retardation and development. Cambridge University Press, Cambridge, pp 80–114

Rahmani Z, Blouin JL, Creau-Goldberg N, Watkins PC, Mattei JR, Poissonnier M, Prieur M, Chettouh Z, Nicole A, Aurias A, et al. (1989) Critical role of the D21S55 region on chromosome 21 in the pathogenesis of Down syndrome. Proc Natl Acad Sci USA 86: 5958–5962

Raz N, Torres IJ, Briggs SD, Spencer WD, Thorton AE, Loken WJ, Gunning FM, Mcquain JD, Driesen NR, Acker JD (1995) Selective neuroanatomic abnormalities in Down's syndrome and their cognitive correlates: evidence from MRI morphometry. Neurology 45: 356–366

Reeves RH, Irving NG, Moran TH, Wohn A, Kitt C, Sisodia SS, Schmidt C, Bronson RT, Davisson MT (1995) A mouse model for Down syndrome exhibits learning and behavior deficits. Nat Genet 11: 177–184

Sambrook J, Fritsch EF, Maniatis E (1989) Molecular cloning: a laboratory manual, 2nd ed. Cold Spring Harbor Press, NY

Sinet PM, Theophile D, Rahmani Z, Chettouh Z, Blouin JL, Prieur M, Noel B, Delabar JM (1994) Mapping of the Down syndrome phenotype on chromosome 21 at the molecular level. Biomed Pharmacother 48: 247–252

Smith DJ, Rubin EM (1997) Functional screening and complex traits: human 21q22.2 sequences affecting learning in mice. Hum Mol Genet 6: 1729–1733

Soeda E, Hou D-X, Osoegawa K, Atsuchi Y, Yamagata T, Shimokawa T, Kishida H, Soeda E, Okano S, Chumakov I, Cohen D, Raff M, Gardiner K, Graw SL, Patterson D, de Jong P, Ashworth LK, Slezak T, Carrano AV (1995) Cosmid assembly and anchoring to human chromosome 21. Genomics 25: 73–84

Sumarsono SH, Wilson TJ, Tymms MJ, Venter DJ, Corrick CM, Kola R, Lahoud MH, Papas TS, Seth A, Kola I (1996) Down's syndrome-like skeletal abnormalities in Ets2 transgenic mice. Nature 379: 534–537

Tassone F, Xu H, Burkin H, Weissman S, Gardiner K (1995) cDNA selection from 10 Mb of chromosome 21 DNA: efficiency in transcriptional mapping and reflections of genome organization. Hum Mol Gen 9: 1509–1518

Villard L, Tassone F, Haymowicz M, Welborn R, Gardiner K (1997) Map location, genomic organization and expression patterns of the human RED1 RNA editase. Somat Cell Mol Genet 23: 135–145

Wisniewski KE (1990) Down syndrome children often have brain with maturation delay, retardation of growth, and cortical dysgenesis. Am J Med Genet [Suppl 7]: 389–403

Authors' address: K. Gardiner, Ph.D., Eleanor Roosevelt Institute, 1899 Gaylord Street, Denver, CO 80206, USA, e-mail: gardiner@eri.uchsc.edu

J Neural Transm (1999) [Suppl] 57: 197–209

Isolation and analysis of chromosome 21 genes potentially involved in Down Syndrome

P. Gosset, G. Ait-Ghezala, P.-M. Sinet, and **N. Créau**

CNRS UMR 8602, Faculté de Médecine Necker Enfants Malades, Paris, France

Down syndrome is the most common birth defect (1 in 700 newborns) and the most important cause of mental retardation. This disease is characterized by a complex phenotype, mainly including morphological abnormalities of the head and limbs, short stature, hypotonia, hyperlaxity of ligaments, visceral malformations (particularly heart defects), and a constant mental retardation. In most cases, it results from the presence of an entire chromosome 21 in excess in all cells of the afflicted individuals. Phenotype-genotype correlation of rare patients with partial trisomy 21 identified a small region of chromosome 21 on 21q22.2, duplication of which is associated with many features of the syndrome (Rahmani et al., 1989, 1990). This region, named Down syndrome Chromosome Region 1 or DCR1, is associated with short stature, hypotonia, joint hyperlaxity, face and limbs dysmorphy, and mental retardation (Delabar et al., 1993).

In order to identify genes potentially involved in the Down syndrome phenotype and particularly in mental retardation, DNA clones overlapping this region were isolated and used to search for transcribed sequences by different methods.

Cloning of the Down syndrome chromosome region 1

Definition of the DCR1

A molecular study of two patients with partial trisomy 21, FGA (Poissonnier et al., 1976) and IGU (Mattei et al., 1981), has highlighted a region around D21S55 marker, which is involved in the pathogenesis of Down syndrome. The proximal boundary of the DCR was mapped between D21S17 and D21S55 by the study of DNA of the patient IGU, and the distal border was localized between D21S55 and the proto-oncogene ETS2, by the study of DNA of the patient FGA (Rahmani et al., 1989, 1990).

Later, more detailed analysis of the genotype-phenotype correlation of other patients with partial trisomy led to more detailed phenotypic maps of chromosome 21 (McKormick, 1989; Korenberg, 1992; Delabar et al., 1993; Korenberg et al., 1994). The main region, named DCR1, is related to several

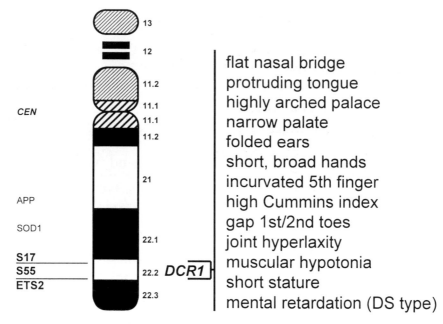

Fig. 1. Cytogenetic and molecular localization of the DCR1 region. The DCR1 region is associated to thirteen features of the Down Syndrome, including mental retardation (see Delabar et al., 1993)

major features of Down syndrome, including mental retardation, muscular hypotonia, joint hyperlaxity, short stature and some other morphological features (Delabar et al., 1993; see Fig. 1).

At the beginning of the study the size of DCR1 was estimated to cover between 400 and 3,000 kilobases and no gene was identified in this region.

YAC cloning of the distal DCR1

The first step for the determination of the DCR1 gene content was the cloning and physical mapping of the region. Regarding the estimated size of the DCR1, the analysis was divided into two parts: proximal to D21S55 (Dufresne-Zacharia et al., 1994), and distal to D21S55 (Crété et al., 1993; Gosset et al., 1995).

For the distal region, between D21S55 and ETS2, three YAC libraries were screened with the available markers D21S55, ETS2 and a new protooncogene ERG, localized near the two other markers. A new marker, 3144C4, was isolated by Alu PCR of YAC 259H11 (Crété et al., 1993) and used to achieve the screening.

33 YACs were isolated from these libraries. Size of each YAC was determined by pulsed-field gel electrophoresis (PFGE).

To establish a physical map of the contig, DNA of each YAC was subjected to restriction with seven enzymes, which recognition site contains one or more "CpG". Size of restriction fragments was analyzed by PFGE and by

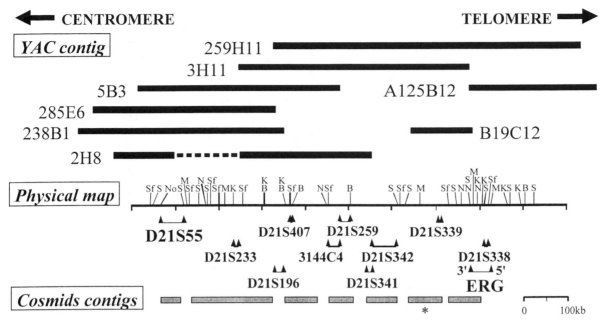

Fig. 2. Cloning of the distal DCR1 between D21S55 and ERG. *Top*: Alignment of 8 YACs of the contig between D21S55 and ERG. Dot-line: deletion in the YAC 2H8. *Center*: Consensus restriction map of the YAC contig established with seven enzymes: *Not*I (No), *Mlu*I (M), *Sfi*I (Sf), *Nru*I (N), *Bss*HII (B), *Sa*lI (S), *Ksp*I (K). *Bottom*: Boxes indicates positions of cosmids contigs along the physical map, contig "E" (KIR4.2) is indicated by an asterisk

hybridization with 17 markers. The restriction map was completed by the study of fragments of YACs partial digestion. This work led to the construction of a precise physical map between D21S55 and ETS2, based on a contig of 12 YACs, showing similarities with restriction fragments of lymphoblasts and fibroblasts genomic DNA. The ERG gene was definitively localized between D21S55 and ETS2 and outside the DCR1 (Gosset et al., 1995) (see Fig. 2).

Cosmids cloning of the distal DCR1

Four YACs of the contig (259H11, 3H11, 5B3 and 285E6), and their inter-Alu PCR products, were used as probes for screening of the ICRF cosmids libraries (Nizetic et al., 1991). A subset of 150 cosmids was isolated from these libraries.

Each cosmid was grown, its DNA was prepared using conventional miniprep technique or with the Qiagen Plasmid Kit (Qiagen Gmbh). DNA of each cosmid was digested by *Eco*RI. Restriction fragments were separated on agarose gels, stained with ethidium bromide, then transferred onto positively charged nylon membranes by the Southern blot method with alkaline buffer.

Each cosmid was hybridized onto the *Eco*RI digests of the other cosmids allowing to establish the overlapping fragments, and the order of cosmids in the contigs. They were also hybridized onto the *Eco*RI digests of the YACs to localize the YAC ends in these contigs and precise the deletion of YAC 2H8.

This approach led to the construction of 7 cosmid contigs between D21S55 and ERG, with persistence of six small gaps (see Fig. 2).

These contigs allowed further investigations in attempting to isolate transcribed sequences, using combination of two molecular biology methods: exon-trapping and cDNA selection.

Isolation of transcribed sequences

Exon-trapping experiment

A subset of cosmids was selected covering the region with a minimum overlap. DNA of these cosmids was digested by *Bam*HI and *Bgl*II, then used for an exon-trapping experiment using the Gibco BRL Exon Trapping System (Life Technologies) with the protocol provided by the manufacturer. An additional step of T_4-ligation was performed for the last cloning into UDG Cloning Vector pAMP10, before electroporation.

Colonies were screened by PCR using a primers pair specific for the vector. PCR products from clones were excised from gel. They were used as probes on membranes containing *Eco*RI digests from the cosmids and from a hybrid cell containing human chromosome 21 to verify their localization as unique probes of chromosome 21.

The resulting positives clones were selected, sequenced, and their sequence compared to databases (Dahmane et al., 1998).

cDNA selection

Three pools of cosmids were used for the experiment as described in Korn et al. (1992). DNA was fully digested by *Eco*RI, biotinylated and immobilized on magnetic beads. cDNAs for the capture were prepared from poly(A+) RNA derived from three human tissues: fetal brain, fetal liver and adult skeletal muscle. Linkers for cloning, containing tissue-specific tags, were ligated to cDNAs and used for amplification. After capture on solid phase, cDNAs were eluted, amplified, and subcloned into pAMP10. cDNA clones were grown, isolated (3 to 4 clones by kb), ordered and spotted onto nylon membranes (Yaspo et al., 1995). These membranes were screened for hybridization with the *Eco*RI fragments of the cosmids isolated from agarose gels, and with exons obtained from exon trapping experiments. Positive clones were used to construct cDNA contigs by hybridization on spotted membranes. These contigs were sequenced and resulting sequences were compared to the databases (Dahmane et al., 1998).

Results

The gene KIR4.2

Isolation of the gene

From these experiments, a group of eight overlapping cDNAs and two exons was identified by cosmids of the contig "E" (see Fig. 2).

The contig led to the identification of a new inwardly rectifier potassium channel gene KIR4.2 (**KCNJ15**) (Gosset et al., 1997). The sequence alignment of exons and cDNAs gave a transcribed sequence of 1,708 bp with an open reading frame of 1,125 bp (EMBL, AC: Y10745), and showed the presence of four different 5′ untranslated regions (EMBL, AC: Y13895 and Y13896) (Gosset et al., 1997) (see Fig. 3). The coding sequence was found similar to a cDNA named IRKK (DBJ AC: D87291) which covers the same gene (Ohira et al., 1997).

Several sequences similar to KIR4.2 were found by comparison with databases: trapped exons 14B1 and HMC20C09 (which is identical to

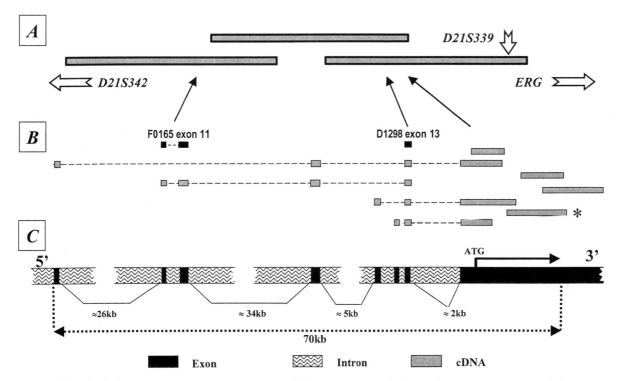

Fig. 3. Schematic representation of the KIR4.2 gene. **A** Schematic representation of the three cosmids of contig E. **B** Respective localization of exon D1298#13 (AC: AJ001891), exon F0165#11 (AC: AJ012369), and cDNAs of KIR4.2 isolated by exon trapping and cDNA selection experiments (not to scale). **C** Exon-intron organization of KIR4.2, comparative to the genomic sequence (AC: AP000001 to AP000022, Hirakawa et al., unpublished). The coding sequence is indicated by an arrow

D1298#13), and ESTs (Expressed Sequence Tagged) ye33c06 and HUMG0008652.

Expression pattern of KIR4.2

Some information concerning tissue expression of KIR4.2, may be obtained from the cDNA selection experiment. Four of the selected cDNAs are derived from skeletal muscle mRNA, two are from fetal brain and two are from fetal liver. One EST is derived from lung (ye33c06). Thus, this gene is expressed, even at a low level, in these tissues.

• *Northern-blot analysis.* Expression analysis of the gene was performed by testing various cDNAs for hybridization on Northern blots from human fetal and adult tissues. Four mRNAs of 9.5, 4.7, 3.2, and 2.2 kb were observed in fetal kidney and fetal lung as well as in adult tissues (Gosset et al., 1997; Ohira et al., 1997). The major transcript, in these tissues, is the 2.2 kb mRNA, which may represent several forms of close size. Two mRNA forms were faintly observed in some brain regions, the 4.7 kb form being predominant (see Fig. 4).

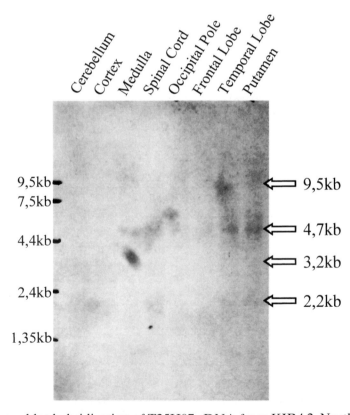

Fig. 4. Northern blot hybridization of T25H07 cDNA from KIR4.2. Northern blot analysis of RNA from adult brain tissues (Clontech, "Human Brain II") hybridized with cDNA clone "T25H07" which is indicated by an asterisk in Fig. 3. Positions and sizes of molecular markers are indicated on the left, sizes of KIR4.2 mRNAs forms observed in kidney are indicated on the right

These results suggest that alternative splicing and expression are regulated in a relative tissue specific manner. Identification of all exons and promoters of the gene will precise these regulations.

• *In situ hybridization.* Riboprobes were prepared from KIR4.2 cDNAs. Antisense riboprobes were used as probes on a 40-days-old embryo and different organs from an 8-weeks-old fetus. Sense probes were used as control on slides. A strong signal was observed in mesonephros and kidney. A lower signal was also observed in the cloaca region, on the epithelium of mandible, and in pancreas and lung. The stage of development at which expression is present in the brain has not yet been determined.

Other genes in the D21S55-ERG part of the DCR1

GIRK2

Several cDNAs corresponding to GIRK2 (or KIR3.2 or **KCNJ6**), which is an other member of the inwardly rectifier potassium channel family, were isolated. This gene is localized near D21S55 (Patil et al., 1995; Tsaur et al., 1995). GIRK2 is covering about 300 kb (Dahmane et al., 1998) and is expressed in brain and in pancreatic beta-cells (Ohira et al., 1997).

Other transcriptional units

cDNA selection and exon-trapping experiments have highlighted seven other highly probable transcription units between D21S55 and ERG (Dahmane et al., 1998). These units are defined as similar to known EST in databases, or containing both exons and cDNAs, or showing a signal by hybridization on Northern blots.

Four units ("CER-19.0" to "CER-22.0") are localized within the large introns of GIRK2. These units are formed by contigs of cDNAs, mainly from fetal brain.

Other transcriptional units are derived from contigs of cDNAs and captured exons from various tissues. Some units showed expression by hybridization on Northern blots: heart and skeletal muscle for CER-19.0, heart, placenta and fetal tissues for CER-27.0.

Function of these new genes is unknown at the present time.

Other genes in the proximal part of the DCR1

Other genes and transcriptional units have been identified and mapped in the proximal part of the DCR1, between CBR and D21S55 (Dahmane et al., 1998):

• *CBR*: CBR is the gene for the carbonyl-reductase (Wermuth et al., 1988; Lemieux et al., 1993).

- *KIAA0136*: This is an ubiquitously expressed gene, with an unknown function (Nagase et al., 1995).
- *CAF1A*: This gene is involved in chromatin assembly (Kaufman et al., 1995; Blouin et al., 1996a; Katsanis et al., 1996).
- *SIM2*: SIM2 is homologous to a transcriptional factor of drosophila, and is expressed in nervous central system, cartilage of face and trunk, muscles of trunk and kidney during development (Chen et al., 1995; Dahmane et al., 1995; Muenke et al., 1995; Yamaki et al., 1996).
- *HCS*: It is the gene for the holocarboxylase-synthetase, involved in metabolism of the biotin (Suzuki et al., 1994; Blouin et al., 1996b; Zhang et al., 1997).
- *TPRD1*: TPRD1 is the gene of a protein containing the tetratricopeptide repeat domain (Ohira et al., 1996), expressed in the mouse during development and in adult brain (Lopes et al., 1999).
- *MNBH*: MNBH is homologous to the drosophila gene *"minibrain"* (Guimera et al., 1996; Song et al., 1996; Shindoh et al., 1996; Chen and Antonarakis, 1997). MNBH is expressed during the development of the nervous central system (Guimera et col., 1996). The mouse homologue is also expressed in adult brain and epithelium tissues (Rahmani et al., 1998).

In addition to these seven known genes, 10 transcriptional units were defined in Dahmane et al. (1998). Two of these units, CER-5.0 and CER-11.0, showed ubiquitous expression by hybridization on Northern blots. Two units showed a tissue-specific expression: skeletal muscle for CER-8.0, liver and kidney for CER-16.0.

Discussion

More than 10% of genes of the chromosome 21 are yet known. Function of most of these genes remain unknown. Nevertheless, involvement of some genes in the pathogenesis of Down Syndrome, particularly mental retardation, may be expected regarding their expression pattern in varied human tissues and during development, or by extrapolation of results obtained for transgenic animal experimentation.

The gene KIR4.2

KIR4.2 is a member of the inwardly rectifier potassium channel family. These channels are concerned in maintenance of resting membrane potential and control excitability. Each KIR subunit is constituted by two membrane spanning segments, separated by a pore-forming region. Four of these subunits (homo or heterotetrameric), are needed to form a functional channel.

Electrophysiologic properties of KIR4.2 (also named Kir1.3) were analyzed by expression in Xenopus oocytes (Shuck et al., 1997). This study failed to show a detectable current by expression of KIR4.2 alone, but co-expression of this gene and Kir1.1 or Kir1.2 sub-units resulted in reduced currents comparing to expression of these sub-units alone, suggesting a dominant negative influence of this channel.

Expression of KIR4.2 is also visualized in brain, as shown by Southern-blot hybridization. In Down Syndrome's brains, effect of a surexpression of KIR4.2 may have heavy consequences, by dominant negative control on other co-expressed inwardly rectifier potassium channel sub-units. Involvement of KIR4.2 in mental retardation of Down syndrome need to be explored by construction and study of transgenic mice over-expressing Kir4.2.

Other genes of the DCR1

Some genes of the DCR1 region are expressed in human nervous system, at different developmental stages, and may be implicated in mental retardation.

MNBH, is expressed in cerebral cortex, cerebellum, hippocampus, spinal cord and retina, in human fetus as in adult (Guimera et al., 1996; Rahmani et al., 1998). Transgenic mice with a YAC containing MNBH were constructed (Smith et al., 1997). These mice show deficit in learning functions. This gene should therefore be considered as a major candidate for the mental retardation when over-expressed in Down Syndrome.

SIM2 is a gene coding for a "helix-loop-helix" protein, homologous to a transcription regulator involved in development of the central nervous system in drosophila. This gene is expressed in rat neuro-epithelium, but also in cartilage of the face and trunk, in muscles of the trunk and in kidney during development. This gene may play an important role in mental retardation and dysmorphy associated to Down Syndrome (Dahmane et al., 1995; Fan et al., 1996; Yamaki et al., 1996).

GIRK2 is an other member of the inwardly rectifier potassium channel family. GIRK2 is expressed in brain and in pancreatic beta-cells. A mutation of this channel sub-unit is associated with the "weaver" phenotype in mouse, where the defect of the GIRK2 channel induces abnormal migration of nervous cells in the cerebellum granular layer, leading to an ataxia (Patil et al., 1995). Knock-out mice for GIRK2, show normal nervous system development but increased sensibility to seizure (Signorini et al., 1997). The effects of GIRK2 over-expression are still to be demonstrated, but may be fundamental in pathogenesis of mental retardation, the results obtained in the mouse suggesting its involvement in neuronal excitability.

Future developments

In the DCR1, the most important region involved in the pathogenesis of Down Syndrome, 18 new potential transcriptional units have been highlighted (Dahmane et al., 1998), in addition to the 9 known genes.

Function of potential genes of DCR1 should be determined as some of them seem to be expressed in brain and may be involved in mechanisms leading to mental retardation when they are present in three copies in trisomy 21.

The level of over-expression of these genes need to be evaluated, for instance by hybridization on Northern blots of Down Syndrome brains, quantitative RT-PCR, or SAGE experiment (Velculescu et al., 1995). Without this step, extrapolation of results observed in transgenic mice must be irrelevant for human pathology.

Acknowledgments

I am grateful to all the staff of UMR 8602 for their help. I am grateful to A.-L. Delezoide for her help for in situ hybridization. This work was supported by CNRS, MNERT, EEC grants, AFM, FRM, Université Paris V, Faculté Necker, and AP-HP.

References

Blouin JL, Duriaux-Sail G, Chen H, Gos A, Morris MA, Rossier C, Antonarakis SE (1996) Mapping of the gene for the p60 subunit of the human chromatin assembly factor (CAF1A) to the Down syndrome region of chromosome 21. Genomics 33(2): 309–312

Blouin JL, Duriaux Saïl G, Antonarakis SE (1996) Mapping of the human holocarboxylase synthetase gene (HCS) to the Down syndrome critical region of chromosome 21q22. Ann Genet 39(3): 185–188

Chen H, Antonarakis SE (1997) Localisation of a human homologue of the Drosophila mnb and rat Dyrk genes to chromosome 21q22.2. Hum Genet 99(2): 262–265

Chen H, Chrast R, Rossier C, Gos A, Antonarakis SE, Kudoh J, Yamaki A, Shindoh N, Maeda H, Minoshima S, Shimizu N (1995) Single-minded and Down Syndrome? Nat Genet 10(1): 9–10

Crété N, Gosset P, Théophile D, Duterque-Coquillaud M, Blouin JL, Vayssettes C, Sinet PM, Créau-Goldberg N (1993) Mapping the Down syndrome chromosome region. Establishment of a YAC contig spanning 1.2 megabases. Eur J Hum Genet 1(1): 51–63

Dahmane N, Charron G, Lopes C, Yaspo ML, Maunoury C, Decorte L, Sinet PM, Bloch B, Delabar JM (1995) Down syndrome-critical region contains a gene homologous to Drosophila sim expressed during rat and human central nervous system development. Proc Natl Acad Sci USA 92(20): 9191–9195

Dahmane N, Ait Ghezala G, Gosset P, Chamoun Z, Dufresne Zacharia MC, Lopes C, Rabatel N, Gassanova Maugenre S, Chettouh Z, Abramowski V, Fayet E, Yaspo ML, Korn B, Blouin JL, Lehrach H, Poutska A, Antonarakis SE, Sinet PM, Créau N, Delabar JM (1998) Transcriptional map of the 2.5-Mb CBR-ERG region of chromosome 21 involved in Down syndrome. Genomics 48(1): 12–23

Delabar JM, Theophile D, Rahmani Z, Chettouh Z, Blouin JL, Prieur M, Noel B, Sinet PM (1993) Molecular mapping of twenty-four features of Down syndrome on chromosome 21. Eur J Hum Genet 1(2): 114–124

Dufresne-Zacharia MC, Dahmane N, Théophile D, Orti R, Chettouh Z, Sinet PM, Delabar JM (1994) 3.6 megabase genomic and YAC physical map of the Down syndrome region on chromosome 21. Genomics 19: 462–469

Fan CM, Kuwana E, Bulfone A, Fletcher CF, Copeland NG, Jenkins NA, Crews S, Martinez S, Puelles L, Rubenstein JL, Tessier-Lavigne M (1996) Expression patterns of two murine homologs of Drosophila single-minded suggest possible roles in embryonic patterning and in the pathogenesis of Down syndrome. Mol Cell Neurosci 7(1): 1–16

Gosset P (1998) Cloning, physical mapping, and identification of genes of the D21S55-ETS2 region, which is involved in the Down Syndrome. [Isolement, cartographie et identification des gènes de la région D21S55-ETS2 impliquée dans la pathogénie de la trisomie 21]. Thesis, Université Paris 7

Gosset P, Crété N, Ait Ghezala G, Théophile D, Van Broeckhoven C, Vayssettes C, Sinet PM, Créau N (1995) A high-resolution map of 1.6 Mb in the Down syndrome region: a new map between D21S55 and ETS2. Mamm Genome 6(2): 127–130

Gosset P, Ait Ghezala G, Korn B, Yaspo ML, Poutska A, Lehrach H, Sinet PM, Créau N (1997) A new inwardly rectifier potassium channel gene (KCNJ15) localized on chromosome 21 in the Down syndrome chromosome region 1 (DCR1). Genomics 44: 237–241

Guimera J, Casas C, Pucharcós C, Solans A, Domènech A, Planas AM, Ashley J, Lovett M, Estivill X, Pritchard MA (1996) A human homologue of Drosophila minibrain (MNB) is expressed in the neuronal regions affected in Down syndrome and maps to the critical region. Hum Mol Genet 5(9): 1305–1310

Katsanis N, Fisher EM (1996) The gene encoding the p60 subunit of chromatin assembly factor I (CAF1P60) maps to human chromosome 21q22.2, a region associated with some of the major features of Down syndrome. Hum Genet 98(4): 497–499

Kaufman PD, Kobayashi R, Kessler N, Stillman B (1995) The p150 and p60 subunits of chromatin assembly factor I: a molecular link between newly synthesized histones and DNA replication. Cell 81(7): 1105–1114

Korenberg JR, Bradley C, Disteche CM (1992) Down syndrome: molecular mapping of the congenital heart disease and duodenal stenosis. Am J Hum Genet 50: 294–302

Korenberg JR, Chen XN, Schipper R, Sun Z, Gonsky R, Gerwehr S, Carpenter N, Daumer C, Dignan P, Disteche C et al (1994) Down syndrome phenotypes: the consequences of chromosomal imbalance. Proc Natl Acad Sci USA 91(11): 4997–5001

Korn B, Sedlacek Z, Manca A, Kioschis P, Konecki D, Lehrach H, Poustka A (1992) A strategy for the selection of transcribed sequences in the Xq28 region. Hum Mol Genet 1(4): 235–242

Lemieux N, Malfoy B, Forrest GL (1993) Human carbonyl reductase (CBR) localized to band 21q22.1 by high-resolution fluorescence in situ hybridization displays gene dosage effects in trisomy 21 cells. Genomics 15(1): 169–172

Lopes C, Rachidi M, Gassanova S, Sinet PM, Delabar JM (1999) Developmentally regulated expression of mtprd, the murine ortholog of tprd, a gene from the Down syndrome chromosomal region 1. Mech Dev 84(1–2): 189–193

Mattei JF, Mattei MG, Beateman MA, Giraud F (1981) Trisomy 21 for the region 21q22.3: identification by high resolution R banding patterns. Hum Genet 56: 409–411

McCormick MK, Schinzel A, Petersen MB, Stetten G, Driscoll DJ, Cantu ES, Tranebjaerg L, Mikkelsen M, Watkins PC, Antonarakis SE (1989) Molecular genetic approach to the characterization of the Down syndrome region of chromosome 21. Genomics 5(2): 325–331

Muenke M, Bone LJ, Mitchell HF, Hart I, Walton K, Hall-Johnson K, Ippel EF, Dietz-Band J, Kvaløy K, Fan CM, et al (1995) Physical mapping of the holoprosencephaly critical region in 21q22.3, exclusion of SIM2 as a candidate gene for holoprosencephaly, and mapping of SIM2 to a region of chromosome 21 important for Down syndrome. Am J Hum Genet 57(5): 1074–1079

Nagase T, Seki N, Tanaka A, Ishikawa K, Nomura N (1995) Prediction of the coding sequences of unidentified human genes. IV. The coding sequences of 40 new genes (KIAA0121-KIAA0160) deduced by analysis of cDNA clones from human cell line KG-1. DNA Res 2(4): 167–174

Nizetic D, Zehetner G, Monaco AP, Gellen L, Young BD, Lehrach H (1991) Construction, arraying, and high-density screening of large insert libraries of human chromosomes X and 21: their potential use as reference libraries. Proc Natl Acad Sci USA 88: 8233–8237

Ohira M, Ootsuyama A, Suzuki E, Ichikawa H, Seki N, Nagase T, Nomura N, Ohki M (1996) Identification of a novel human gene containing the tetratricopeptide repeat domain from the Down syndrome region of chromosome 21. DNA Res 3(1): 9–16

Ohira M, Seki N, Nagase T, Suzuki E, Nomura N, Ohara O, Hattori M, Sakaki Y, Eki T, Murakami Y, Saito T, Ichikawa H, Ohki M (1997) Gene identification in 1.6 Mb region of the Down syndrome region on chromosome 21. Genome Res 7: 47–58

Patil N, Cox DR, Bhat D, Faham M, Myers RM, Peterson AS (1995) A potassium channel mutation in weaver mice implicates membrane excitability in granule cell differentiation. Nat Genet 11(2): 126–129

Poissonnier M, Saint-Paul B, Dutrillaux B, Chassaigne M, Gruyer P, de Blignières-Strouk G (1976) Trisomie 21 partielle (21q21–21q22.2). Ann Genet 19(1): 69–73

Rahmani Z, Blouin JL, Créau-Goldberg N, Watkins PC, Mattei JF, Poissonnier M, Prieur M, Chettouh Z, Nicole A, Aurias A, Sinet PM, Delabar JM (1989) Critical role of the D21S55 region on chromosome 21 in the pathogenesis of Down syndrome. Proc Natl Acad Sci USA 86: 5958–5962

Rahmani Z, Blouin JL, Créau-Goldberg N, Watkins PC, Mattei JF, Poissonnier M, Prieur M, Chettouh Z, Nicole A, Aurias A, et al (1990) Down syndrome critical region around D21S55 on proximal 21q22.3. Am J Med Genet [Suppl] 7: 98–103

Rahmani Z, Lopes C, Rachidi M, Delabar JM (1998) Expression of the mnb (dyrk) protein in adult and embryonic mouse tissues. Biochem Biophys Res Commun 253(2): 514–518

Shindoh N, Kudoh J, Maeda H, Yamaki A, Minoshima S, Shimizu Y, Shimizu N (1996) Cloning of a human homolog of the Drosophila minibrain/rat Dyrk gene from the Down syndrome critical region of chromosome 21. Biochem Biophys Res Commun 225(1): 92–99

Shuck ME, Piser TM, Bock JH, Slightom JL, Lee KS, Bienkowski MJ (1997) Cloning and characterization of two K+ inward rectifier (Kir) 1.1 potassium channel homologs from human kidney (Kir1.2 and Kir1.3). J Biol Chem 272(1): 586–593

Signorini S, Liao YJ, Duncan SA, Jan LY, Stoffel M (1997) Normal cerebellar development but susceptibility to seizures in mice lacking G protein-coupled, inwardly rectifying K+ channel GIRK2. Proc Natl Acad Sci USA 94(3): 923–927

Smith DJ, Stevens ME, Sudanagunta SP, Bronson RT, Makhinson M, Watabe AM, O'Dell TJ, Fung J, Weier HU, Cheng JF, Rubin EM (1997) Functional screening of 2 Mb of human chromosome 21q22.2 in transgenic mice implicates minibrain in learning defects associated with Down syndrome. Nat Genet 16(1): 28–36

Song WJ, Sternberg LR, Kasten-Sportès C, Keuren ML, Chung SH, Slack AC, Miller DE, Glover TW, Chiang PW, Lou L, Kurnit DM (1996) Isolation of human and murine homologues of the Drosophila minibrain gene: human homologue maps to 21q22.2 in the Down syndrome critical region. Genomics 38(3): 331–339

Suzuki Y, Aoki Y, Ishida Y, Chiba Y, Iwamatsu A, Kishino T, Niikawa N, Matsubara Y, Narisawa K (1994) Isolation and characterization of mutations in the human holocarboxylase synthetase cDNA. Nat Genet 8(2): 122–128 (Abstract)

Velculescu VE, Zhang L, Vogelstein B, Kinzler KW (1995) Serial analysis of gene expression. Science 270(5235): 484–487

Wermuth B, Bohren KM, Heinemann G, von Wartburg JP, Gabbay KH (1988) Human carbonyl reductase. Nucleotide sequence analysis of a cDNA and amino acid sequence of the encoded protein. J Biol Chem 263(31): 16185–16188

Yamaki A, Noda S, Kudoh J, Shindoh N, Maeda H, Minoshima S, Kawasaki K, Shimizu Y, Shimizu N (1996) The mammalian single-minded (SIM) gene: mouse cDNA structure and diencephalic expression indicate a candidate gene for Down syndrome. Genomics 35(1): 136–143

Yaspo ML, Gellen L, Mott R, Korn B, Nizetic D, Poustka AM, Lehrach H (1995) Model for a transcript map of human chromosome 21: isolation of new coding sequences from exon and enriched cDNA libraries. Hum Mol Genet 4(8): 1291–1304

Zhang XX, León-Del-Rio A, Gravel RA, Eydoux P (1997) Assignment of holocarboxylase synthetase gene (HLCS) to human chromosome band 21q22.1 and to mouse chromosome band 16C4 by in situ hybridization. Cytogenet Cell Genet 76(3–4): 179

Authors' address: Dr. P. Gosset, Laboratoire d'Histo-Embryologie-Cytogénétique, Hôpital Necker Enfants-Malades, 149 rue de Sèvres, F-75743 Paris Cedex 15, France, e-mail: philippe.gosset@nck.ap-hop-paris.fr

J Neural Transm (1999) [Suppl] 57: 211–220
© Springer-Verlag 1999

Differential display reveals deteriorated mRNA levels of NADH3 (complex I) in cerebellum of patients with Down Syndrome

K. Krapfenbauer[1], **B. Chul Yoo**[1], **N. Cairns**[2], and **G. Lubec**[1]

[1] Department of Pediatrics University of Vienna, Austria
[2] Institute of Psychiatry, Department of Neuropathology, MRC Brain Bank, University of London, United Kingdom

Summary. Although gene hunting has been carried out in Down Syndrome (DS) cells, information on expressional differences in DS brain is limited. We have recently described expressional differences in fetal DS brain but cannot assign these findings to "DS per" se or simply to "neurodegeneration".

We therefore performed gene hunting in cerebellum of adult patients with DS and Alzheimer's disease (AD) neuropathology, AD and controls.

The gene hunting method used was differential display and pools of the individual groups were examined to rule out allelic differences.

Differential display revealed the absence of a band, identified by sequencing and gene bank work as matching the NADH3 gene (99.1% identity) in cerebellum of DS patients. Dot blots showed the presence of NADH3 signals in only two out of 7 DS patients.

We show at the transcriptional level that a mitochondrial enzyme, the complex I, NADH3, is significantly downregulated in DS cerebellum. This extends previous work on deficiencies of the electron transport chain in platelets of patients with DS.

Introduction

Gene hunting in Down Syndrome (DS) tissues and cells is an innovative approach to study expressional differences. Using subtractive hybridization we were the first to report a changed expressional pattern at the mRNA level in DS brain of sex and age matched fetuses at the 23rd week of gestation (Labudova et al., 1999a; Kitzmueller et al., 1999). A first series of genes was subsequently determined at the protein and/or activity level in brains of adult patients with DS (Labudova et al., 1999b, 1998) confirming semiquantitative mRNA results obtained in fetal brain. In order to rule out expressional changes by neurodegeneration per se rather than by trisomy 21 we decided to perform gene hunting in brains of adult patients with AD and with DS based upon the fact that all DS patients from 40 years of age present with AD neuropathology (Burger et al., 1973; Wisniewski et al., 1985). Differential display was selected for gene hunting as this method was sensitive and highly reproducible in our laboratory. Cerebellum was selected as it is regularly affected in DS. Using this technique we detected a band in controls and brains of patients with AD, missing in DS brain, representing a fragment, which was amplified, sequenced and identified by gene bank work as NADH3.

Methods

Principle of differential display

The general strategy of differential display is presented in Fig. 1. It depends on the combination of three techniques brought together by a concept of (a) reverse transcription by anchored primers (Downstream-primer) (b) choice of arbitrary primer for setting lengths of cDNA to be amplified by the polymerase chain reaction (PCR), each of the cDNA corresponding to a part of a m-RNA and (c) separation of the amplified cDNA by gel sequencing gel electrophoreses.

Samples

Briefly, post mortem brain samples were obtained from the MRC London Brain Bank for Neurodegenerative Disease. In all DS brain samples there was evidence of abundant beta A plaques and neurofibrillary tangles. The AD patients fulfilled the National Institute of neurological and Communicative Disorders and Stroke, Alzheimer Disease and related Disorders Association (NINCDS/ADRDA) criteria for probable AD (Tierney et al., 1988). The histological diagnosis of AD was established and consistent with the CERAD criteria for definite diagnosis of AD (Mirra et al., 1991). The control brains were from individuals with no history of neurological or psychiatric illness. The major causes of death was bronchopneumonia in DS and AD patients and heart disease in controls. Post mortem interval and age of AD, DS patients and controls was comparable (Seidl et al., 1998).

Tissue samples were stored at −80°C and the freezing chain was never interrupted. Cerebellum (C) of patients with DS (n = 7), AD (n = 9) and controls (n = 7) was used for our studies. Cerebellum samples of male patients in each group were pooled in order to rule out allelic differences and used for isolation of RNA.

Isolation of RNAs

Isolation of RNA was carried out using RNazol-B method, obtained from Molecular Research Centre (cat.: RN-160), and was performed following the manufacturer's protocols. The integrity of the RNA-samples was checked by running 1–3 μg of each RNA on a 7% (v/v) formaldehyde agarose gel. After preparation of total RNA-pools the mRNA was isolated by using Quick Prep Micro mRNA purification kit (Pharmacia Biotech, Uppsala, Sweden; cat. 27-92-55-01). For reverse transcription 1 μg of the cleaned mRNA was diluted to a concentration of 0,1 μg/μl. For each $HT_{11}M$-downstream primer (with M = G,A and C, s. Table 1), three reverse transcription reactions were set up: 65°C for 5 min., 37°C for 60 min., 95°C for 5 min.

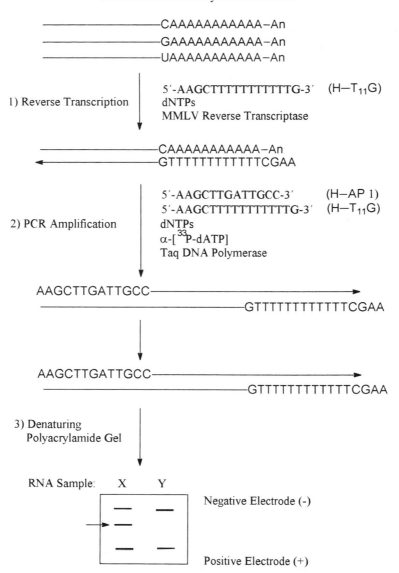

Fig. 1. Schematic presentation of the differential display principle

PCR-Amplification of 3' Termini of mRNA

After reversed transcription the cDNA-samples were directly used for PCR-amplification. For each downstream primer the following combination for the PCR was set: 8.2 μl dH$_2$O, 2 μl of 10xPCR buffer (100 mM Tris.HCl, pH 8.4, 500 mM KCl, 15 mM MgCl$_2$, 0.01% gelatine), 1.6 μl of 25 μM dNTP, 2 μl of 2 μM arbitrary primer (upstream-primer), 2 μl of 2 μM HT$_{11}$M — primer, 0.2 μl (1 unit) of AmpliTaq-Polymerase (Perkin Elmer, Norwalk, CT). The upstream primer set (HAP-1, HAP-2, HAP-3 and HAP-4, s. Table 1) was purchased from GenHunter Corp. (RNA Image kit; Nashville, TN, cat: G501). The tubes were mixed well by pipetting up and down and centrifuged to collect the probes. The PCR reactions were carried out as follows: 4°C for 1 min., 94°C for 30 sec, 40°C for 2 min., 72°C for 20 sec, for 40 cycles, then 72°C for 5 min.

Table 1. Differential display primers used in the present study

No	Primer	Conc.	Sequence
1	H-T$_{11}$G	2 μM	5′-AAGCT G-3′
2	H-T$_{11}$A	2 μM	5′-AAGCT A-3′
3	H-T$_{11}$C	2 μM	5′-AAGCT C-3′
4	H-AP1	2 μM	5′-AAGCTTGATTGCC-3′
5	H-AP2	2 μM	5′-AAGCTTCGACTGT-3′
6	H-AP3	2 μM	5′-AAGCTTTGGTCAG-3′
7	H-AP4	2 μM	5′-AAGCTTCTCAACG-3
8	H-AP5	2 μM	5′-AAGCTTAGTAGGC-3′
9	H-AP6	2 μM	5′-AAGCTTGCACCAT-3′
10	Lgh	2 μM	5′-GACGCGAACGAAGCAAC-3′
11	Rgh	2 μM	5′-CGACAACACCGATAATC-3′

Denaturing Polyacrylamide Gel-Electrophoresis

3.5 μl of the PCR-samples plus 2 μl of loading dye (9.5% formamide, 0.09% bromophenolblue and 0.09% xylene cyanol FF) were incubated at 80°C for 2 min. before loading onto a 6% denatured DNA-sequencing gel. Electrophoresis were carried out for about 3 hr at 60 W constant power. After electrophoresis the gel was directly transferred to a piece of Whatman 3 MM paper (Clifton, NJ, cat: 3030-917), covered with a sheet of "Saran wrap paper" and dried under vacuum at 80°C for 1 hr. After complete drying the Saran wrap paper was removed and exposed to a X-ray film at room temperature for 24–72 hr.

Reamplification and cloning of the cDNA-probe

The band of interest, i.e. the band present in controls and AD and absent in DS, was cut out from the dried gel after alignment with the autoradiography. The DNA was extracted along with the Whatmann 3 MM paper in 50 μl of 4x distilled water for 10 min. at room temperature and boiled for 15 min. The tubes were spun down and reamplification initiated using the same primer set and PCR-conditions except that the dNTP concentrations were 20 μM instead of 2 μM and no isotope was added. To check the quality of the samples, 10 μL of the reamplified PCR-samples were run on a 6% acrylamide-gel with 9.5% formamide, 0.09% bromophenolblue and 0.09% xylene cyanol FF as a loading dye and stained with ethidium bromide. Reamplified cDNAs were directly cloned into a PCR-Trap vector using the cloning system, version 3.0 from GenHunter (Nashville, TN, cat: P404) and performed as given in the manufacturer's protocols. Recombinant colonies were picked up with sterile micropipets tips and deposited in a lysis buffer. The tubes were centrifuged at 12,000 × g for 1 min. and 2 μl of the supernatant was used for PCR with vector, derived Lgh- and Rgh-Primer (s. Table 1). The integrity of the insert was checked by running 10 μL on a 6% (v/v) acrylamide gel and subsequently sequenced by MWG (Ebersberg, Germany) The sequence of the PCR-product was aligned with the EMBL — gene bank database using BLAST and FASTA programs.

Verification of NADH-RNA levels by Dot-Blot-Analysis

An aliquot of the total RNAs from DS, AD and controls was used for dot blot analysis according to the method given previously (Hardmeier et al., 1997). RNA was transferred

PCR TRAP Cloning Vector (before cloning)

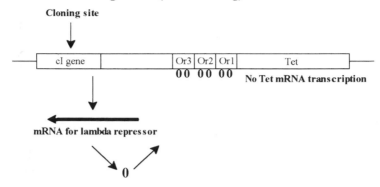

PCR TRAP Cloning Vector (after cloning of PCR product)

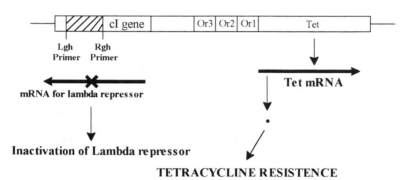

Fig. 2. Schematic presentation of the vector used

to a positively charged membrane (Gene Screen Plus, DuPont, cat: NEF-986) by dot-blotting, fixed with 0,05 M sodium hydroxide (NaOH) for 5 min at room temperature, and equilibration was allowed at pH = 7.0 with three washes in 2× standard saline citrate.

After fixation of bound RNA, the nylon membrane was incubated in pre-hybridization solution [0,25 M phosphate buffer, pH = 7.2, containing 5% (wt/vol.) SDS, 1 mM EDTA and 0,5% blocking reagent (from DuPont, cat: NEL-203)] for 4 hr. at 65°C in a hybridization oven. The cDNA probe for NADH3 prepared from our differential display system was denatured before labelling by boiling for 10 min and subsequent cooling on ice, and labelled with fluorescein-12-dUTP using the Renaissance Random Primer Fluorescein-12-dUTP Labelling Kit (DuPont, cat: NEL-203).

Blots were hybridized overnight at 65°C with the denatured probes (100 ng/ml of the prehybridization buffer). After hybridization, non-specifically bound material was removed by posthybridizing washes with 0.5× and 0.1× prehybridization buffer (10 min each at 65°C). The 0.5× and 0.1× prehybridization buffers were brought to 65°C before use and the second wash was performed at room temperature.

Hybridized blots were blocked with 0.5% blocking reagent in 0.1 M Tris·HCl (pH = 7.2) and 0.15 M NaCl for 1 hr at room temperature. Membranes were then incubated with antifluorescein horseradish peroxidase antibody (DuPont cat: NEL-203) at a 1 : 500 dilution in the solution given above for 1 hr under constant shaking. Membranes were washed four times for 5 min each in the solution given above.

The Nucleic Acid Chemiluminescence Reagent (DuPont, cat: NEL-201) was added to the membrane and incubated for 1 min. Excess detection reagent was removed by use of filter papers, and the membrane was exposed to autoradiography Reflection films (DuPont, cat: NEF-496) for 15 min at 4°C.

Results

Differential display revealed a series of expressional differences. The most intriguing finding was a band (Fig. 3A and insert B) obtained from clone CM9 present in AD and controls which was not detectable in DS. The sequence and length is given in Fig. 4 and the alignment allowed the assignment to human mitochondrial DNA (Arnason et al., 1996), 16,579 nt with 99.1% identity in 223 nt, representing the NADH3 gene (Fig. 4).

Dot blot analysis revealed the presence of NADH3 RNA in all 7 control samples, 7 out of 9 AD samples but in only 2 out of 7 DS samples NADH3 RNA was detectable. The chi square test showed that this finding was significant at the $P < 0.05$ level.

Discussion

As shown in the Results the NADH3 transcript was found in cerebellum of two out of seven patients with DS only, thus clearly indicating downregulation of this mitochondrial enzyme of complex I in the majority of patients with trisomy 21.

Due to the heterogeneity of trisomy 21 patients we cannot expect impairment of NADH3 in all seven patients studied (Epstein, 1992).

Our finding is contradicting the gene dosage effect hypothesis making overexpression of genes encoded on chromosome 21 responsible for DS phenotype and pathology (Epstein, 1992): de Coo and coworkers (1997) performed molecular cloning and characterization of the human mitochondrial complex I and assigned this gene to a single location on chromosome 21q22.3. According to this hypothesis NADH3 should be overexpressed in DS, but this hypothesis has been already challenged (Greber-Platzer et al., 1999).

Mitochondrial enzyme deficiencies have been already reported in DS (Prince et al., 1994): significant reductions of mitochondrial monoamine oxidase, cytochrome oxidase and isocitrate dehydrogenase were shown to be reeduced in platelets; so far no data on DS brain were available.

Our data, however preliminary they are, point to an involvement of a mitochondrial key enzyme. The probability that reduced NADH3 RNA was reduced due to neuronal cell loss in patients with DS is highly unlikely as we have been applying identical RNA amounts and, furthermore, normalization versus the housekeeping gene beta-actin (data not shown) confirmed our results.

Moreover, studying the role of mitochondrial enzymes in DS becomes even more attractive as other mitochondrial enzymes have been localized on chromosome 21 (Chen et al., 1995). Impaired mitochondrial redox systems

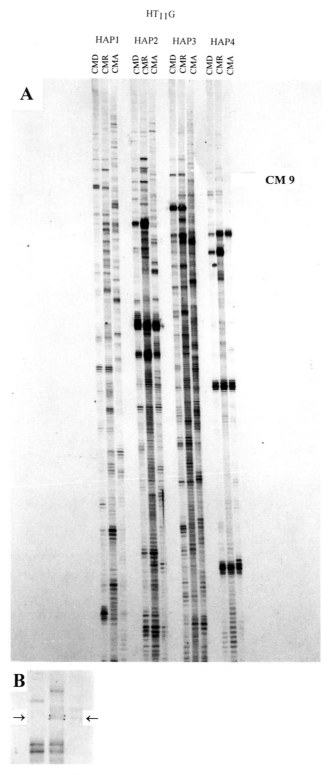

Fig. 3. Presentation of the differential display pattern obtained in cerebellum with different primer pairs revealing clear separation of cDNAs Triplets of lanes correspond to each primer (I–IV). The first lane in the individual triplets shows DS, the second controls, the third AD samples. The lane at the very left represents the control treated in the same manner as samples, in which no band was found. As shown in the insert, the band of NADH3 (magnification showing the pattern resulting from primer HAP4) was absent in DS patients, but present in controls and AD patients

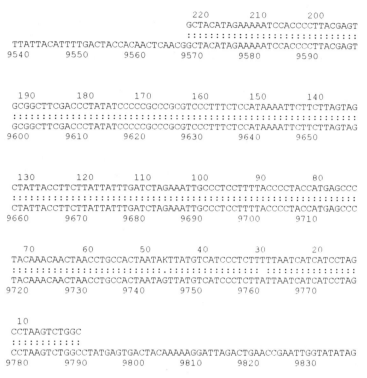

Fig. 4. The alignment of the sequence of clone CM9 with the sequence obtained from EMBL showed high identity of 99.1% in 223 nt overlap

would also be compatible with observations on increased oxidative stress in DS (Busciglio et al., 1995).

Mitochondrial enzymes have been shown to be deranged in patients with a wide range of neurodegenerative disorders including Parkinson's disease, Amyotrophic lateral sclerosis and AD (Cassarino and Bennett Jr, 1999).

It has been hypothesized that the etiology of sporadic AD may involve mutations in the mitochondrial genome in the genes encoding complex IV of the electron transport chain (Parker et al., 1994; Swerdlow et al., 1997). Inhibition of electron transport, in turn, can generate damaging active oxygen species and their presence in AD has been suggested by several groups (see reviews: Benzi et al., 1995; Markesberry et al., 1997), however, not unequivocal (Hayn et al., 1996; Seidl et al., 1997, 1998). There is some biological evidence that inhibition of the complex IV inhibitor sodium azide generates active oxygen species directly linking deficiency of mitochondrial enzyme with oxidative stress (Smith et al., 1994). Additional evidence for a tentative role of deficient complex IV in AD is provided by the observation that inhibition of complex IV resulted in an increase of amyloid precursor protein (Gabuzda et al., 1994). Data from this study do not indicate the involvement of complex I — NADH3 in the pathogenesis of AD, although it is intriguing that 2 patients with AD did not show any visible dots of NADH3 RNA in contrast to controls which were positive in all cases. Also in AD heterogeneity has to be taken into account when results are interpreted.

In conclusion, we detected NADH3 deficiency at the transcriptional level in cerebellum of patients with DS, a finding that maybe relevant for the development of brain damage extending knowledge on mitochondrial pathobiochemistry. We are currently studying activity levels of mitochondrial enzymes of the electron transport chain in the brain of patients with DS and AD, which is hampered by the fact that the mitochondrial systems are disrupted in postmortem brain, thus requiring cumbersome adaptation of technology.

Acknowledgement

We (G. L.) are highly indebted to the Red Bull Company, Salzburg, Austria, for generous support of the study.

References

Arnason U, Xu X, Gullberg A (1996) Comparison between the complete DNA sequences of homo and the common chimpanzee based on non-chimeric sequences. J Mol Evol 42: 145–152

Benzi G, Moretti A (1995) Are reactive oxygen species involved in Alzheimer's disease? Neurobiol Aging 16: 661–674

Burger PC, Vogel FS (1973) The development of pathologic changes of Alzheimer's disease and senile dementia in patients with Down's syndrome. Am J Pathol 73: 457–476

Busciglio J, Yankner BA (1995) Apoptosis and increased generation of reactive oxygen species in Down's syndrome neurons in vitro. Nature 378: 776–779

Casserino DS, Bennett JP (1999) An evaluation of the role of mitochondria in neurodegenerative diseases: mitochondrial mutations and oxidative pathology, protective nuclear Responses and cell death in neurodegeneration. Brain Res Rev 29: 1–25

Chen H, Morris MA, Rossier C, Blouin JL, Antonarakis SE (1995) Cloning of the cDNA for the human ATP synthase OSCP subunit (ATP50) by exon trapping and mapping to chromosome 21q22.1–q22.2. Genomics 28: 470–476

De Coo RFM, Buddiger P, Smeets HJM, van Oost BA (1997) Molecular cloning and characterization of the human mitochondrial NADH:oxidoreductase 10kDa gene (NDUFV3). Genomics 45: 434–437

Epstein CJ (1992) Down Syndrome (Trisomy 21) In: Scriver CR, Beaudet AL, Sly WS, Valle D (eds) The metabolic basis of inherited disease. McGraw Hill, New York, pp 749–794

Gabuzda D, Busciglio J, Chen LB, Matsudiara P, Yankner B (1994) Inhibition of energy metabolism alters the processing of the amyloid precursor protein and indeuces a potentialoly amyloidogenic derivative. J Biol Chem 269: 13623–13628

Greber-Platzer S, Schatzmann-Turhani D, Wollenek G, Lubec G (1999) Evidence against the current hypothesis of "gene dosage effects" of trisomy 21: ets2, encoded on chromosome 21 is not overexpressed in hearts of patienjts with Down Syndrome. Biochem Biophys Res Commun 254: 395–399

Hardmeier R, Hoeger H, Khoshsorur A, Lubec G (1997) Transcription and activity of superoxide dismutase, catalase and glutathione peroxidase following irradiation in radiation resistant and radiation sensitive mice. Proc Natl Acad Sci USA 94: 7572–7576

Hayn M, Kremser K, Singewald N, Nemethova M, Lubec G (1996) Evidence against the involvement of active oxygen species in the pathogenesis of Down Syndrome and Alzheimer Disease. Life Sci 59: 537–544

Kitzmueller E, Labudova O, Cairns N, Lubec G (1999) Differences in gene expression in fetal Down Syndrome brain. J Neural Transm (this volume)

Labudova O, Krapfenbauer K, Moenkemann H, Rink H, Kitzmüller E, Cairns N, Lubec G (1998) Decreased transcription factor junD in brains of patients with Down Syndrome. Neurosci Lett 252: 159–162

Labudova O, Kitzmueller E, Rink H, Cairns N, Lubec G (1999) Gene expression in fetal Down Syndrome brain as revealed by subtractive hybridization. J Neural Transm (this volume)

Labudova O, Kitzmüller E, Rink H, Cairns N, Lubec G (1999) Increased phosphoglycerate kinase in brains of patients with Down's syndrome but not with Alzheimer's disease. Clin Sci 96: 279–285

Markesberry WR (1997) Oxidative stress hypothesis in Alzheimer's disease. Free Radic Biol Med 23: 134–147

Mirra SS, Heyman A, McKeel D, Sumi S, Crain BJ (1991) The consortium to establish a registry for Alzheimer's disease (CERAD). II. Standardization of the neuropathological assessment of Alzheimer's disease. Neurology 41: 479–486

Parker WD, Parks JK, Filley CM, Kleinschmidt-DeMasters (1994) Electron transport defects in Alzheimer's disease braqin. Neurology 44: 1090–1096

Prince J, Jia S, Bave U, Anneren G, Oreland L (1994) Mitochondrial enzyme deficiencies in Down's syndrome. J Neural Transm [PD-Sect] 8: 171–181

Seidl R, Schuller E, Cairns N, Lubec G (1997) Evidence against increased glycoxidation in Alzheimer's disease. Neurosci Lett 232: 49–52

Seidl R, Greber S, Schuller E, Bernert G, Cairns N, Lubec G (1997) Evidence against increased oxidative DANN-damage in Down Syndrome. Neurosci Lett 235: 137–140

Smith TS, Bennett Jr JP (1997) Mitochondrial toxins in neurodegenerative diseases. I. In vivo brain hydroxyl radical production during systemic MPTP treatment or following microdialysis infusion of methylpyridinium or azide ions. Brain Res 765: 183–186

Swerdlow RH, Parks JK, Cassarino DS, Maguire DJ, Maguire RS, Bennett Jr JP, Davis RE, Parker Jr WD (1997) Cybrids in Alzheimer's disease: a cellular model of the disease? Neurology 49: 918–925

Tierney MC, Fisher RH, Lewis AJ, Torzitto ML, Snow WG, Reid DW, Nieuwstraten P, Van Rooijen LAA, Derks HJMG, Van Wijk R, Bischop A (1988) The NINCDA-ADRDA work group criteria for the clinical diagnosis of probable Alzheimer's disease. Neurology 38: 359–364

Wisniewski KE, Wisniewski HM, Wen GY (1985) Occurrence of neuropathological changes and dementia of Alzheimer's disease in Down's syndrome. Ann Neurol 17: 278–282

Authors' address: Prof. Dr. G. Lubec, Department of Pediatrics, University of Vienna, Währinger Gürtel 18, A-1090 Vienna, Austria, e-mail: gert.lubec@akh-wien.ac.at

J Neural Transm (1999) [Suppl] 57: 221–232

Serotonin (5-HT) in brains of adult patients with Down Syndrome

R. Seidl[1], S. T. Kaehler[3], H. Prast[3], N. Singewald[3], N. Cairns[2], M. Gratzer[1], and G. Lubec[1]

[1] Department of Pediatrics, University of Vienna, Austria
[2] Brain Bank, Institute of Psychiatry, University of London, London, United Kingdom
[3] Department of Pharmacology and Toxicology, University of Innsbruck, Austria

Summary. Down syndrome (DS) is a genetic disease with developmental brain abnormalities resulting in early mental retardation and precocious, age dependent Alzheimer-type neurodegeneration. Furthermore, non-cognitive symptoms may be a cardinal feature of functional decline in adults with DS. As the serotonergic system plays a well known role in integrating emotion, cognition and motor function, serotonin (5-HT) and its main metabolite, 5-hydroxyindol-3-acetic acid (5-HIAA) were investigated in post-mortem tissue samples from temporal cortex, thalamus, caudate nucleus, occipital cortex and cerebellum of adult patients with DS, Alzheimer's disease (AD) and controls by use of high performance liquid chromatography (HPLC). In DS, 5-HT was found to be age-dependent significantly decreased in caudate nucleus by 60% (DS: mean ± SD 58.6 ± 28.2 vs. Co: 151.7 ± 58.4 pmol/g wet tissue weight) and in temporal cortex by about 40% (196.8 ± 108.5 vs. 352.5 ± 183.0 pmol/g), insignificantly reduced in the thalamus, comparable to controls in cerebellum, whereas occipital cortex showed increased levels (204.5 ± 138.0 vs. 82.1 ± 39.1 pmol/g). In all regions of DS samples, alterations of 5-HT were paralleled by levels of 5-HIAA, reaching significance compared to controls in thalamus and caudate nucleus. In AD, 5-HT was insignificantly reduced in temporal cortex and thalamus, unchanged in cerebellum, but significantly elevated in caudate nucleus (414.3 ± 273.7 vs. 151.7 ± 58.4 pmol/g) and occipital cortex (146.5 ± 76.1 vs. 82.1 ± 39.1 pmol/g). The results of this study confirm and extend putatively specific 5-HT dysfunction in basal ganglia (caudate nucleus) of adult DS, which is not present in AD. These findings may be relevant to the pathogenesis and treatment of cognitive and non-cognitive (behavioral) features in DS.

Introduction

Down syndrome (DS) is a genetic disease with developmental brain abnormalities (Wisniewski and Kida, 1994) resulting in early mental retardation and precocious, age-dependent Alzheimer-type neurodegeneration (Nadel and

Epstein 1992; Wisniewski et al., 1985). Virtually all DS individuals manifest progressive accumulation of beta-amyloid plaques and neurofibrillary tangles beyond 25 years of age, clinical dementia however occurs only in one third beyond 40 years of age (Wisniewski et al., 1985). Although beta-amyloid plaques and neurofibrillary tangles are reliable semiquantitative markers for the presence of AD pathology, there is no compelling evidence that they, by themselves, cause dementia (Mann et al., 1990). Nevertheless, irrespective of whether dementia occurs, a progressive loss of intellectual functions, as diminished long-term memory and impaired visuospatial construction, appears in many older individuals with DS (Haxby, 1989). In addition, non-cognitive, behavioral symptoms are a cardinal feature of functional decline in adults with DS (Geldmacher et al., 1997) and some patients with DS who are thought to be demented are suffering from major depression (Warren et al., 1989).

Besides consistent findings of cholinergic deficits, neurochemical studies in DS revealed other neurotransmitter systems affected similar to those in AD including monoaminergic and serotonergic system (see Epstein, 1995; Godridge et al., 1987; Yates, 1980, 1983, 1986; Mann and Yates, 1986). Of all neurotransmitter networks in the mammalian brain, the serotonin (5-HT) system is the phylogenetically oldest and most expansive (Jacobs and Fornal, 1995). Serotonergic neurons originate in the dorsal and median raphe nuclei of the brain stem and project nerve terminals to virtually every regions of the brain with primary targets including substantia nigra, hypothalamus, thalamus, amygdaloid-hippocampal area, caudate, putamen and nucleus accumbens, and cortical areas including the frontal, occipital, insular, parietal, temporal and cerebellar cortices (Azmitia and Whitaker-Azmitia, 1991). 5-HT modulates a wide variety of physiological functions by activating an ever-expanding list of families of 5-HT receptors. In the human brain serotonergic neurons play known roles in integrating emotion, cognition and learning, motor function (sensorimotor reactivity), and pain as well as circadian and neuroendocrine functions including food intake, sleep and sexual activity. The uneven neuronal localization of the receptor types and subtypes may explain why serotonin has been implicated in so many clinical conditions: affective disorder (depression), schizophrenia, anxiety disorders, anorexia nervosa, sleep-wake disturbances, obsessive-compulsive disorders, eating disorders and impulse-related disorders or personality features including aggression, substance abuse, gambling, obsessive control, and attention-deficit disorders (Casper, 1998; Lesch and Mössner, 1998; Lucki, 1998). The hypothesis of aberrant 5-HT neurotransmission in the pathogenesis of depression is supported by a multitude of studies that have demonstrated blunted 5-HT function (Casper, 1998). Findings included basal CSF 5-hydroxyindol-3-acetic acid (5-HIAA) levels, plasma tryptophan levels, blood platelet 5-HT function, hormonal response to 5-HT specific challenge tests and recently, decreased 5-HT transporters in midbrain regions as measured by SPECT and 5-HT2 receptor increase in response to antidepressant treatment (as evaluated by PET) (see review Staley et al., 1998). Finally, it has been suggested that 5-HT receptors may also be targeted to treat neurodegenerative disorders (Bowen et al., 1994; Lesch and Mössner, 1998).

In adult DS, significant reductions of 5-HT, partly paralleled by decreased levels of 5-HIAA, have been reported in most telencephalic brain regions (Reynolds and Godridge, 1985; Yates et al., 1986, Godridge et al., 1987; Risser et al., 1997) and neuronal loss in 5-HT projection side, midbrain dorsal raphe nuclei has also been identified (Mann et al., 1985).

However, the "serotonin story" in DS goes back to the mid-sixties. Following first report of blood-serotonin deficiency in DS (Tu and Zellweger, 1965) several studies appeared, showing reduced 5-HT levels in blood, cerebrospinal fluid and platelets (for review see Epstein, 1995). In particular, decreased 5-HT platelet uptake was suggested to serve as a model for neurotransmitter uptake and storage in CNS synaptosomes (see Epstein, 1995). Subsequently, reversal of hypotonia by administration of 5-hydroxytryptophan (Bazelon et al., 1967) and reduction of self-injurious behavior by dietary increase in serotonin (Gedye, 1990, 1991) have been reported. In other trials, 5-hydroxytryptophan showed either no beneficial effect on intelligence development (Weise et al., 1974) or lead to an increased risk for infantile spasms (Coleman, 1971).

The aim of this study was to delineate and/or confirm relevant differences of 5-HT function between AD and aged DS patients with AD-like pathology. In order to differentiate whether changes observed are likely to be secondary to the prominent and characteristic cortical pathology, subcortical, primary 5-HT target brain regions were included.

We report the results of determination of 5-HT and its main metabolite 5-HIAA in brain homogenates of five different brain areas in DS subjects with AD-like neuropathologic lesions compared to AD and control subjects.

Materials and methods

The levels of 5-HT and 5-HIAA were quantified by HPLC in 5 different brain regions of 9 adult DS individuals with AD-like neuropathologic lesions (3 females, 6 males; mean age of 55.8 ± 7.5 years), 10 patients with Alzheimer disease (2 females, 8 males; mean age 59.0 ± 8.0 years) and 12 controls (3 females, 7 males; mean age 47.3 ± 19.2 years). Postmortem brain samples (temporal lobe — superior temporal gyrus, thalamus, caudate nucleus, occipital lobe — Brodmann area 17/18, and cerebellar cortex) were obtained from the Medical Research Council's Brain Bank for Neurodegenerative Diseases, Institute of Psychiatry, London, UK. Corresponding brain regions of the contralateral hemisphere were used for histopathological analysis. As there was evidence of abundant β-amyloid plaques and neurofibrillary tangles in all DS brains, the patients included were considered to represent a homogeneous population of adult DS patients with Alzheimer-like neuropathological lesions. As these patients had not been systematically assessed with psychometric tests, it is impossible to state whether they represent an homogenous population also from psychological point of view. AD patients fulfilled the National Institute of Neurological and Communicative Disorders and Stroke and the Alzheimer Disease and Related Disorders Association (NINCDS/ADRDA) criteria for probable AD (Tierney et al., 1988). The histological diagnosis of AD was established consistent with the CERAD criteria (Mirra et al., 1991) for a "definite" diagnosis of AD. The controls were brains from individuals with no history of neurological or psychiatric illness.

The major cause of death was bronchopneumonia in DS and AD patients and heart disease in the controls. Post-mortem interval of brain dissection in DS, AD and controls was 28.8 ± 18.6, 24.9 ± 19.8 and 33.5 ± 14.7 hours, respectively.

Determination of serotonin (5-HT) and 5-hydroxyindolyl-acetic acid (5-HIAA) was performed as published previously (Singewald et al., 1997). Tissue was taken into 0.1 N HCl, weighed, homogenized for 30 sec by the use of a Branson Sonifier 250, a duty cycle of 75% and a micro tip limit 3.5. Samples were thawed prior to use at 4°C and spun down in a vacuum centrifuge 20 min at 193,000 g. The supernatant was used for HPLC. The HPLC system used consisted of a Jasco 880 PU pump (Jasco, Tokyo, Japan) operating at a flow rate of 1.5 ml/min, a CMA 260 degaser (CMA, Stockholm, Sweden), and a electro-chemical detector (LC-4B, BAS, West Lafayette, USA) set at +600 mV. Using a flow splitter, the flow rate through the analytical column (StepStik microbore column, 150 × 1 mm, 5 micrometer C18, BAS) was 70 microliters per min. The analytical column was protected by a guard column (SepStik, 14 × 1 mm, 5 micrometer C8, BAS). The SepStik column was directly coupled to a BAS Unijet 3 mm glassy carbon electrode MF-1003. Samples of 50 µl were automatically injected by a CMA 200 Refrigerated Microsampler. The injected port and guard column were interconnected by a short piece of peek tubing with an internal diameter of 0.005 inches. The mobile phase consisted of 88% phosphate buffer (0.1 M NaH2PO4, 1 mM sodium octanesulphonic acid, 10 mM NaCl and 0.5 mM Na2-EDTA; pH was adjusted to 3.5 with o-phosphoric acid), 6% acetonitrile and 6% methanol. Evaluation of 5-HT and 5-HIAA was carried out by comparing peak heights of samples with external standard solutions containing various concentrations of 5-HT and 5-HIAA by using an integrator (SIC Chromatocoder 12, System Instruments, Tokyo, Japan). Retention time of serotonin was 8.5 min and the minimum detection limit was 0.3 pg per sample at a signal: noise ratio of 3.

Statistics

Results were expressed as means ± standard deviation (SD). Between-group differences were investigated by non-parametric Mann-Whitney U test. Within-group correlations were done with non-parametric Spearman rank-correlation coefficient procedure. The level of significance was set at P < 0.05. All analyses were run with the SPSS-PC V 8.0 statistical software package.

Results

The mean transmitter concentrations of DS, AD and control brain tissue are given in Table 1. The thalamus showed highest 5-HT and 5-HIAA concentrations in all three groups. In DS, 5-HT was found to be significantly decreased in caudate nucleus by 60% and in temporal cortex by 40%, insignificantly reduced in the thalamus, comparable to controls in cerebellum, whereas occipital cortex showed increased levels. In all regions of DS samples, alterations of 5-HT were paralleled by levels of 5-HIAA, reaching significance compared to controls in thalamus and caudate nucleus. In AD, 5-HT was insignificantly reduced in temporal cortex and thalamus, 5-HIAA in both regions were unchanged. Caudate nucleus and occipital cortex in AD exhibited significantly elevated 5-HT levels, also not paralleled by 5-HIAA. Cerebellar levels of 5-HT and 5-HIAA were comparable to controls. Compared to AD, DS 5-HT and 5-HIAA of caudate nucleus were significantly decreased

225

Table 1. Concentration of 5-HT and 5-HIAA (HPLC) in Down syndrome, Alzheimer's disease and Control brain regions (in pmol/g wet tissue weight; data are shown as mean ± SD)

	Down syndrome		Alzheimer's disease		Control	
	5-HT	5-HIAA	5-HT	5-HIAA	5-HT	5-HIAA
Temporal cortex	196.8 ± 108.8* (n = 7)	296.9 ± 152.7 (n = 7)	218.5 ± 116.8 (n = 9)	519.0 ± 230.2 (n = 9)	352.5 ± 183 (n = 11)	532.1 ± 273.4 (n = 10)
Thalamus	550.8 ± 508.1 (n = 6)	521.4 ± 328.9* (n = 6)	660.5 ± 284.8 (n = 8)	963.6 ± 447.9 (n = 8)	734.9 ± 405.6 (n = 9)	998.0 ± 300.2 (n = 9)
Caudate nucleus	58.6 ± 28.2* (n = 6)	275.0 ± 128.7* (n = 6)	414.3 ± 273.7* (n = 5)	601.1 ± 85.6 (n = 5)	151.7 ± 58.45 (n = 5)	1,044.9 ± 452.2 (n = 5)
Occipital cortex	204.5 ± 138* (n = 9)	497.1 ± 342.9 (n = 7)	146.5 ± 76.1* (n = 10)	289.4 ± 149.4 (n = 10)	82.1 ± 39.1 (n = 10)	304.1 ± 278.8 (n = 10)
Cerebellum	178.4 ± 110.5 (n = 5)	426.5 ± 496.5 (n = 5)	204.2 ± 145.7 (n = 7)	751.0 ± 556.6 (n = 7)	173.1 ± 35.6 (n = 5)	516.5 ± 251.1 (n = 5)

Statistics: Mann-Whitney-U test: *: P < 0.05, Down syndrome, Alzheimer disease vs. Controls

(P = 0.014, 5-HT and 5-HIAA). Age, post-mortem interval, sex and agonal state revealed no influence on 5-HT and 5-HIAA in AD and controls, whereas in DS 5-HT concentration in caudate nucleus of DS exhibited a significantly negative correlation with age (Spearman's rho — 0.9, P = 0.037).

Discussion

The results reveal major differences of alterations and regional distribution of these alterations of 5-HT levels and its main metabolite 5-HIAA between DS and AD compared to control samples. Findings in caudate nucleus seem to be most interesting. In DS, 5-HT and 5-HIAA levels in caudate nucleus exhibited an age-dependent decrease, whereas in AD 5-HT was significantly elevated.

In previous studies, Reynolds and Godridge (1985) reported losses of 5-HT and 5-HIAA in temporal, frontal cortex and hippocampus of postmortem tissue taken from 5 Down syndrome patients, aged 34–64 years of age. Extended measurements in the same patients (Godridge et al., 1987) revealed significant reductions of 5-HT in most cortical and subcortical regions, reaching highest significance in temporal cortex and caudate nucleus, closely resembling the results of the present study. In a study by Yates et al. (1986), 5-HT was markedly reduced in the amygdala, cingulate cortex and caudate nucleus of DS brains with neuropathological features of Alzheimer's disease, unaltered in one DS brain with no Alzheimer pathology and reduced only in the amygdala and cingulate cortex but not in caudate nucleus of the AD brains. Forebrain 5-HT is derived nearly entirely from neurons located in the dorsal and median raphe nuclei of the midbrain. Prominent forebrain terminal regions include the hypothalamus, cortex, hippocampus, amygdala and striatum. Furthermore, 5-HT neurons are highly bifurcated, indicating that they are structured ideally for influencing the function of several regions of the central nervous system simultaneously. However, ascending projections from the raphe nuclei are organized in a topographical manner. The median raphe projects heavily to hippocampus and hypothalamus, whereas the striatum is innervated predominantly by the dorsal raphe. Interestingly, dorsal raphe axons appear to be more vulnerable to neurotoxic amphetamine derivatives, while median raphe axons appear to be more resistant. The raphe nuclei in turn receive input from dopaminergic (substantia nigra, ventral tegmental area), cholinergic (superior vestibular nucleus) and noradrenergic (locus coeruleus) neurons. Other afferents include neurons from the hypothalamus, thalamus and limbic forebrain structures (Jacobs and Azmitia, 1992; Lucki, 1998; Frazer and Hensler, 1999).

The questions are whether consistently found alterations of 5-HT in DS caudate nucleus reflect selective cell loss of dorsal raphe nuclei neurons and/ or indicate a probably disease-related serotonergic striatal dysfunction, or whether they are likely to be secondary to prominent forebrain, especially cortical Alzheimer-like neurodegeneration?

Although losses are more pronounced in older DS cases, a direct relation-ship between 5-HT loss and plaques and tangles seems unlikely since areas such as caudate nucleus in DS contain few or none of these features (Yates et al., 1986; Godridge et al., 1987). In line with this view, the youngest case (age 34 years) in the study by Godridge et al. (1987), in which plaques and tangles were absent, showed substantial 5-HT and noradrenaline deficits and normal choline acetyltransferase. In contrast, postmortem 5-HT of a 27 years old DS subject was comparable to control values in caudate nucleus (Yates et al., 1986). Neuropathologically, number of nerve cells and the volume of nucleo-lus in the dorsal tegmental nucleus in one young DS patient showed no substantial alterations, whereas six middle-aged patients with cortical and hippocampal AD-pathology showed significant loss of cells from nucleus basalis, locus coeruleus, dorsal motor vagus and dorsal tegmental nucleus (Mann et al., 1985). In the present study, comprising only DS patients with abundant AD-pathology, 5-HT in caudate nucleus significantly decreased with age. Therefore, although both, serotonergic abnormalities and accumula-tion of plaques and tangles increase with advancing age in DS, a causative relationship is not yet proven. This view is further supported by increased 5-HT levels in caudate nucleus of AD brains in the present study. No clinical, biochemical and/or neuropathological data exist supporting a serotonergic dysfunction in infant and childhood DS brain. However knowledge would be of particular interest, as the early expression of the 5-HT in developing midbrain raphe neurons and their projecting terminals prior to synapto-genesis indicates that it is an important regulator of morphogenetic activities during early embryonic development, including cell proliferation, migration and differentiation (Lauder, 1993). Reductions of 5-HT in caudate nucleus, thalamus and temporal cortex of DS brains were paralleled by decreased levels of 5-HIAA, elevated 5-HT in occipital cortex paralleled by increased 5-HIAA. The concentrations of 5-HIAA in terminal serotonergic regions are believed to derive from 5-HT released and metabolized outside the sero-tonergic neuron (Celeda and Artigas, 1993). Hence, increased 5-HIAA con-tent and an increased 5-HIAA/5-HT ratio have been employed as index of increased functional activity of serotonergic neurons and increased 5-HT turnover. An increase in 5-HT turnover might indicate a possible functional compensation for serotoninergic fiber loss or a possible decrease in 5-HT binding (Rodriguez-Gomez et al., 1995). In normal humans, a stability of 5-HT terminals with increasing age has been found (Andersson et al., 1992). Results of the present study suggest that the age dependent 5-HT decrease in DS caudate nucleus is not compensated by an increase in 5-HT turnover. Is there further evidence for abnormalities of caudate nucleus with respect to other neurotransmitter systems? Similar to serotonergic system, caudate nucleus exhibited unchanged noradrenaline, dopamine, HVA and choline acetyltransferase in one 27 year old DS subject, whereas in 4 DS subjects with AD-pathology dopamine and choline acetyltransferase but not noradrenaline showed significant deficits (Yates et al., 1983). In the study by Godridge et al. (1987) 5-HT deficits in caudate and putamen contrasted to unaltered dopam-ine and HVA levels. The choline acetyltransferase activity was not determined in this study, but was found to be reduced in caudate nucleus of 3 middle-aged DS subjects in the same order of magnitude as in amygdala and parahippocampal gyrus (Yates et al., 1980). Another issue to delineate whether a DS-specific 5-HT dysfunction of caudate nucleus can be assumed or not might be to review dopaminergic findings. Comparing senile AD, presenile AD and adult DS subjects, dopamine was found to be decreased in caudate nucleus of DS, whereas it was unaltered in senile and presenile AD (Yates et al., 1983). Furthermore, most workers agree that the dopaminergic system is only minimally and perhaps merely secondary affected in AD (for review see Mann and Yates, 1986; Procter et al., 1995), although some studies found decreased dopamine and HVA concentrations in the caudate nucleus (Gottfries et al., 1983; Procter et al., 1995). Interestingly, biochemical disturbances were more pronounced in patients with early onset of dementia. Thus, in late-onset AD there is less evidence that caudate nucleus pathology is a necessary part of the AD pathogenesis (Procter et al., 1995).

Concepts of basal ganglia organization suggest structurally and functionally segregated pathways that link putamen function to motor but caudate function to cognitive performance and PET studies in Parkinson disease, another neurodegenerative condition, provided evidence that memory impairment may be assigned to altered dopamine neuronal integrity in the caudate nucleus (Holthoff-Detto et al., 1997). In summary, low content of 5-HT and its metabolite 5-HIAA is suggestive of loss of serotonergic axon terminals within the caudate nucleus of adult DS subjects. As also other (dopaminergic and cholinergic) but not all neurotransmitter systems are involved, these findings might not be attributed to general neuronal loss.

Recently, treatment with selective serotonin-reuptake inhibitor (SSRI) medication in six DS patients, aged 23 to 63 years, showed improvement in behaviors and on objective measures, such as workplace productivity. These DS patients presented with functional decline in adult life, but non-cognitive symptoms were prominent and included aggression, social withdrawal, compulsive behaviors and cognitive dysfunction at varying degrees. Authors concluded, that treatment trials with SSRIs may, therefore, be warranted in such cases (Geldmacher et al., 1997). In an earlier study, low dose antidepressant treatment combined with a serotonin-enhancing diet in a mentally handicapped adult with Down syndrome showing signs of Alzheimer-type dementia decreased aggressive behavior (Gedye, 1991).

In contrast to the findings in DS, 5-HT and 5-HIAA levels in AD in the present study were slightly decreased only in the thalamus and temporal cortex (insignificantly), but elevated in caudate nucleus and occipital cortex. In AD, the situation with respect to 5-HT is complex. Although reduced 5-HT and 5-HIAA concentrations have been found in cortical regions, hippocampus and basal ganglia, this was by no means a consistent finding (for review see Mann and Yates, 1986; Procter et al., 1995; Bierer et al., 1997). In addition, cell loss in the dorsal raphe region does not seem to be as great as that within the nucleus basalis complex (review Mann and Yates, 1986; Palmer et al., 1988). Recently, Chen et al. (1996) reported unchanged 5-HT and 5-HIAA concentrations in frontal and temporal cortex, and suggest that previous studies may have been subject to unintentional selection of behaviorally disturbed AD patients or those receiving neuroleptic medication (Chen et al., 1996). Furthermore it is clear at present, that structural pathology of raphe neurons is not necessarily reflected in reduced 5-HT innervation, as increased turnover of transmitter and/or synaptic plasticity within remaining 5-HT terminals maintains this activity (Procter et al., 1995; Chen et al., 1996).

Despite these inconsistent findings in AD and lack of knowledge in early DS, disturbances in functional 5-HT transporter (5-HTT) expression are discussed to play a role in the pathophysiology of neurodegenerative disorders (Lesch and Mössner, 1998). In a population-based association study the frequency of the low-activity allele of the 5-HTTLPR (5-HTT gene-linked polymorphic region, modulating the transcriptional activity of 5-HTT promoter) was increased in patients with late-onset Alzheimer disease (Li et al., 1997). 5-HTT knock-out mice showed marked 5-HT reductions in brain stem, frontal cortex, hippocampus and striatum, altered somatosensory cortex formation, premature cognitive deficits and alterations of myelinated fiber tracts. As neuronal cell damage and damage in myelinated fiber tracts were also observed in heterozygous knockout animals, the possibility that subtle changes in the dynamics of 5-HT transport exert acute or long-term effects on neurodevelopment, adult brain plasticity and neurodegenerative processes has been raised (Lesch and Mössner, 1998). Limitation (limits) of postmortem neurochemical studies as mentioned by Chen (1996) have precisely been summarized by Staley et al. (1998). These limits are: — inaccurate retrospective case histories, — lack of diagnosis of comorbid neuropsychiatric disorders, — insufficient knowledge of drug treatment histories, — inability to control interindividual clinical parameters that might alter brain neurochemical measurements, — and a single measurement at the end stage of the disease.

In conclusion, studies including the present one, carried out in postmortem tissue specimen, have identified a putatively specific 5-HT dysfunction in basal ganglia (caudate nucleus), which is not present in AD. These findings may be relevant for the pathogenesis of cognitive and non-cognitive (behavioral) features in DS. As little is known about the status of 5-HT synaptic markers in the DS brain, in vivo brain imaging modalities such as PET and SPECT may in the future offer the opportunity to investigate the state of the 5-HT synapse in living DS patients. In addition, studies of allelic variation of 5-HT transporter gene expression may contribute to understand the relevance of adaptive 5-HT uptake function in DS brain development, plasticity and neurodegeneration.

Acknowledgement

We (G. L.) are highly indebted to the Red Bull Company, Salzburg, Austria, for generous support of the study.

References

Andersson A, Sundman I, Marcusson J (1992) Age stability of human brain 5-HT terminals studied with 3H paroxetine binding. Gerontol 38: 127–132

Azmitia E, Whitaker-Azmitia P (1991) Awakening the sleeping giant: anatomy and plasticity of the brain serotonergic system. J Clin Psychiatry 52 [Suppl 12]: 4–16

Bazelon M, Paine RS, Coeiw VA, Hunt P, Houck JC, Mahanand D (1967) Reversal of hypotonia in infants with Down's syndrome by administration of 5-hydroxytryptophan. Lancet i: 1130–1133

Becker LE, Mito T, Takashima S, Onodera K (1991) Growth and development of the brain in Down syndrome. Prog Clin Biol Res 373: 133–152

Bierer LM, Haroutunian V, Gabriel S, Knott PJ, Carlin LS, Purohit DP, Perl DP, Schmeidler J, Kanof P, Davis KL (1995) Neurochemical correlates of dementia severity in Alzheimer's disease: relative importance of the cholinergic deficits. J Neurochem 64: 749–760

Bowen BB, Francis PT, Chessel IP, Webster MT (1994) Neurotransmission-the link integrating Alzheimer's disease? Trends Neurosci 17: 149–150

Casper RC (1998) Serotonin, a major player in regulation of feeding and affect. Biol Psychiatry 44: 795–797

Celeda P, Artigas F (1993) Effects of local and systemic MAO inhibitors on extracellular brain 5-hydroxytryptamine and 5-hydroxyindoleacetic acid in the frontal cortex and raphe nuclei of freely moving rats. An in vivo microdialysis study. Naunyn Schmiedebergs Arch Pharmacol 347: 583–590

Chen CP, Alder JT, Bowen DM, Esiri MM, McDonald B, Hope T, Jobst KA, Francis PT (1996) Presynaptic serotonergic markers in community-acquired cases of Alzheimer's disease: correlations with depression and neuroleptic medication. J Neurochem 66: 1592–1598

Coleman M (1971) Infantile spasms associated with 5-hydroxytryptophan administered in patients with Down's syndrome. Neurol 21: 911

Epstein CJ (1995) Down Syndrome (Trisomy 21). In: Scriver SR, Beaudet AL, Sly WS, Valle D (eds) The metabolic and molecular bases of inherited disease. McGraw-Hill, New York, pp 749–794

Frazer A, Hensler JG (1999) Serotonin. In: Siegel GJ, Agranoff BW, Albers RW, Fisher SK, Uhler MD (eds) Basic neurochemistry, molecular, cellular and medical aspects, 6th edn. Lippincott Raven, Philadelphia New York, pp 263–293

Gedye A (1990) Dietary increase in serotonin reduces self-injurious behaviour in a Down's syndrome adult. J Ment Defic Res 34: 195

Gedye A (1991) Serotonergic treatment for aggression in a Down's syndrome adult showing signs of Alzheimer's disease. J Ment Defic Res 35: 247–258

Geldmacher DS, Lerner AJ, Voci JM, Noelker EA, Somple LC, Whitehouse PJ (1997) Treatment of functional decline in adults with Down syndrome using seletive serotonin-reuptake inhibitor drugs. J Geriatr Psychiat Neurol 10: 99–104

Godridge H, Reynolds GP, Czudek C, Calcutt NA, Benton M (1987) Alzheimer-like neurotransmitter deficits in adult Down's syndrome brain tissue. J Neurol Neurosurg Psychiatry 50: 775–778

Gottfries CG, Adolfsson R, Aquilonius SM, Carlsson A, Eckernas SA, Nordberg A, Oreland L, Svennerholm L, Wiberg A, Winblad B (1983) Biochemical changes in dementia disorders of Alzheimer type (AD/SDAT). Neurobiol Aging 4: 261–271

Haxby JV (1989) Neuropsychological evaluation of adults with Down's syndrome: patterns of selective impairment in nondemented old adults. J Ment Defic Res 33: 193–197

Holthoff-Detto K, Kessler J, Herholz K, Bonner H, Pietrzyk U, Wurker M, Ghaemi M, Wienhard K, Wagner R, Heiss WD (1997) Functional effects of striatal dysfunction in Parkinson disease. Arch Neurol 54: 145–150

Jacobs BL, Azmitia EC (1992) Structure and function of the brain serotonin system. Physiol Rev 72: 165–229

Jacobs B, Fornal C (1995) Serotonin and behavior, a general hypothesis. In: Bloom F, Kupfer D (eds) Psychopharmacology: the fourth generation of progess. Raven Press, New York, pp 461–469

Lauder JM (1993) Neurotransmitters as growth regulatory signals: role of receptors and second messengers. Trends Neurosci 16: 233–240

Lesch KP, Mössner R (1998) Genetically driven variation in serotonin uptake: is there a link to affective spectrum, neurodevelopmental and neurodgenerative disorders? Biol Psychiatry 44: 179–192

Li T, Holmes C, Sham PC, Vallada H, Birkett J, Kirov G (1997) Allelic functional variation of serotonin transporter expression is a susceptibility factor for late-onset Alzheimer's disease. Neuroreport 8: 683–686

Lucki I (1998) The spectrum of behaviors influenced by serotonin. Biol Psychiatry 44: 151–162

Mann DMA, Yates PO (1986) Neurotransmitter deficits in Alzheimer's disease and in other dementing disorders. Hum Neurobiol 5: 147–158

Mann DMA, Yates PO, Marcyniuk B, Ravindra CR (1985) Pathological evidence for neurotransmitter deficits in Down's syndrome of middle age. J Ment Defic Res 29: 125–135

Mann DMA, Royston MC, Ravindra CR (1990) Some morphological observations on the brains of patients with Down's syndrome: their relationship to age and dementia. J Neurol Sci 99: 153

Mirra SS, Heyman A, McKeel D, Sumi SM, Crain BJ (1991) The consortium to establish a registry for Alzheimer's disease (CERAD). II. Standardisation of the neuropathological assessment of Alzheimer's Disease. Neurol 41: 479–486

Nadel L, Epstein CJ (eds) (1992) Down Syndrome and Alzheimer disease. Wiley-Liss, New York (Prog Clin Biol Res 379)

Palmer AM, Stratmann GC, Procter AW, Bowen DM (1988) Possible neurotransmitter basis of behavioral changes in Alzheimer's disease. Ann Neurol 23: 616–620

Procter AW, Francis PT, Chen CPLH, Chessel IP, Dijk S, Clarke NA, Webster MT, Bowen DM (1995) The neurochemical pathology of Alzheimer's disease. In: Allen SJ, Dawbarn D (eds) Neurobiolgy of Alzheimer disease. BIOS Scientific Publ, Oxford, pp 193–221

Reynolds GP, Godridge H (1985) Alzheimer-like brain monoamine deficits in adults with Down's syndrome. Lancet ii: 1368–1369

Risser D, Lubec G, Cairns N, Herrera-Marschitz M (1997) Excitatory amino acids and monoamines in parahippocampal gyrus and frontal cortical pole of adults with Down syndrome. Life Sci 60: 1231–1237

Rodriguez-Gomez JA, de la Roza C, Machado A, Cano J (1995) The effect of age on the monoamines of the hypothalamus. Mech Ageing Dev 77: 185–195

Singewald N, Kaehler S, Hemeida R, Philippu A (1997) Release of serotonin in the rat locus coeruleus: effects of cardiovascular, stressful and noxious stimuli. Eur J Neurosci 9: 556–562

Staley JK, Malison RT, Innis RB (1998) Imaging of the serotonergic system: interactions of neuroanatomical and functional abnormalities of depression. Biol Psychiatry 44: 534–549

Tierney MC, Fisher RH, Lewis AJ, Torzitto ML, Snow WG, Reid DW, Nieuwstraten P, Van Rooijen LAA, Derks HJGM, Van Wijk R, Bischop A (1998) The NINCDA-ADRDA work group criteria for the clinical diagnosis of probable Alzheimer's disease. Neurol 38: 359–364

Tu JB, Zellweger H (1965) Blood-serotonin deficiency in Down's syndrome. Lancet ii(7415): 715–716

Warren AC, Holroyd S, Folstein P (1989) Major depression in Down's syndrome. Br J Psychiatry 155: 202–207

Weise P, Koch R, Shaw KNF, Rosenfeld MJ (1974) The use of 5-HTP in the treatment of Down's syndrome. Pediatr 54: 165–167

Wisniewski KE, Kida E (1994) Abnormal neurogenesis and synaptogenesis in Down syndrome brain. Dev Brain Dysfunct 7: 289–301

Wisniewski KE, Wisniewski HM, Wen GY (1985) Occurrence of neuropathological changes and dementia of Alzheimer's disease in Down syndrome. Ann Neurol 17: 278–282

Yates CM, Simpson J, Maloney AFJ, Gordon A, Reid AH (1980) Alzheimer-like cholinergic deficiency in Down syndrome. Lancet Nov 1st: 979

Yates CM, Simpson J, Gordon A, Maloney AFJ, Allison Y, Ritchie IM, Urquhart A (1983) Catecholamines and cholinergic enzymes in pre-senile and senile Alzheimer-type dementia and Down's syndrome. Brain Res 280: 119–126

Yates CM, Simpson J, Gordon A (1986) Regional brain 5-hydroxytryptamine levels are reduced in senile Down's syndrome as in Alzheimer's disease. Neurosci Lett 65: 189–192

Authors' address: Prof. Dr. G. Lubec, Department of Pediatrics, University of Vienna, Währinger Gürtel 18-20, A-1090 Vienna, Austria, e-mail: gert.lubec@akh-wien.ac.at

J Neural Transm (1999) [Suppl] 57: 233–245

Mechanisms of neuronal death in Down's syndrome

Zs. Nagy

OPTIMA, Departments of Neuropathology and Pharmacology, University of Oxford,
Oxford, United Kingdom

Summary. There is growing evidence that neuronal death in Down's syndrome is due to apoptotic mechanisms. The phenomena, however, that trigger and regulate programmed cell death in the Down's syndrome-related neurodegeneration are still much debated.

In vitro evidence has suggested that the main factor responsible for neuronal death in this condition is the accumulation of β-amyloid, due to the overexpression of its precursor protein. Another hypothesis argues for the importance of reactive oxygen species in neuronal death. However, the in vivo findings do not entirely support either theories.

We propose that neuronal apoptosis, as well as the formation of Alzheimer-type pathology, in Down's syndrome is due to an aberrant re-entry of neurones into the cell division cycle. Due to the simultaneous overexpression of conflicting cell cycle regulatory signals, the mitogenic amyloid precursor and the differentiation factor S100, the cell cycle is abandoned. Subsequently the cell cycle arrest may lead to either the formation of Alzheimer-related pathology or to apoptotic cell death.

Introduction

Down's syndrome (DS) is the most common genetic cause of mental retardation. Most patients (95%) have trisomy 21, while in the remaining patients the syndrome is due to chromosomal translocation.

The brain pathology most commonly associated with Downs syndrome consists of decreased cell numbers, altered neuronal differentiation (Mann, 1997), altered synaptic connectivity (Mann, 1997; Prinz et al., 1997; Takashima et al., 1994) and an early, inevitable, development of amyloid plaques (AP) and neurofibrillary tangles (NFT) similar to those seen in sporadic Alzheimer's disease (SAD). It is generally accepted that the regional distribution and evolution of cell loss and AD-type changes are similar to that seen in sporadic AD (Braak et al., 1991; Hyman, 1992). Although there are differences between the two diseases, indicating that their evolution is not entirely parallel in all respects, the comparable development of NFT pathology suggests that there is a common pathway that leads to the worsening functional deficit and cell death in both sporadic and DS-related AD.

Alzheimer-type pathology in Down's syndrome

The main constituent of AD-type plaques is the β-amyloid protein, formed through an alternative cleavage of the amyloid precursor protein (APP). This β-amyloid is believed to be partly produced by neurones and deposited as a diffuse extracellular mass (Armstrong, 1996). The transformation of diffuse amyloid into primitive plaques is aided by glia cells (Armstrong, 1996).

The paired helical filaments (PHF) of NFTs consist mainly of the hyperphosphorylated form of the microtubule associated protein tau. The antigenic properties of PHFs and the phosphorylation pattern of tau were found to be identical in DS-related and sporadic AD (Mann, 1997).

In DS, as in SAD, the primary focus of plaque and tangle formation is the amygdala, layer II of the entorhinal cortex, the subiculum and CA1 regions of the hippocampus proper (Hof et al., 1995; Hyman, 1992; Hyman et al., 1995; Mann et al., 1986b). As the disease evolves the pathology spreads out to the neocortical association areas and to the olfactory bulb and tract (Hof et al., 1995; Hyman, 1992; Hyman et al., 1995; Mann et al., 1986b). However, quantitative analyses show that NFTs are more numerous in the hippocampi of DS patients than in SAD sufferers (Mann et al., 1987). While the mean density of β-amyloid deposits is greater in DS, the formation of large clusters of amyloid deposits predominate in sporadic AD (Armstrong, 1996; Hof et al., 1995; Hyman et al., 1995). This finding, consistent with the gene-dose effect, points to the fact that while the high density of diffuse amyloid deposits may be important in the development of DS, the clustering tendency of amyloid is important in SAD (Armstrong, 1996).

Cell loss in Down's syndrome

The observation of lower neuronal counts in the hippocampus proper of foetuses with trisomy 21 relative to age-matched controls led to the hypothesis that the brain atrophy seen in DS is mainly due to a congenital malformation (Sylvester, 1984). However, later studies indicate that an actual loss of nerve cells does occur as these patients develop AD-type pathology (Mann, 1997).

Mapping of cell loss in different regions of the brain indicates that, similar to plaque and tangle pathologies, the neuronal populations vulnerable to cell death in DS patients are identical to those identified in SAD, namely the pyramidal and non-pyramidal cells of the entorhinal cortex, hippocampus and association neocortical regions (Mann, 1997). There is a major loss in subcortical afferent connections based on acetylcholine, noradrenalin and serotonin (Mann et al., 1986a). In the cortex somatostatin containing large pyramidal neurones (using glutamate and aspartate) are also affected (Mann et al.,

1986a). The cells affected in both conditions represent an interconnected neuronal network (Braak et al., 1991; Mann et al., 1986a). However, when the two patient populations were compared, the hippocampal cell loss was found to be greater in DS patients, while neocortical cell loss was more pronounced in SAD sufferers (Mann et al., 1987).

The greater numbers of amyloid plaques, NFTs and more pronounced cell death in the hippocampus in DS-related AD, as compared to SAD, is probably due to the fact that while in DS perhaps 20–30 years elapse until the disease progresses from mild to severe, sporadic AD is only a 10–15 year process (Hyman, 1992). On the other hand, the relatively less severe pathology in the neocortex of DS patients may account for the absence of clinically detectable dementia in some elderly DS sufferers despite severe AD-related brain pathology (Mann, 1997). Alternatively, it is possible that the compensatory mechanisms, that appear in response to increased functional demands on healthy remaining neurones, are different in elderly patients suffering from AD than those activated in Down's syndrome (Nagy et al., 1999).

In summary, it is likely that multiple factors and their interactions are responsible for the neuronal death and the development of dementia in DS patients. However, the mechanisms that lead to the development of AD-related pathology and neuronal death in DS are still debated.

Apoptosis in the developing nervous system

Apoptosis, or programmed cell death, was first described as a phenomenon associated with developmental events. It is widely accepted that after the last division cycle is completed by neuronal progenitors and synaptogenesis, the final stage of neuronal differentiation, begins the redundant neurones are eliminated via an apoptotic pathway. Apoptosis therefore, contributes to the final patterning of the brain playing a major role in the final stages of its development.

The chain of morphological changes associated with apoptosis start with the condensation of the nucleus and lead to the fragmentation of the cell (Majno et al., 1995). It was initially believed that the execution of the "death programme" required de novo protein synthesis by the cells. It was, however, found recently that the effector proteins of apoptosis, the caspases, are constitutively present in every cell and that the "death programme" can be carried out by cells even if protein synthesis was previously blocked (Steller, 1995).

The main regulatory proteins that prevent or precipitate the apoptotic programme are part of the Bcl-2 protein family (Adams et al., 1998). The first identified member of the family was the Bcl-2 protein that was found to prevent apoptosis. The effects of another member of the family, the Bcl-x protein, depend on its alternative splicing. The long form, Bcl-x_L, similar to Bcl-2, is able to prevent apoptosis, while the short form, Bcl-x_S, precipitates cell death. The recently discovered Bax, similar to Bcl-x_S, is able to initiate the cell death programme. The different members of the Bcl-2 family form heterodimers with one another and block each other's actions. Therefore, whether a cell survives or undergoes apoptosis depends on the expression and ratio of these proteins (Adams et al., 1998). Factors that ensure neuronal survival (i.e. target derived growth factors, establishment of functional synapses) or conversely trigger apoptosis (i.e. growth factor or target deprivation) in the developing nervous system do so by influencing the expression pattern of these apoptosis regulatory proteins.

Apoptosis and the cell division cycle

Apoptosis is also known as a regular feature of neoplasms. Extensive cancer research has shed further light on the regulatory pathways of programmed cell death and its relationship with cell division (Meikrantz et al., 1995; Pines, 1995).

The cell division cycle is divided into four phases. The first gap phase (G_1) precedes DNA replication (S phase), while the second gap phase (G_2) represents the preparation of the cell for mitosis (M phase).

The re-entry of a cell into the cell division cycle from the quiescent (G_0) phase requires the presence of mitogenic stimuli. The unperturbed progression of the cycle through its four phases depends on the sequential expression, activation and degradation of the cyclin/cyclin-dependent kinase (cyclin/cdk) complexes (Grana et al., 1995; Pines, 1994a,b).

The presence of mitogenic growth factors, that initiate the transition from G_0 to G_1, trigger the expression/activation of the cyclin D/cdk4,6 complex (Grana et al., 1995). If mitogenic growth factors are not sufficient to support cell division cyclin D expression will cease immediately (Grana et al., 1995). Up to the mid-late stages of G_1 this will lead to cell cycle arrest and subsequent re-differentiation. However, if the cycle is well advanced into the late G_1 phase the absence of cyclin D will have no effect and the cell will remain committed to the division cycle (Chiarugi et al., 1994; Grana et al., 1995). The transition from G_1 to S phase is regulated by the cyclin E/cdk2 complex. If the transition is not successful DNA replication cannot begin and the cell cycle will be arrested at the G_1/S transition point (Grana et al., 1995; Pines, 1994a,b). The G_1 arrest can have two alternative outcomes: re-differentiation or apoptotic cell death. If cyclin A production, necessary for DNA replication, has already started the G_1 arrest results in the activation of a Bax-dependent apoptotic pathway (Chiarugi et al., 1994). However, in the absence of cyclin A re-differentiation is possible and it is aided by the presence of the Bcl-2 protein (Chiarugi et al., 1994).

After the G_1/S transition is successfully completed, cyclin E is degraded and cdk2 associates with cyclin A (Grana et al., 1995; Pines, 1994a,b). The normal progression of DNA replication (S phase) requires, among others, the presence and activation of the cyclin A/cdk2 complex. Once DNA replication is completed cyclin A is required for the activation of the cdc2 kinase to allow the progression of the cell into the G_2 phase. The degradation of cyclin A, in early G_2, is followed by the activation of the cyclin B/cdc2 complex which is needed for the transition into mitosis (M phase). In genetically unstable cell populations (unreplicated DNA, errors in DNA replication, incorrect assembly of the mitotic apparatus) the cyclin B/cdc2 complex is inhibited and the G_2/M transition is blocked. At this stage the cell cycle arrest will be followed by apoptosis (Grana et al., 1995; Pines, 1994a,b). However, if the G_2/M transition is successful cyclin B is rapidly degraded to allow the completion of mitosis. The continued presence of cyclin B leads to cell death due to abortive mitosis (Grana et al., 1995; Pines, 1994a,b).

In summary, the progression of the cell division cycle is a tightly controlled process orchestrated by the successive expression/activation/degradation of cyclin/cdk complexes. These in turn are regulated by extrinsic and intrinsic factors that ensure that the cell cycle is arrested in an environment that is not able to support a healthy cell division. If the cell cycle arrest occurs in the early G_1 phase re-differentiation is possible, however, in more advanced stages the cell cycle arrest will lead to apoptotic cell death.

Apoptotic mechanisms in Down's syndrome brain

It is becoming widely accepted that apoptosis may be responsible for neuronal death in degenerative diseases of the adult and ageing brain (Cotman, 1998; Desjardins et al., 1998). There is also evidence to indicate that the expression of apoptosis regulatory proteins is altered in the brain of DS sufferers (Busciglio et al., 1995; de la Monte et al., 1998; Nagy et al., 1997a; Sawa et al., 1997). These findings lead to the hypothesis that neuronal death in DS occurs via an apoptotic mechanism. Although several theories have been proposed recently to explain the massive neuronal loss in DS the cellular phenomena involved in triggering apoptosis remain elusive (Fig. 1).

From SAD studies it is plausible to envisage that the accumulation of PHFs is the main cause of neuronal death by impeding neuronal transport which in turn would affect growth factor dependent neuronal survival (Trojanowski et al., 1995). On the other hand the reduced synaptic plasticity caused by NFT accumulation could lead to further synaptic loss and subsequent worsening of the dementing process (Masliah et al., 1993a,b). This reasoning, however, does not explain the excessive cell death and synaptic loss that occurs very early in the DS sufferers and clearly precedes NFT formation.

B-amyloid toxicity

Since the main feature of trisomy 21 is the overproduction of APP, with subsequent β-amyloid deposition, it is tempting to hypothesise that this gradual accumulation of β-amyloid is responsible for the dementing process in DS. It has been proposed that amyloid deposition disrupts the neuropil affecting synaptic integrity that was found to be the strongest correlate of dementia severity in SAD (Masliah et al., 1993a,b).

Other studies indicate that β-amyloid may be directly toxic to neurones in culture (Mattson, 1997). It has been shown that the presence of β-amyloid also

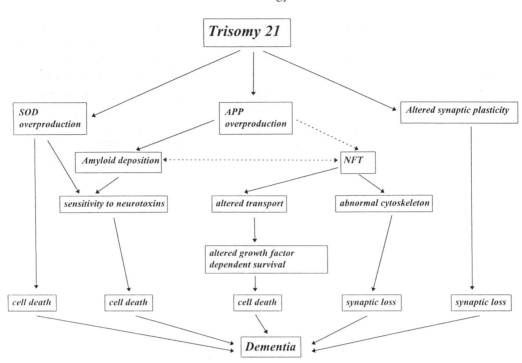

Fig. 1. Cellular mechanisms leading to dementia in Down's syndrome

increases the sensitivity of neurones to neurotoxic insults in vitro (Mattson, 1997).

Recent experiments have indicated that the presence of β-amyloid is able to induce glycogen synthase kinase (GSK-3β) activation in primary neuronal cultures (Imahori et al., 1997). In turn GSK-3β activation leads to tau hyperphoshorylation then apoptotic cell death in these neurones (Imahori et al., 1997). The possibility of in vivo detrimental effects of β-amyloid on neurones is also supported by the finding that ubiquitin-positive degenerating neurite aggregation appears much earlier in DS brains than in age matched controls and it was found to be accentuated in the vicinity of amyloid deposits, giving rise to senile plaques (Mattiace et al., 1991). It seems, therefore, plausible that the increased β-amyloid deposition, due to overproduction of APP, could lead to apoptotic neuronal death in DS brains.

However, it is also known from extensive studies on SAD that amyloid deposition does not relate to the development of dementia and it is not directly related to NFT formation which is the main pathological feature related to dementia severity (Nagy et al., 1995).

The free radical theory

Another pathway implicated in apoptotic mechanisms in neurodgenerative diseases is oxidative stress due to the accumulation of reactive oxygen species

(ROS). Since the superoxide dismutase (SOD) is overexpressed in Down's syndrome it is possible that oxidative stress may be involved in neuronal death in trisomy 21. It was indeed found that overexpression of SOD leads to an increased sensitivity of neurones to neurotoxins (Bar Peled et al., 1996). The finding that neurones from DS foetuses, when cultured, undergo apoptotic cell death due to oxidative stress provided further strength to the hypothesis that a defective metabolism of ROS is the cause of neuronal cell death in Down's syndrome (Busciglio et al., 1995). Evidence from animal models also support the possibility of abnormal neuronal differentiation and connectivity due to oxidative stress since it was found that SOD transgenic mice have defective mossy fibre inervation in the hippocampus (Barkats et al., 1993). However, a recent study on SAD and DS brains found no evidence for a pathogenic role of SOD, lipid peroxidation or ROS in either conditions (Hayn et al., 1996).

Abortive mitosis theory

The role of cell cycle-related phenomena in cell death and the development of AD-related plaque and tangle pathology in sporadic AD is increasingly accepted (Nagy et al., 1998; Vincent et al., 1998a).

The first indications of cell division related phenomena being involved in AD emerged with the identification of the protein kinases responsible for tau phosphorylation. The MAP kinases, GSK-3β, cdk2 and cdk5 are all kinases expressed during cell division (Lovestone et al., 1997). It is also known that neuroblastoma cells are capable of producing AD-type NFTs that, similar to those found in AD brains, were found to be immunopositive for cell division-related proteins (Smith et al., 1995). In vitro studies also indicated that during the G$_2$ phase of the cell cycle APP is processed into amyloidogenic fragments (Suzuki et al., 1997). It was also found that while APP itself constitutes a mitogenic signal for several cell types in culture (Alvarez et al., 1995) the diffuse amyloid deposits of AD also accumulate potentially mitogenic growth factors, such as bFGF and EGF (Araujo et al., 1992; Birecree et al., 1988). The possibility of a genetically determined failure of cell cycle regulation in DS is further supported by the finding that these patients are significantly more cancer prone that the age-matched control population (Altmann et al., 1998; Dieckmann et al., 1997; Satge et al., 1997). Furthermore, DNA repair mechanisms were found to be malfunctioning in DS patients (Au et al., 1996).

All thesm led us to search for cell cycle-related proteins in the hippocampi of DS sufferers and sporadic AD patients in relation to cell death and the accumulation of AD-type pathology.

Our preliminary findings of nuclear Ki-67 expression in DS hippocampi indicated that the neurones of the postnatal and adult hippocampus arc not necessarily quiescent and that they are able to re-enter the cell division cycle (Nagy et al., 1997c). It was also evident that this cell cycle re-entry precedes the development of AD-related pathology and massive cell death in the hippocampi of DS sufferers. However, although nuclear Ki-67 expression clearly indicates cell cycle re-entry it does not provide any information on the different phases of the cell division cycle. To gain more detailed information about the neuronal cell cycle in DS we examined the expression pattern of cyclins A, B, D and E (Nagy et al., 1997b). The nuclear expression of these proteins is specific for the different phases of the cell division cycle, and can be used to identify these phases. While there was no evidence of nuclear cyclin A or cyclin D we found that nuclear cyclin E and cyclin B were present in neurones in all hippocampal subregions in DS patients. Furthermore, we, and others, have found that in pyramidal cells cyclin B was co-localised with the AD-type hyperphosphorylated tau protein (Nagy et al., 1997b; Vincent et al., 1997). The finding of cyclins E and B expressed in nuclei of pyramidal and granule cells of the hippocampus strengthened our hypothesis that neurones of the postnatal and adult nervous system are not necessarily quiescent. The distribution pattern, however, of nuclear cyclin expression was intriguing. It was apparent that that cell cycle re-entry of neurones of the subiculum and CA1 regions of the hippocampus precedes the development of the AD-type pathology (Nagy et al., 1997b,c; Vincent et al., 1997, 1996, 1998b) while in other hippocampal regions it seemed to predict massive cell death (Nagy et al., 1997b). The factor responsible for this differential effect was identified later when we analysed the expression pattern of cell death-related proteins in DS hippocampi (Nagy et al., 1997a). It emerged that the apoptosis-promoting Bax protein was highly expressed in DS hippocampi and its expression was mutually exclusive with that of the AD-type hyperphosphorylated tau (Nagy et al., 1997a).

The "death cascade" in Down's syndrome

Based on evidence from several studies we can conclude that neurones of the adult nervous system are capable of re-entering the cell division cycle. The cell cycle re-entry, depending on the extrinsic signals and intrinsic regulatory mechanisms may have different outcomes. Where normal regulatory mechanisms are in place the cell cycle progression is probably halted at the G_1/S transition point and cells may re-differentiate. However, the same G_1/S transition arrest may lead to apoptotic cell death in neuronal populations via a p53/Bax-dependent pathway, that was found to be activated in DS neurones (de la Monte et al., 1998). If cell cycle regulatory mechanisms fail the cell cycle is allowed to progress into the G_2 phase. Since there is no indication of mitosis in DS brains (no mitotic figures) we can safely assume that the cell cycle in these neurones is arrested at the G_2/M transition point. At this stage re-differentiation is not possible. The cells arrested at this transition point of the cycle may undergo apoptosis via the p53/Bax related pathway. However, if the cell death programme is not initiated at this stage cytoskeletal microtubules are destabilised and numerous protein kinases, MAP kinases, GSK-3β, cyclin A/cdk2, cdk5, are activated (Lew et al., 1995a,b). These are the kinases that

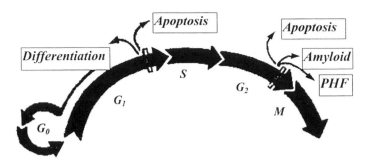

Fig. 2. Abortive cell division cycle in Down's syndrome neurones

were found to be able to hyperphosphorylate the microtubule associated tau protein into an AD-type state (Lovestone et al., 1997). On the other hand the G_2 phase-specific processing of the APP protein will lead to the accumulation of amyloidogenic fragments resulting in extracellular amyloid deposition (Suzuki et al., 1997). All these led to the hypothesis that cell cycle re-entry in neurones may be followed either by the development of AD-related pathology or by apoptotic cell death (Fig. 2) (Nagy et al., 1998).

The mitogenic stimulus responsible for this aberrant cell cycle re-entry may be the overexpression of APP in DS neurones and glia (Alvarez et al., 1995). Additionally the process may be facilitated by the synaptic disgenesis (Ferrer et al., 1990). Another protein, elevated in DS, (on chromosome 21 on the region obligatory for DS) is S-100. This is a differentiation factor that was shown to be capable of triggering apoptosis (Whitaker Azmitia et al., 1997). It is possible that while the elevated APP levels act as the mitogenic stimulus, the effects of increased amounts of S100 may contribute to the cell cycle arrest and precipitate the death programme via a Bax-dependent pathway in neurones. In neurones that do not express the Bax protein, the G_2 arrest may not necessarily result in cell death, but could lead to the formation of AD-related pathology. The amyloid deposits, by accumulating mitogenic growth factors, would further accelerate cell cycle re-entry of neurones, while the formation of neuritic pathology would lead to synaptic loss and subsequent neuronal de-differentiation in the interconnected areas of the brain. Thus the process could quickly become a self perpetuating vicious circle that gradually affects interconnected brain regions (Fig. 3).

In summary, we believe that in DS patients the genetically determined failure of cell cycle regulation is at the root of brain pathology. This hypothesis would provide the "missing link" between the different pathological features, excessive cell death and AD-type plaques and tangles, and explain their gradual development in specific, interconnected, areas of the brain. The genetically determined nature of the main factors maintaining this "death cascade", APP and S100 overexpression, could also explain why cell death and AD-related pathology appear very early and are more severe in DS patients than in sporadic AD sufferers.

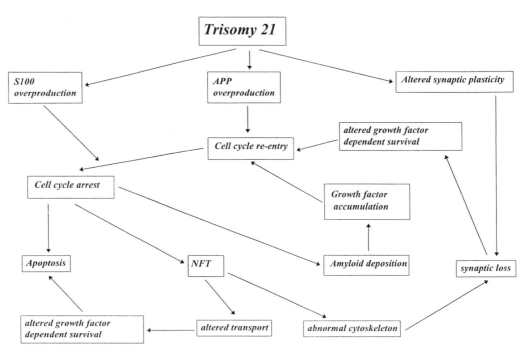

Fig. 3. The "death cascade" in Down's syndrome

Acknowledgements

I would like to thank the all members of OPTIMA who made my research into Alzheimer's disease possible and Dr. B. McDonald who kindly provided the autopsy material from Down's syndrome patients. I am also grateful to Mrs. S. Litchfield and Miss M. Reading for their valuable technical contribution to all my studies.

I would like to take this opportunity to thank my father, Mr. S. Nagy, for the illustrations used in this review.

All our studies in Alzheimer's disease were supported by grants from Bristol-Myers Squibb and Research into Ageing.

References

Adams JM, Cory S (1998) The Bcl-2 protein family: arbiters of cell survival. Science 281: 1322–1326

Altmann AE, Halliday JL, Giles GG (1998) Associations between congenital malformations and childhood cancer. A register-based case-control study. Br J Cancer 78: 1244–1249

Alvarez J, Moreno RD, Inestrosa NC (1995) Mitosis of Schwann cells and demyelination are induced by the amyloid precursor protein and other protease inhibitors in the rat sciatic nerve. Eur J Neurosci 7: 152–159

Araujo DM, Cotman CW (1992) Beta-amyloid stimulates glial cells in vitro to produce growth factors that accumulate in senile plaques in Alzheimer's disease. Brain Res 569: 141–145

Armstrong RA (1996) Correlations between the morphology of diffuse and primitive beta-amyloid (A beta) deposits and the frequency of associated cells in Down's syndrome. Neuropathol Appl Neurobiol 22: 527–530

Au WW, Wilkinson GS, Tyring SK, Legator MS, el Zein R, Hallberg L, Heo MY (1996) Monitoring populations for DNA repair deficiency and for cancer susceptibility. Environ Health Perspect 3: 579–584

Bar Peled O, Korkotian E, Segal M, Groner Y (1996) Constitutive overexpression of Cu/Zn superoxide dismutase exacerbates kainic acid-induced apoptosis of transgenic-Cu/Zn superoxide dismutase neurons. Proc Natl Acad Sci USA 93: 8530–8535

Barkats M, Bertholet JY, Venault P, Ceballos Picot I, Nicole A, Phillips J, Moutier R, Roubertoux P, Sinet PM, Cohen Salmon C (1993) Hippocampal mossy fiber changes in mice transgenic for the human copper-zinc superoxide dismutase gene. Neurosci Lett 160: 24–28

Birecree E, Whetsell WO, Jr, Stoscheck C, King LE, Jr, Nanney LB (1988) Immunoreactive epidermal growth factor receptors in neuritic plaques from patients with Alzheimer's disease. J Neuropathol Exp Neurol 47: 549–560

Braak H, Braak E (1991) Neuropathological stageing of Alzheimer-related changes. Acta Neuropathol Berl 82: 239–259

Busciglio J, Yankner BA (1995) Apoptosis and increased generation of reactive oxygen species in Down's syndrome neurons in vitro. Nature 378: 776–779

Chiarugi V, Magnelli L, Cinelli M, Basi G (1994) Apoptosis and the cell cycle. Cell Mol Biol Res 40: 603–612

Cotman CW (1998) Apoptosis decision cascades and neuronal degeneration in Alzheimer's disease. Neurobiol Aging 19: S29–S32

de la Monte SM, Sohn YK, Ganju N, Wands JR (1998) P53- and CD95-associated apoptosis in neurodegenerative diseases. Lab Invest 78: 401–411

Desjardins P, Ledoux S (1998) The role of apoptosis in neurodegenerative diseases. Metab Brain Dis 13: 79–96

Dieckmann KP, Rube C, Henke RP (1997) Association of Down's syndrome and testicular cancer. J Urol 157: 1701–1704

Ferrer I, Gullotta F (1990) Down's syndrome and Alzheimer's disease: dendritic spine counts in the hippocampus. Acta Neuropathol Berl 79: 680–685

Grana X, Reddy EP (1995) Cell cycle control in mammalian cells: role of cyclins, cyclin dependent kinases (CDKs), growth suppressor genes and cyclin-dependent kinase inhibitors (CKIs). Oncogene 11: 211–219

Hayn M, Kremser K, Singewald N, Cairns N, Nemethova M, Lubec B, Lubec G (1996) Evidence against the involvement of reactive oxygen species in the pathogenesis of neuronal death in Down's syndrome and Alzheimer's disease. Life Sci 59: 537–544

Hof PR, Bouras C, Perl DP, Sparks DL, Mehta N, Morrison JH (1995) Age-related distribution of neuropathologic changes in the cerebral cortex of patients with Down's syndrome. Quantitative regional analysis and comparison with Alzheimer's disease. Arch Neurol 52: 379–391

Hyman BT (1992) Down syndrome and Alzheimer disease. Prog Clin Biol Res 379: 123–142

Hyman BT, West HL, Rebeck GW, Lai F, Mann DM (1995) Neuropathological changes in Down's syndrome hippocampal formation. Effect of age and apolipoprotein E genotype. Arch Neurol 52: 373–378

Imahori K, Uchida T (1997) Physiology and pathology of tau protein kinases in relation to Alzheimer's disease. J Biochem Tokyo 121: 179–188

Lew J, Qi Z, Huang QQ, Paudel H, Matsuura I, Matsushita M, Zhu X, Wang JH (1995a) Structure, function, and regulation of neuronal Cdc2-like protein kinase. Neurobiol Aging 16: 263–268

Lew J, Wang JH (1995b) Neuronal cdc2-like kinase. Trends Biochem Sci 20: 33–37

Lovestone S, Reynolds CH (1997) The phosphorylation of tau: a critical stage in neurodevelopment and neurodegenerative processes. Neuroscience 78: 309–324

Majno G, Joris I (1995) Apoptosis, oncosis, and necrosis. An overview of cell death. Am J Pathol 146: 3–15

Mann DM, Yates PO (1986a) Neurotransmitter deficits in Alzheimer's disease and in other dementing disorders. Hum Neurobiol 5: 147–158

Mann DM, Yates PO, Marcyniuk B, Ravindra CR (1986b) The topography of plaques and tangles in Down's syndrome patients of different ages. Neuropathol Appl Neurobiol 12: 447–457

Mann DM, Yates PO, Marcyniuk B, Ravindra CR (1987) Loss of neurones from cortical and subcortical areas in Down's syndrome patients at middle age. Quantitative comparisons with younger Down's patients and patients with Alzheimer's disease. J Neurol Sci 80: 79–89

Mann DMA (1997) Neuropathological changes of Alzheimer's disease in persons with Down's syndrome. In: Esiri MM, Morris JH (eds) The neuropathology of dementia. Cambridge University Press, Cambridge, pp 122–136

Masliah E, Mallory M, DeTeresa R, Alford M, Hansen L (1993a) Differing patterns of aberrant neuronal sprouting in Alzheimer's disease with and without Lewy bodies. Brain Res 617: 258–266

Masliah E, Miller A, Terry RD (1993b) The synaptic organization of the neocortex in Alzheimer's disease. Med Hypotheses 41: 334–340

Mattiace LA, Kress Y, Davies P, Ksiezak Reding H, Yen SH, Dickson DW (1991) Ubiquitin-immunoreactive dystrophic neurites in Down's syndrome brains. J Neuropathol Exp Neurol 50: 547–559

Mattson MP (1997) Cellular actions fo beta-amyloid precursor protein and its soluble and fibrillogenic derivatives. Physiol Rev 77: 1081–1132

Meikrantz W, Schlegel R (1995) Apoptosis and the cell cycle. J Cell Biochem 58: 160–174

Nagy Zs, Esiri MM, Jobst KA, Morris JH, King EM, McDonald B, Litchfield S, Smith A, Barnetson L, Smith AD (1995) Relative roles of plaques and tangles in the dementia of Alzheimer's disease: correlations using three sets of neuropathological criteria. Dementia 6: 21–31

Nagy Zs, Esiri MM (1997a) Apoptosis-related protein expression in the hippocampus in Alzheimer's disease. Neurobiol Aging 18: 565–571

Nagy Zs, Esiri MM, Cato AM, Smith AD (1997b) Cell cycle markers in the hippocampus in Alzheimer's disease. Acta Neuropathol Berl 94: 6–15

Nagy Zs, Esiri MM, Smith AD (1997c) Expression of cell division markers in the hippocampus in Alzheimer's disease and other neurodegenerative conditions. Acta Neuropathol Berl 93: 294–300

Nagy Zs, Esiri MM, Smith AD (1998) The cell division cycle and the pathophysiology of Alzheimer's disease. Neuroscience 84: 731–739

Nagy Zs, Esiri MM, LeGris M, Matthews PM (1999) Mitochondrial enzyme expression in the hippocamppus in relation to Alzheimer-type pathology. Acta Neuropathol Berl 97: 346–354

Pines J (1994a) The cell cycle kinases. Semin Cancer Biol 5: 305–313

Pines J (1994b) Protein kinases and cell cycle control. Semin Cell Biol 5: 399–408

Pines J (1995) Cyclins, CDKs and cancer. Semin Cancer Biol 6: 63–72

Prinz M, Prinz B, Schulz E (1997) The growth of non-pyramidal neurons in the primary motor cortex of man: a Golgi study. Histol Histopathol 12: 895–900

Satge D, Sasco AJ, Cure H, Leduc B, Sommelet D, Vekemans MJ (1997) An excess of testicular germ cell tumors in Down's syndrome: three case reports and a review of the literature. Cancer 80: 929–935

Sawa A, Oyama F, Cairns NJ, Amano N, Matsushita M (1997) Aberrant expression of bcl-2 gene family in Down's syndrome brains. Brain Res Mol Brain Res 48: 53–59

Smith TW, Lippa CF (1995) Ki-67 Immunoreactivity in Alzheimer's disease and other neurodegenerative disorders. J Neuropathol Exp Neurol 54: 297–303

Steller H (1995) Mechanisms and genes of cellular suicide. Science 267: 1445–1449

Suzuki T, Ando K, Isohara T, Oishi M, Lim GS, Satoh Y, Wasco W, Tanzi RE, Nairn AC, Greengard P, Gandy SE, Kirino Y (1997) Phosphorylation of Alzheimer beta-amyloid precursor-like proteins. Biochemistry 36: 4643–4649

Sylvester PE (1984) Ammon's horn or hippocampal sclerosis without epilepsy in mental handicap. Br J Psychiatry 144: 538–541

Takashima S, Iida K, Mito T, Arima M (1994) Dendritic and histochemical development and ageing in patients with Down's syndrome. J Intellect Disabil Res 38: 265–273

Trojanowski JQ, Lee VM (1995) Phosphorylation of paired helical filament tau in Alzheimer's disease neurofibrillary lesions: focusing on phosphatases. Faseb J 9: 1570–1576

Vincent I, Rosado M, Davies P (1996) Mitotic mechanisms in Alzheimer's disease? J Cell Biol 132: 413–425

Vincent I, Jicha G, Rosado M, Dickson DW (1997) Aberrant expression of mitotic cdc2/cyclin B1 kinase in degenerating neurons of Alzheimer's disease brain. J Neurosci 17: 3588–3598

Vincent I, Ding XL, Zheng JH, Zhang SH (1998a) Mitotic kinases: initiators and mediators of neurodegeneration? Neurobiol Aging [Suppl] 4S

Vincent I, Zheng JH, Dickson DW, Kress Y, Davies P (1998b) Mitotic phosphoepitopes precede paired helical filaments in Alzheimer's disease. Neurobiol Aging 19: 287–296

Whitaker Azmitia PM, Wingate M, Borella A, Gerlai R, Roder J, Azmitia EC (1997) Transgenic mice overexpressing the neurotrophic factor S-100 beta show neuronal cytoskeletal and behavioral signs of altered aging processes: implications for Alzheimer's disease and Down's syndrome. Brain Res 776: 51–60

Author's address: Dr. Zs. Nagy, OPTIMA, Radcliffe Infirmary NHS Trust, Woodstock Road, Oxford OX2 6HE, United Kingdom, e-mail: zsuzsa.nagy@pharm.ox.ac.uk

J Neural Transm (1999) [Suppl] 57: 247–256

Impaired brain glucose metabolism in patients with Down Syndrome

O. Labudova[1], **N. Cairns**[2], **E. Kitzmüller**[1], and **G. Lubec**[1]

[1] Department of Pediatrics, University of Vienna, Austria
[2] Institute of Psychiatry, Brain Bank, London, United Kingdom

Summary. A series of impaired metabolic functions in Down Syndrome (DS) including glucose handling has been described. Recent information from positron emission tomography studies in DS patients and our finding of downregulated phosphoglucose isomerase (PGI) in fetal brain with DS by gene hunting using subtractive hybridization, made us investigate PGI, a key enzyme of glucose metabolism, in brain of patients with DS, Alzheimer's disease (AD) and controls. PGI and phosphofructokinase (PFK) activities were determined in frontal, parietal, temporal, occipital lobe and cerebellum of 9 controls, 9 patients with DS and 9 patients with AD. PGI activity in DS brain was significantly decreased in frontal, temporal lobe and cerebellum, comparable to controls in parietal lobe and elevated in occipital lobe. Brain PGI activity of patients with AD was comparable to controls in all regions tested. PFK, a rate limiting enzyme of glucose metabolism, was comparable between all brain regions of all three groups. Data of this study confirm impaired glucose metabolism in DS proposed in literature and found by positron emission tomography (PET) studies. We show that changes in glucose handling in patients with AD as evaluated by PET studies are not supported by our data, although not contradictory, as determinants other than glucose metabolizing enzymes as e.g. vascular factors and glucose transport may account for these findings. Changes of downregulated PGI found by subtractive hybridization at the transcriptional level in fetal DS brain along with our findings in DS brain regions suggest a strong specific link between glucose metabolism and DS rather than AD.

Introduction

Down Syndrome (DS) is the most frequent genetic cause of dementia and although the trisomic state can be incriminated for the phenotype, the underlying pathomechanisms are far from being understood. All subjects with DS over age 40 years show neuropathological and neurochemical abnormalities on postmortem brain examinations that are indistinguishable from those seen in Alzheimer's disease (AD), Burger (1973) and Wisniewski

(1985), and 75% of persons with DS over age 60 are demented (Haxby, 1988; Lai, 1989).

A long list of impaired metabolic functions in DS has been reported, Epstein (1992) including deteriorated glucose metabolism. Increased phosphofructokinase has been demonstrated in red blood cells as well as in fibroblasts (Baikie, 1965; Bartels, 1968; Layzer, 1972; Anneren, 1987; Vora, 1981) and glucose-6-phosphate dehydrogenase was found to be increased in blood cells (Hsia, 1971).

Glucose metabolism has been evaluated in brain of DS patients in vivo using positron emission tomography (PET) with [^{18}F] fluorodeoxyglucose, a well-established method for the measurement of regional cerebral glucose metabolism rates in humans and older demented adults with DS showed lower than normal rates of absolute glucose metabolism in temporal and parietal association neocortical areas (Schapiro, 1992), resembling a pattern seen in AD (Haxby, 1985). Dani (1996) and coworkers showed in a longitudinal study that older subjects with DS without cognitive impairment maintain regional cerebral glucose metabolism within the normal range for many years, although repeated measurements showed a progressive decline of glucose metabolism over time and after the onset of dementia.

In a recent publication Pietrini (1997) and coworkers demonstrated by a PET study that abnormalities in cerebral glucose metabolism of patients with DS at risk for AD prior to dementia, appeared during stimulation in the first cortical regions typically affected in AD. Their data indicate that a stress paradigm can detect metabolic abnormalities in the preclinical stages of AD despite normal cerebral metabolism at rest.

Performing gene hunting in DS using the principle of subtractive hybridization (Labudova, 1998), we detected a sequence with strong homology to phosphoglucose isomerase (PGI), key enzyme of glucose metabolism, which was significantly downregulated in temporal lobe of fetal DS brain as compared to normal fetal brain. We therefore decided to study the activity of PGI in several regions of postmortem brains of patients with DS, AD and controls.

Materials and methods

Subtractive hybridization protocol (Labudova, 1998)

Fetal brain samples of two fetuses with DS and two age and sex matched controls, 23rd week of gestation, were obtained from the Brain Bank of the Institute of Psychiatry (Denmark Hill, London, UK). Hippocampus (gyrus parahippocampalis) was taken into liquid nitrogen and ground for the isolation of mRNA. Isolation of mRNA was performed using the Quick Prep Micro mRNA purification kit (Pharmacia Biotech Inc., Uppsala, Sweden, cat. 27 92 55 01). 1 μg of mRNA from each (of the two) preparation was quality — checked by cDNA cloning kit (Gibco, Life Technologies, Eggenstein, Germany, cat. 18248-013) using the incorporation of [alpha — ^{32}P] dATP (Amersham, Buckinghamshire, UK, cat AA0004)) with subsequent electrophoresis on 1% agarose followed by autoradiography. The Reflection film (Dupont NEF 496) was exposed to the gel for a period of two hours at room temperature.

Construction of the subtractive library: 10 μg each of mRNA from brain of DS and control were biotinilated by UV irradiation at 360 nm according to the instructions supplied in the subtractor kit (Invitrogen, Leek, Netherlands, cat K4320-01). 1 μg of mRNA — pools each from the DS brain sample was subject to reverse transcriptase reaction (subtractor kit, Invitrogen) and the cDNA — pools were hybridized with the corresponding biotinilated mRNAs from controls. The substractive hybridization mixture was incubated with streptavidin according to the subtractor kit given above and thus the biotinilated molecules (non-induced biotinilated mRNAs and the hybrid [biotinilated mRNAs/cDNAs]) complexed. The streptavidine complexes were removed by repeated phenol-chloroform extraction and subtracted cDNAs were separated from the aquous phase by alcohol precipitation (subtractor kit).

In order to amplify and clone subtracted cDNAs, they were ligated with Not I — linkers followed by Not I — digestion. These Not I linked cDNAs were ligated to Not I site of sPORT 1 cloning vector (cDNA cloning kit, Gibco). To enable visualization of subtracted cDNAs the cloned cDNAs were amplified using universal primers

I   5'-GTAAAACGACGGCCAGT-3'
II 5'-ACAGCTATGACCATG-3'

from multiple cloning site of the sPORT-1 vector (cDNA cloning kit, Gibco). Amplified cDNAs were analysed on 1% agarose electrophoresis.

Cloning of subtracted Not I — linked cDNAs: Not I linked cDNAs ligated with sPORT 1 vector were used for the transformation of highly competent INFalpha F' E. coli cells (Invitrogen, Leek, Netherlands, cat C2020-03) and plated clones were analysed by plasmid isolation kit (Quiagen, Hilden, Germany, cat 12245) and digestion with Eco RI/ Hind III. Recombinant clones were sequenced by K. Granderath, MWG — Biotech (Ebersberg, Germany). Homologies were determined by computer assisted comparison of data from the genebank sequence library: fastA@ebi.ac.uk (GBALL, Gen Bank Heidelberg, Germany).

Subtractive hybridization was performed cross-wise i.e. DS sample mRNA subtraction from control and vice versa at the 1:3 level (DSmRNA: control mRNA).

Determination of glucose metabolising enzymes

The brain regions temporal, frontal, occipital, parietal cortex and cerebellum of patients with DS (n = 9; 3 females, 6 males; 56.1 ± 7.1 years old), AD (n = 9; 6 females, 3 males; 72.3 ± 7.6 years old) and controls (n = 9; 5 females, 4 males; 72.6 ± 9.6 years old) characterized in a previous publication (Seidl, 1997), were used for the studies at the activity level. Briefly, post mortem brain samples were obtained from the MRC London Brain Bank for Neurodegenerative Diseases, Institute of Psychiatry). In all DS brains there was evidence of abundant beta A plaques and neurofibrillary tangles. The AD patients fulfilled the National Institute of neurological and Communicative Disorders and Stroke and Alzheimer Disease and Related Disorders Association (NINCDS/ADRDA) criteria for probable AD (Mirra, 1991). The histological diagnosis of AD was established and was consistent with the CERAD criteria (Tierney, 1988), for a "definite" diagnosis of AD. The controls were brains from individuals with no history of neurological or psychiatric illness. The major cause of death was bronchopneumonia in DS and AD patients and heart disease in controls. Post mortem interval of brain dissection in AD, DS and controls was 34.1 ± 13.7, 30.6 ± 17.5 and 34.8 ± 15.0 hrs. Tissue samples were stored at −70°C and the freezing chain was never interrupted.

Shock frozen brains of patients with DS, AD and controls were thawed on ice. Samples were homogenized by twenty strokes in a Potter-Elvehjem homogenizer on ice (Schmidt, 1977), and the homogenates were centrifuged for 10 min at 4000 g at 4°C. The supernatant was used for the determination of protein (Bradford, 1976), and enzyme determinations.

Determination of PGI

The principle and procedure given by Beutler (1975) was followed. The principle of the assay is given by the formula:

glucose-6-phosphate $\xrightarrow{\text{PGI}}$ fructose-6-phosphate

Fructose-6-phosphate serves as a substrate in this assay, and the glucose-6-phosphate formed is measured by linking it to the reduction of NADP through glucose-6-phosphate dehydrogenase.

The procedure was as follows: chemicals listed in Table 1 were added to cuvettes with a critical volume of less than 1 ml. As a large number of assays were carried out, multiple of the indicated volumes of tris buffer, $MgCl_2$, NADP, fructose-6-phosphate, and water were mixed and appropriate quantities of the mixture distributed into cuvettes. Increase of optical density of the system is measured against that of the blank at 340 nm at 37°C for 10–20 min. A recorder expansion giving a full-scale reading of 1.0 OD units was used. A blank was carried out to be certain that the glucose-6-phosphate dehydrogenase was free of PGI activity. Calculations were performed exactly as given in the reference and enzyme activity was expressed as U/mg protein (Beutler, 1975).

Determination of phosphofructokinase (PFK)

The principle and procedure given by Beutler (1975) was followed. The principle is that PFK catalyzes the phosphorylation of fructose-6-phosphate by ATP to fructose-1,6-diphosphate:

Fructose-6-phosphate + ATP $\xrightarrow{\text{PFK}}$ fructose-1,6-diphosphate + ADP

In this assay the fructose-1,6-diphosphate formed is measured by the conversion to dihydroxyacetone phosphate (DHAP) through the aldolase and triose phosphate isomerase (TPI) reactions. The DHAP is then reduced by the action of alpha-glycerophosphate dehydrogenase, oxidizing NADH to NAD. The sequence of reactions involved is as follows:

Table 1. Sequence of enzyme reactions

	blank	system
	microliters	
1 M Tris-HCl pH 8.0 cont. 5 mM EDTA	100	100
0.1 M $MgCl_2$	100	100
2 mM NADP	100	100
Glucose-6-phosphate dehydrogenase 10 U/ml (diluted in beta-mercaptoethanol-EDTA stabilizingsolution)	10	10
0.02 M fructose-6-phosphate	—	100
H_2O	685	585
Incubate at 37°C for 1 hour		
Supernatant of brain tissue homogenate	5	5

Table 2. Sequence of enzyme reactions

	blank	system	low system + ADP
		microliters	
1 M Tris-HCl pH 8.0 cont. 5 mM EDTA	100	100	100
0.03 M ADP	—	—	10
0.1 M MgCl$_2$	200	200	200
2 mM NADP	100	100	100
auxilliary enzyme solution*	100	100	100
0.02 M fructose-6-phosphate	—	100	5
H$_2$O	390	290	375
Supernatant of brain tissue homogenate	10	10	10
Incubate at 37°C for 1 hour 0.02 M ATP	100	100	100

* Auxiliary enzyme solution: (NH4)2 SO4 suspension of 50 units of each enzyme per ml is made by adding 100 units of aldolase, 100 units of TPI and 100 U of alpha-glycerophosphate dehydrogenase to a tube and bringing to a total vol of 2 ml with saturated (NH4)2SO4 solution. The suspension is stable at 4°C

$$\text{fructose-1,6-diphosphate} \xrightarrow{\text{aldolase}} \text{glyceraldehyde-3-phosphate} + \text{DHAP}$$

$$\text{Glyceraldehyde-3-phosphate} \xrightarrow{\text{TPI}} \text{DHAP}$$

$$\text{DHAP} + \text{NADH} + \text{H}^+ \xrightarrow{\text{alpha-glycerophosphate dehydrogenase}} \text{alpha-glycerophosphate} + \text{NAD}^+$$

The oxidation of NADH is measured at 340 nm. The procedure was carried out as shown in Table 2.

The chemicals are added to cuvettes with a critical volume of less than 1 ml: The decrease in optical density of the system and the low system + ADP is measured against that of the blank for 10–20 min. A recorder expansion giving a full — scale reading of 0.4 OD units was suitable for our samples. A blank assay was carried out to be certain that the aldolase, TPI and alpha-glycerophosphate dehydrogenase used were free of PFK activity. Calculations were carried out as described in the reference and U of activity were expressed per mg protein.

Statistical methods

The ANOVA with subsequent Kruskal-Wallis test was applied and a level of $p < 0.05$ was considered significant.

Results

Subtractive hybridization

The corresponding downregulated sequence of our clone 274 found by subtractive hybridization is given in Fig. 1 and reveals 100% homology of our sequence of 308 bp with phosphoglucose isomerase from humans and high

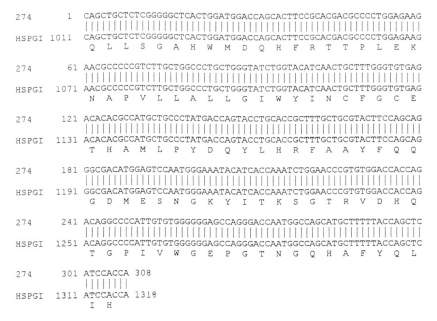

Fig. 1. Alignment of the nucleic acid and deduced amino acid sequence found by subtractive hybridization with the sequence obtained from the GenBank Heidelberg (HSPGI). 100% of homology was found when comparing the 308 bp sequence expressed in the clone 274 with the open reading frame of human PGI (accession number X51346)

homology to PGIs from several species, accession numbers X15196; and U54764, U54765, U54766, M37267, X13977 (Katz, 1996).

Results of PGI and PFK

PGI in DS patients brain was significantly decreased in frontal, temporal lobe and cerebellum whereas significantly increased in occipital lobe and comparable to controls in parietal lobe (Fig. 2).

PGI in patients with AD was comparable to controls in all brain regions tested (Fig. 2).

PFK determinations were comparable between all brain regions from controls, DS and AD patients (Fig. 3).

Discussion

High tech instrumentation as PET has shown impaired in vivo glucose metabolism in brain regions of patients with both, DS and AD, the underlying enzymatic deterioration in brain has not been shown before and changes in glucose handling enzymes could be suggested from in vitro tests in DS cells but not from brain tissue. The availability of a potent gene hunting technique, subtractive hybridization, and of fetal brain tissue, enabled us to show that

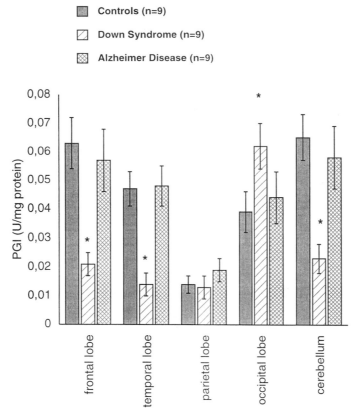

Controls (n=9)

Down Syndrome (n=9)

Alzheimer Disease (n=9)

Fig. 2. PGI activity (U/mg protein) ind different brain regions of Controls, Down Syndrome and Alzheimer's disease

PGI, a key enzyme for glucose metabolism, was downregulated this early in life. This along with observations of dysregulated glycolytic enzymes in children with DS (Baikie, 1965; Bartels, 1968; Layzer, 1972; Anneren, 1987; Vora, 1981; Hsia, 1971), makes us suggest that impaired glucose metabolism is linked to the disease per se rather than to secondary changes occuring in DS brain, AD pathology. And indeed, no PGI activity changes were found in all brain regions tested in AD patients (Fig. 2).

Our result do not contradict previous findings of changes of glycolytic enzymes in DS, as no studies on brain tissue and on PGI were carried out in the past. Data from our study, however, are clearly indicating that impaired glucose handling is linked to DS rather than to AD, although all DS patients showed neuropathological changes indistinguishable from AD. Even this observation may not be in disagreement with PET studies (Schapiro, 1992; Haxby, 1985; Dani, 1996; Pietrini, 1997), as altered rates of glucose (i.e. [^{18}F] fluorodeoxyglucose-) metabolism seen by this technique may not be reflecting biochemical-enzymatic changes only, but also other factors as e.g. vascular factors and glucose transport.

Another glucose metabolism — rate limiting enzyme, PFK which is encoded on chromosome 21, was unchanged in all brain samples (Fig. 3) and this

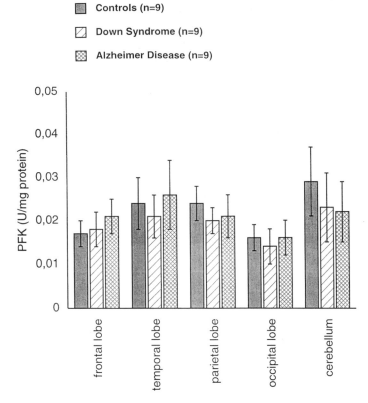

Fig. 3. PFK activity (U/mg protein) ind different brain regions of Controls, Down Syndrome and Alzheimer's disease

is in contradiction with previous reports of increased PFK in fibroblasts and blood cells of DS patients. Again, no brain tissue was examined for glucose metabolizing enzymes so far, but one current concept for the development of the DS phenotype is "gene dosage effects": trisomic chromosome 21 should lead to overexpression of genes encoded on this chromosome and there is some support for this hypothesis from studies on brain tissue (Epstein, 1992, 1986; Pash, 1991).

In conclusion, we describe decreased PGI in fetal brain with DS at the transcriptional level by subtractive hybridization and decreased PGI activity in several brain regions of patients with DS but not with AD. These data suggest a link of this enzyme deficiency to DS per se rather than to neuropathology common to aged patients with DS and AD (Wisniewski, 1985; Haxby, 1988). Data presented in this study may serve as the underlying biochemical mechanism for in vivo findings of glucose metabolic rates on DS brain by PET studies.

Acknowledgements

We are highly indebted to the Red Bull Company, Salzburg, Austria, for generous financial support and to Ing. E. Piegler for excellent technical assistance.

References

Anneren KG, Korenberg JR, Epstein CJ (1987) Phosphofructokinase activity in fibroblasts aneuploid for chromosome 2. Hum Genet 76: 63–65

Baikie AG, Loder PB, de Grouchy GC, Pitt DB (1965) Phosphohexokinase activity in erythrocytes in mongolism: another possible marker for chromosome 21. Lancet i: 412–414

Bartels H, Kruse K (1968) Enzymbestimmungen in Erythrozyten bei Kindern mit Down Syndrome. Humangenetik 5: 305–309

Beutler E (1975) Red cell metabolism. A manual of biochemical methods, 2nd ed. Grune and Stratton, NY, pp 40–45

Bradford M (1976) A rapid and sensitive method for the quantitation of microgram quantities of protein utilizing the principle of protein- dye binding. Anal Biochem 72: 248–254

Burger PC, Vogel FS (1973) The development of pathologic changes of Alzheimer's disease and senile dementia in patients with Down's syndrome. Am J Pathol 73: 457–476

Dani A, Pietrini P, Furey M, McIntosh CL, Horwitz B, Freo U, Alexander GE, Schapiro MB (1996) Brain cognition and metabolism of Down syndrome adults in association with development of dementia. Neuroreport 7: 2933–2936

Epstein CJ (1986) The consequences of chromosomal imbalance. Principles, mechanisms, and models. Cambridge University Press, New York

Epstein CJ (1992) Down Syndrome (Trisomy 21). In: Scriver CR, Beaudet AL, Sly WS, Valle D (eds) The metabolic and molecular basis of inherited disease. McGraw Hill, New York, pp 749–794

Haxby JV, Schapiro MB (1988) Longitudinal study of neuropsychological function in older adults with Down syndrome. In: Epstein C, Nadel L (eds) Down Syndrome and Alzheimer disease. Wiley-Liss, New York, pp 35–50

Haxby JV, Duara R, Grady CL, Cutler NR, Rapoport SI (1985) Relations between neuropsychological and cerebral metabolic asymmetries in early Alzheimer's disease. J Cereb Blood Flow Metab 5: 193–200

Hsia DY, Justice P, Smith GF, Dowben RM (1971) Down's syndrome. A critical review of the biochemical and immunological data. Am J Dis Child 121: 153–163

Katz LA (1996) Transkingdom transfer of the phosphoglucose isomerase gene. J Mol Evol 43: 453–459

Labudova O, Lubec G (1998) cAMP upregulates the transposable element mys-1: a possible link between signaling and mobile DNA. Life Sci 62: 431–437

Lai F, Williams RS (1989) A prospective study of Alzheimer's disease in Down syndrome. Arch Neurol 46: 849–853

Layzer RB, Epstein CJ (1972) Phosphofructokinase and chromosome 21. Am J Hum Genet 24: 533–543

Mirra SS, Heyman A, McKeel D, Sumi S, Crain BJ (1991) The consortium to establish a registry for Alzheimer disease (CERAD). II. Standardisation of the neuropathological assessment of Alzheimer's disease. Neurology 41: 479–486

Pash J, Smithgall T, Bustin M (1991) Chromosomal protein HMG — 14 is overexpressed in Down syndrome. Exp Cell Res 193: 232–236

Pietrini P, Dani A, Furey ML, Alexander GE, Freo U, Grady CL, Mentis MJ, Mangot D, Simon EW, Horwitz B, Hazby JV, Schapiro MB (1997) Low glucose metabolism during brain stimulation in older Down's Syndrome subjects at risk for Alzheimer's disease prior to dementia. Am J Psychiatry 154: 1063–1069

Schapiro MB, Grady CL, Haxby JV (1992) Nature of mental retardation and dementia in Down's syndrome: study with PET, CT and neuropsychology. Neurobiol Aging 13: 723–734

Schmidt HJ, Schaum U, Pichotka JP (1977) The influence of mode and intensity of homogenization on the absolute value and stability of oxygen consumption of guinea pig liver homogenates. Z Naturforsch 32: 908–912

Seidl R, Greber S, Schuller E, Bernert G, Cairns N, Lubec G (1997) Evidence against increased oxidative DNA damage in Down Syndrome. Neurosci Lett 235: 137–140

Tierney MC, Fisher RH, Lewis AJ, Torzitto ML, Snow WG, Reid DW, Nieuwstraaten P, Van Rooijen LAA, Derks HJGM, Van Wijk R, Bischop A (1988) The NINCDA-ADRDA work group criteria for the clinical diagnosis of probable Alzheimer's disease. Neurology 38: 359–364

Vora S, Franke U (1981) Assignment of the human gene for liver-type 6-phosphofruc-tokinase isoenzyme (PFK-L) to chromosome 21 by using somatic cell hybrids and monoclonal anti-L-antibody. Proc Natl Acad Sci USA 78: 3738–3742

Wisniewski KE, Wisniewski HM, Wen GY (1985) Occurrence of neuropathological changes and dementia of Alzheimer's disease in Down's syndrome. Ann Neurol 17: 278–282

Authors' address: Prof. Dr. G. Lubec, Department of Pediatrics, University of Vienna, Währinger Gürtel 18, A-1090 Vienna, Austria, e-mail: gert.lubec@akh-wien.ac.at

J Neural Transm (1999) [Suppl] 57: 257–267

Oxidative stress and neural dysfunction in Down Syndrome

R. C. Iannello, P. J. Crack, J. B. de Haan, and **I. Kola**

Centre for Functional Genomics and Human Disease, Institute of Reproduction
and Development, Monash Medical Centre, Clayton, Australia

Summary. Total or partial trisomy of chromosome 21 occurs with relatively high frequency and is responsible for the occurrence of Down syndrome. Phenotypically, individuals with Down syndrome display characteristic morphological features and a variety of clinical disorders. One of the challenges for researchers in this field has been to ascertain and understand the relationship between the Down syndrome phenotype with the gene dosage effect resulting from trisomy of chromosome 21. Much attention therefore, has been given towards investigating the consequences of overexpressing chromosome 21-linked genes. In particular, an extensive analysis of SOD1 and APP have provided important insights as to how perturbations in the expression of their respective genes may contribute to the Down syndrome phenotype. In this review we will highlight studies which support a key role for SOD1 and APP in the pathogenesis of neural abnormalities observed in individuals with Down syndrome. Central to this relationship is how the redox state of the cell is affected and its consequences to neural function and integrity.

Introduction

One of the most prevalent genetic disorders in our community today is Down syndrome (DS). It occurs at a frequency of one in 700–1,000 live births and results as a consequence of either a total or partial trisomy of chromosome 21 (Lejeune et al., 1959). Although patients with DS are characterised by very specific morphological features (Patterson, 1987), they exhibit a broad and varied range of clinical pathologies including mental retardation, microcephaly, bone and skeletal abnormalities, congenital malformations, and increased incidence of leukemia and diabetes (Epstein, 1986; Pueschel, 1990; Kola, 1997; Kola and Hertzog, 1997). One of the most obvious features of DS is premature aging, with many DS individuals often developing an early onset of Alzheimer's type dementia (Kesslak et al., 1994; Korenberg, 1995).

A central approach towards elucidating the mechanisms which contribute to the different pathologies in DS is based on the favoured hypothesis that DS is a gene dosage disorder (Anneren and Edman, 1993; Groner, 1995; Sumarsono et al., 1996; Kola, 1997). As such, considerable effort and progress is being made towards the identification and determination of the biological roles of chromosome 21-linked genes, several of which have already provided insights into the molecular processes underlying this disorder (Kola and Hertzog, 1997). One of these processes implicates reactive oxygen species (ROS) as important contributors in the pathogenesis of the DS phenotype. ROS are molecules which are formed spontaneously as part of the natural cellular processes involving oxygen, and when present at relatively low concentrations, ROS serve as important signaling molecules and regulators of gene expression (Khan and Wilson, 1995; Winyard and Blake, 1997). However, at high concentrations, they have the potential to cause cellular damage either directly or by serving as substrates for the formation of other ROS (Kehrer, 1993).

A number of chromosome 21-linked genes have been implicated in modulating the steady state levels of cellular ROS. Consequently, oxidative stress resulting from the disruption of this steady state is thought to play a prominent role in several of the clinical features of DS. The aim of this review, is to highlight the relationship between genes located on chromosome 21 and their influence on cellular antioxidant pathways. In particular, we will focus on how perturbations to these pathways may account for some of the clinical outcomes associated with DS such as Alzheimer's-type dementia and premature aging. There are a number of genes located on chromosome 21 which have been shown or suspected of playing a role in regulating the redox state of the cell. These include the *CBS* gene which codes for the enzyme cystathione-β-synthase (Kraus et al., 1998), the gene coding for carbonyl reductase (*CBR*) (Wermuth, 1982; Kelner, 1997), the heatshock protein family member, *STCH* (Brodsky et al., 1995), and a newly described gene, *DSCR1* (Fuentes et al., 1995). However, further work on these genes is required before their contribution to perturbations in ROS metabolism in the context of DS is revealed. Therefore, for the purposes of this review, much of the discussion will centered around two extensively studied chromosome 21-linked genes, *SOD1* and *APP*.

Cu/Zn-superoxide dismutase-1

The gene coding for the antioxidant enzyme, Cu/Zn-superoxide dismutase-1 (SOD1), is localized to 21q22.1. This enzyme plays a key role in the metabolism of ROS (Fridovich, 1986) by catalysing the dismutation of superoxide anions to hydrogen peroxide. Hydrogen peroxide is then neutralized to water and oxygen through the actions of either glutathione peroxidase (GPX) and/or catalase (Yu, 1994; de Haan et al., 1995, 1997). This combined enzymatic action also prevents the subsequent formation of the more noxious and reactive hydroxyl radicals (Baker and Gebicki, 1984; Kappus, 1985; Yim et al., 1990). A number of studies have demonstrated that in individuals with DS, the levels of SOD1 are elevated in a variety of cell types and organs (Feaster et al., 1977; Brooksbank and Balazs, 1984; Anneren and Epstein, 1987; de Haan et al., 1995, 1997; Cristiano et al., 1995). It has therefore been speculated that the production of hydrogen peroxide in these tissues is increased causing a shift in the steady state level of ROS, thus triggering a series of molecular events leading to metabolic impairment and loss of cellular function.

Of particular significance is the observation that in most diploid tissues, GPX and/or catalase activities increase to compensate elevated levels of SOD1 as a function of age. However, no such adaptation by either GPX or catalase was observed in brain (de Haan et al., 1992; Cristiano et al., 1995). This has important implications, particularly with respect to the neuropathologies associated with DS since its suggests that the brain is more prone to oxidative stress. Indeed, a number of studies have already demonstrated that DS neurons generate increased levels of ROS. In one study, primary cultures of DS cortical neurons from aborted conceptuses differentiate normally in culture but subsequently degenerate and undergo apoptosis whereas control neurons remain viable (Busciglio and Yankner, 1995). More importantly, the extent of degeneration and apoptosis of DS neurons could be significantly reduced by free radical scavengers. Finally, further support for the involvement of increased SOD1 activity in neurological perturbations is derived from Gahtan et al. (1998). In their study, *SOD1* transgenic mice were shown to have an increase in tetanic stimulation-evoked formation of hydrogen peroxide resulting in cognitive deficits and impaired hippocampal long term potentiation.

The importance of SOD1 in chronic neurodegeneration is highlighted in familial amyotrophic lateral sclerosis (FALS). In this disorder, approximately 15–25% of FALS patients have mutations in Cu/Zn superoxide dismutase. These mutations result in a gain of SOD1 function and links FALS to toxicity by reactive oxygen species. Transgenic mice carrying human SOD1 mutations develop selective motor neuron impairment that strongly resembles the pathology seen in FALS patients (Gurney et al., 1994). Evidence from these transgenic models suggests that the mutant enzymes in SOD1-linked FALS enhance the production of hydroxyl radical and hydrogen peroxide thereby resulting in oxidative damage of vital cellular targets (Bruijn et al., 1997; Bogdanov et al., 1998). Interestingly, it has been recently reported that recapitulating SOD1 mutations such as those found in FALS selectively inactivate the glial glutamate transporter GLT1 (Trotti et al., 1999). GLT1 is the predominant mechanism for the clearing of glutamate from the synaptic cleft and protecting neurons from glutamate toxicity. Inactivation of GLT1 is known to result in neuronal degeneration via excitotoxic mechanisms (Rothstein et al., 1996). Although there are no known reports of DS individuals exhibiting FALS-like pathology, this study provides the first hard evidence that oxidative stress, through SOD1, can lead directly to molecular perturbations of specific cellular targets and subsequent neuronal death.

In the past, one of the more extensively investigated areas relating to DS research has been the phenomenon of premature aging. The proposal that SOD1 may have a strong pathophysiological involvement in this process comes from research investigating the normal aging process. In those studies a strong correlation was demonstrated between increased generation of ROS

(Harman, 1994), an altered antioxidant defense and increased oxidative stress and cellular damage (Matsuo, 1993) as a function of age. As expected from a gene dosage effect, an analysis of antioxidant enzymes in DS patients revealed that SOD1 activity was elevated in various cell types and organs, including fetal brain (Sinet et al., 1975; Anneren and Epstein, 1987; Feaster et al., 1977; Brooksbank and Balazs, 1984). Interestingly however, the levels of catalase or GPX activity in many of these tissues and cells were relatively unchanged compared with non-DS controls. The implication here being that the failure of GPX and catalase to adapt to the increased levels of SOD1 activity would result in the generation of ROS, oxidative stress and cellular damage. Indeed, in DS tissues where the SOD1/GPX ratio was skewed, lipid peroxidation was observed. As mentioned earlier, studies undertaken by de Haan et al. (1992) and Cristiano et al. (1995) demonstrated that there was a concomitant increase in GPX/catalase activity with an increase in SOD1 activity in most organs tested as a function of age. Interestingly, the brain was the only organ in which this adaptation did not occur and where lipid peroxidation increased with age. Furthermore, analysis of cell lines which overexpress SOD1 (as a consequence of *SOD1* transfection), but not a concomitant increase in either GPX or catalase activity, were demonstrated to contain higher intracellular levels of hydrogen peroxide and displayed a senescence phenotype indicated by a slower growth rate, altered morphology and increased expression of the senescence marker, Cip1 (de Haan et al., 1996) . Conversely, cell lines which adapted to the overexpression of SOD1 by elevating cellular GPX activity were biochemically and morphologically indistinguishable from their untransfected parental controls. The study therefore provided evidence to support a role for SOD1, alterations in the SOD1/GPX ratio and oxidative stress in the aging process.

Additional support for the hypothesis that an altered antioxidant balance contributes to oxidative damage and aging is derived from the analysis of *SOD1* transgenic mice. Avraham et al. (1988, 1991) and Yarom et al. (1988) reported that *SOD1* transgenic mice undergo premature aging with respect to neuromuscular junction morphology (NMJ). In these mice, the degeneration and withdrawal of terminal axons in tongue muscle along with the degeneration of endplate structures of the leg muscle were similar to the changes observed in muscles of aging rats and mice (Fahim and Robbins, 1982; Cardasis, 1983) and tongue muscles of individuals with DS (Yarom et al., 1986, 1987). Furthermore, neuromorphological abnormalities usually associated with premature aging such as the diminution of Mossy-fiber innervation in the hippocampus, was also reported to occur (Barkats et al., 1993) as was an increase in thiobarbituric acid-reactive material in the brains of these mice, suggesting increased levels of peroxidation and oxidative damage (Ceballot-Picot et al., 1992).

Together, the studies presented here, provide evidence supporting the notion that ROS and oxidative stress may be implicated in neuronal abnormalities associated with DS. Central to this is the influence SOD1 exerts on cellular antioxidant pathways, and significantly, its consequences in the brain which appears to be particularly susceptible to oxidative stress.

Amyloid precursor protein

In addition to a high incident of mental retardation there are also several neuroanatomical abnormalities which associated with the DS phenotype. These include decreased brain weight, diminished gyral size and a simplified convolutional pattern (Crome and Stern, 1972; Urich, 1976). At the microscopic level DS brains present decreased neuronal number/density (Colon, 1972; Galaburda and Kemper, 1979; Ross et al., 1984), trans-synaptic degeneration (Marin-Padilla, 1976) and alterations in dendritic spine structure (Marin-Padilla, 1972). Several neurotransmitters are also depleted in the DS brain (Coyle et al., 1986) and electrophysiological alterations have been described in dorsal root ganglion cells in culture (Scott et al., 1981; Caviedes et al., 1990, 1991). Notably however, most DS patients exhibit dementia and Alzheimer's like pathology by middle age. Consequently, researchers have investigated a number of parallels between DS and Alzheimer's disease (AD). In both AD and DS, intracellular and extracellular deposits of proteins in tangles, neuropil threads, and neuritic plaques are correlated with neuronal dysfunction leading to dementia (Hardy, 1997). Of particular significance is the identification of the genetic basis underlying the familial types of AD of which mutations in the genes responsible have been mapped to chromosomes 1, 14, 19 and 21.

There is now mounting evidence that both amyloid β-peptide (Aβ) and oxidative stress are implicated in the pathogenesis of AD. It has been shown that Aβ, which is derived from the amyloid precursor protein (APP) is overproduced in the brain of individuals with AD and forms insoluble fibrillar aggregates (senile or neuritic plaques) that promote neuronal degeneration (Hardy, 1997). Since the APP gene resides within chromosome 21, DS patients would have an increased production of this protein. Indeed, historically, it was this linkage of APP to chromosome 21 and the correlation that AD is present in a large majority of the DS population that gave enormous impetus for the amyloid theory of Alzheimer's disease.

A possible neurotoxic effect of Aβ deposits may involve the generation of reactive oxygen species (ROS) which disrupts cellular calcium homeostasis and render neurons more susceptible to excitotoxicity and apoptosis. In vitro studies have shown that the Aβ peptides can be toxic to a wide variety of neuronal cell types through the disruption of Ca2+ homeostasis (Mattson et al., 1992, 1993). Under this model, disturbances in Ca^{2+} homeostasis, through the accumulation of Aβ, could lead to increased intracellular Ca^{2+} (Koizumi et al., 1998; Silei et al., 1999). This would cause depolarization of the mitochondrial membrane, thereby disturbing the respiratory chain and production of ATP (Cassarino and Bennet, 1998). In turn, ATP deficiency could interfere with the function of plasma membrane Ca2+ ATPase resulting in a further elevation of cytosolic Ca2+ and subsequent cell death. It is interesting to note that cortical and hippocampal neurons from trisomy 16 mice, TS16 (a mouse model for DS), also display elevated levels of basal Ca2+ (Schuchmann et al., 1998) and that glutamate augmented cell death in TS16 neuronal cultures. While it is unclear from this study if susceptibility to glutamate induced excitotoxicity is linked to Ca2+ homeostasis or increased mitochondrial ROS production, other studies have clearly demonstrated a relationship between high concentrations of Aβ and the production of ROS (Hensley et al., 1994; Behl, 1994). In addition, Yan et al. (1996) showed that Aβ can bind to RAGE (a receptor for advanced-glycosylation end products) at concentrations which generate markers of oxidative stress. Finally, complimenting these studies is the correlation which exists between increased oxidative stress on neurons and with plaques and tangles in Alzheimer's disease (Yan et al., 1996), Lewy bodies in Parkinson's disease and Lewy body associated dementia (Lyras et al., 1998).

Interestingly, some experimental evidence suggests that oxidative stress may actually exacerbate Aβ aggregation (Multhaup et al., 1997) while fibrillar Aβ may increase intraneuronal generation of free radicals (Yan et al., 1995). This in turn, could provide a trigger for an inflammatory reaction causing further oxidative damage. Several studies have implicated hydrogen peroxide as being the major source of oxidative stress generated by Aβ peptides in in vitro systems (Behl et al., 1994; Yan et al., 1995). Lending support to this notion are in vivo, studies which show that SOD1 levels in the brains of aged APP mice are elevated, especially around Aβ deposits (Papolla et al., 1998).

Finally, oxidative stress impairs mitochondrial function, resulting in energy depletion. Impaired cellular energy metabolism in turn increases the amyloidogenic processing of Aβ precursor protein (APP) potentially leading to increased Aβ production and subsequent oxidative stress (Gabuzda et al., 1994). What this leads to is a pathological vicious cycle, which amplifies itself to increase oxidative stress and thereby neuronal damage. These studies described above not only strengthen support for the involvement of these two chromosome 21-linked genes in the neuropathology observed in DS but also shed some important insights as to the molecular pathways which may be operating in this disorder.

In familiar AD, mutations involving genes coding for APP, presenilin 1 (PS1) and presenilin (PS2) result in increased production of the 42(43) Aβ peptide via the abnormal processing of APP. However, in DS, these familial mutations are not common (Hardy 1997; Tanzi et al., 1996). In a recent article, van Leewen and colleagues (1998) presented evidence of novel transcript frameshift mutations of APP and Ubiquitin-B to account for some non-familial types of AD. These mutations were found to occur at the transcript level and not the genomic level implying that these mutations were generated via a transcriptional or post-transcriptional RNA editing process. It is hypothesized that postmitotic neurons are less capable of compensating for transcript-modifying activity are thus particularly sensitive to the accumulation of frameshifted proteins. Thus, during aging, neurons may generate and accumulate abnormal proteins leading to cellular disturbances and neurodegeneration. It is tempting to speculate that transcript mutations in APP (or PS1 and PS2) caused by increased oxidative stress may be one mechanism by which Alzheimer's is propagated in the DS population.

Conclusion

There is now strong experimental evidence to support a prominent role for ROS and oxidative stress in the neuropathogenesis of DS. The imbalance of ROS homeostasis resulting from elevated levels of SOD1 and overexpression of APP, has been eloquently demonstrated through a number of investigative avenues. While the analysis of DS tissue has provided useful insights, a more fundamental understanding of this disorder has come from the use of transgenic mouse models and an analysis of ROS involvement in other neurological disorders.

The approach towards the delineation of the molecular pathways involved in the DS phenotype mostly centers around the investigation of single chromosome 21 genes and the consequences of their overexpression. However, as more genes become available for analysis and their biological functions better understood, the challenge for researchers in this field will be to understand the compounding effects of multiple genes and how the sum of these perturbations give rise to the DS phenotype.

References

Anneren G, Epstein CJ (1987) Lipid peroxidation and superoxide dismutase-1 and glutathione peroxidase activities in trisomy 16 fetal mice and human trisomy 21 fibroblasts. Pediatr Res 21: 88–92

Anneren G, Edman B (1993) Down syndrome- a gene dosage disease caused by trisomy of genes within a small segment of the long arm of chromosome 21, exemplified by the study of effects from the superoxide-dismutase type 1 (SOD-1) gene. APMIS [Suppl] 40: 71–79

Avraham KB, Schickler M, Sapoznikov D, Yarom R, Groner Y (1988) Down's syndrome: abnormal neuromuscular junction in tongue of transgenic mice with elevated levels of human Cu/Zn-superoxide dismutase. Cell 54(6): 823–829

Avraham KB, Sugarman H, Rotshenker S, Groner Y (1991) Down's syndrome: morphological remodeling and increased complexity in the neuromuscular junction of transgenic CuZn-superoxide dismutase mice. J Neurocytol 20(3): 208–215

Baker MS, Gebicki JM (1984) The effect of pH on the conversion of superoxide to hydroxyl free radicals. Arch Biochem Biophys 234: 258–262

Barkats M, Bertholet JY, Venault P, Ceballos-Picot I, Nicole A, Phillips J, Moutier R, Roubertoux P, Sinet PM, Cohen-Salmon C (1993) Hippocampal mossy fiber changes in mice transgenic for the human copper-zinc superoxide dismutase gene. Neurosci Lett 160(1): 24–28

Behl C, Davis JB, Lesley R, Schubert D (1994) Hydrogen peroxide mediates amyloid beta protein toxicity. Cell 77(6): 817–827

Bogdanov MB, Ramos LE, Xu Z, Beal MF (1998) Elevated "hydroxyl radical" generation in vivo in an animal model ofamyotrophic lateral sclerosis. J Neurochem 71(3): 1321–1324

Brodsky G, Otterson GA, Parry BB, Hart I, Patterson D, Kaye FJ (1995) Localization of STCH to human chromosome 21q11.1. Genomics 30(3): 627–628

Brooksbank BWL, Balazs R (1984) Superoxide dismutase, glutathione peroxidase and lipid peroxidation in Down's syndrome fetal brain. Dev Brain Res 16: 37–44

Bruijn LI, Beal MF, Becher MW, Schulz JB, Wong PC, Price DL, Cleveland DW (1997) Elevated free nitrotyrosine levels, but not protein-bound nitrotyrosine or hydroxyl radicals, throughout amyotrophic lateral sclerosis (ALS)-like disease implicate tyrosine nitration as an aberrant in vivo property of one familial ALS-linked superoxide dismutase 1 mutant. PNAS 94(14): 7606–7611

Busciglio J, Yankner BA (1995) Apoptosis and increased generation of reactive oxygen species in Down's syndrome neurons in vitro. Nature 378(6559): 776–779

Cardasis CA (1983) Ultrastructural evidence of continued reorganization at the aging (11–26 months) rat soleus neuromuscular junction. Anat Rec 207: 399–415

Cassarino DS, Bennett JP Jr (1999) An evaluation of the role of mitochondria in neurodegenerative diseases: mitochondrial mutations and oxidative pathology, protective nuclear responses, and cell death in neurodegeneration. Brain Res Brain Res Rev 29(1): 1–25

Caviedes P, Ault B, Rapoport SI (1990) The role of altered sodium currents in action potential abnormalities of cultured dorsal root ganglion neurons from trisomy 21 (Down syndrome) human fetuses. Brain Res 510(2): 229–236

Caviedes P, Koistinaho J, Ault B, Rapoport SI (1991) Effects of nerve growth factor on electrical membrane properties of cultured dorsal root ganglia neurons from normal and trisomy 21 human fetuses. Brain Res 556(2): 285–291

Ceballos-Picot I, Nicole A, Clement M, Bourre JM, Sinet PM (1992) Age-related changes in antioxidant enzymes and lipid peroxidation in brains of control and transgenic mice overexpressing copper-zinc superoxide dismutase. Mutat Res 275(3–6): 281–293

Cristiano F, de Haan JB, Iannello RC, Kola I (1995) Changes in the levels of enzymes which modulate the antioxidant balance occur during aging and correlate with cellular damage. Mech Ageing Dev 80(2): 93–105

Colon EJ (1972) The structure of the cerebral cortex in Down's syndrome: a quantitative analysis. Neuropadiatrie 3: 362–376

Coyle JT, Oster-Granite ML, Gearhart JD (1986) The neurobiologic consequences of Down syndrome. Brain Res Bull 16(6): 773–787

Crome L, Stern J (1972) Pathology of mental retardation. In: Crome L, Stern J (eds) Down syndrome. Churchill Livingstone, Edinburgh, pp 200–224

de Haan JB, Newman JD, Kola I (1992) Cu/Zn superoxide dismutase mRNA and enzyme activity, and susceptibility to lipid peroxidation, increases with aging in murine brains. Brain Res Mol Brain Res 13(3): 179–187

de Haan JB, Cristiano F, Iannello RC, Kola I (1995) Cu/Zn-superoxide dismutase and glutathione peroxidase during aging. Biochem Mol Biol Int 35(6): 1281–1297

de Haan JB, Cristiano F, Iannello R, Bladier C, Kelner MJ, Kola I (1996) Elevation in the ratio of Cu/Zn-superoxide dismutase to glutathione peroxidase activity induces features of cellular senescence and this effect is mediated by hydrogen peroxide. Hum Mol Genet 5(2): 283–292

de Haan JB, Wolvetang EJ, Cristiano F, Iannello R, Bladier C, Kelner MJ, Kola I (1997) Reactive oxygen species and their contribution to pathology in Down syndrome. Adv Pharmacol 38: 379–402

Epstein CJ (1986) The consequences of chromosome imbalance: principles, mechanisms and models. Cambridge University Press, New York

Fahim MA, Robbins N (1982) Ultrastructural studies of young and old mouse neuromuscular junctions. J Neurocytol 11(4): 641–656

Feaster WW, Kwok LW, Epstein CJ (1977) Dosage effects for superoxide dismutase-1 on nucleated cells aneuploid for chromosome 21. Am J Hum Genet 29: 563–570

Fridovich I (1986) Superoxide dismutases. Adv Enzymol Rel Areas Mol Biol 58: 61–97

Fuentes JJ, Pritchard MA, Planas AM, Bosch A, Ferrer I, Estivill X (1995) A new human gene from the Down syndrome critical region encodes a proline-rich protein highly expressed in fetal brain and heart. Hum Mol Genet 4(10): 1935–1944

Gabuzda D, Busciglio J, Chen LB, Matsudaira P, Yankner BA (1994) Inhibition of energy metabolism alters the processing of amyloid precursor protein and induces a potentially amyloidogenic derivative. J Biol Chem 269(18): 13623–13628

Gahtan E, Auerbach JM, Groner Y, Segal M (1998) Reversible impairment of long-term potentiation in transgenic Cu/Zn-SOD mice. Eur J Neurosci 10(2): 538–544

Galaburda AM, Kemper TL (1979) Cytoarchitectonic abnormalities in developmental dyslexia: a case study. Ann Neurol 6(2): 94–100

Groner Y (1995) Transgenic models for chromosome 21 gene dosage effects. Prog Clin Biol Res 393: 193–212

Gurney ME, Pu H, Chiu AY, Dal Canto MC, Polchow CY, Alexander DD, Caliendo J, Hentati A, Kwon YW, Deng HX, et al (1994) Motor neuron degeneration in mice that express a human Cu,Zn superoxide dismutase mutation. Science 264(5166): 1772–1775

Hardy J (1997) Amyloid, the presenilins and Alzheimer's disease. TINS 20: 154–159

Harman D (1994) Free-radical theory of aging: increasing the functional life span. Ann NY Acad Sci 717: 1–15

Hensley K, Carney JM, Mattson MP, Aksenova M, Harris M, Wu JF, Floyd RA, Butterfield DA (1994) A model for Beta-amyloid aggregation and neurotoxicity based on free radical generation by the Peptide — relevance to alzheimer disease. PNAS 91(8): 3270–3274

Kappas H (1985) Lipid peroxidation: mechanism, analysis, enzymology and biological relevance. In: Sies H (ed) Oxidative stress. Academic Press, London, pp 273–311

Kehrer JP (1993) Free radicals, mediators of tissue injury and disease. Crit Rev Toxicol 23: 21–48

Kelner MJ, Estes L, Rutherford M, Uglik SF, Peitzke JA (1997) Heterologous expression of carbonyl reductase: demonstration of prostaglandin 9-ketoreductase activity and paraquat resistance. Life Sci 61: 2317–2322

Kesslak JP, Nagata SF, Lott I, Nalcioglu O (1994) Magnetic resonance imaging analysis of age-related changes in the brains of individuals with Down's syndrome. Neurology 44(6): 1039–1045

Khan AU, Wilson T (1995) Reactive oxygen species as cellular messengers. Chem Biol 2: 437–445

Koizumi S, Ishiguro M, Ohsawa I, Morimoto T, Takamura C, Inoue K, Kohsaka S (1998) The effect of a secreted form of beta-amyloid-precursor protein on intracellular Ca2+ increase in rat cultured hippocampal neurons. Br J Pharmacol 123(8): 1483–1489

Kola I (1997) Simple minded mice from "in vivo" libraries. Nature Genet 16: 8–9

Kola I, Hertzog PJ (1997) Animal models in the study of the biological function of genes on human chromosome 21 and their role in the pathophysiology of Down syndrome. Hum Mol Genet 6: 1713–1727

Korenberg JP (1995) Mental modeling. Nature Genet 11: 109–111

Kraus JP, Oliveriusova J, Sokolova J, Kraus E, Vlciek C, de Franchis R, Maclean KN, Bao L, Bukovska G, Patterson D, Pacies V, Ansorge W, Koziich V (1998) The human cystathionine β-synthase (CBS) gene: complete sequence, alternative splicing, and polymorphisms. Genomics 52: 313–324

Lejeune J, Gauthier M, Turpin R (1959) Etude des chromosomes somatiques de neufs enfants mongoliens. CR Acad Sci (Paris) 248: 1721–1722

Lyras L, Perry RH, Perry EK, Ince PG, Jenner A, Jenner P, Halliwell B (1998) Oxidative damage to proteins, lipids, and DNA in cortical brain regions from patients with dementia with Lewy bodies. J Neurochem 71(1): 302–312

Marin-Padilla M (1972) Pyramidal cell abnormalities in the motor cortex of a child with Down's aberrations: a Golgi study. Brain Res 44(2): 625–629

Marin-Padilla M (1972) Structural abnormalities of the cerebral cortex in human chromosomal aberrations: a Golgi study. Brain Res 44(2): 625–629

Mattson MP, Cheng B, Davis D, Bryant K, Lieberburg I, Rydel RE (1992) beta-Amyloid peptides destabilize calcium homeostasis and render human cortical neurons vulnerable to excitotoxicity. J Neurosci 12(2): 376–389

Mattson MP, Cheng B, Culwell AR, Esch FS, Lieberburg I, Rydel RE (1993) Evidence for excitoprotective and intraneuronal calcium-regulating roles for secreted forms of the beta-amyloid precursor protein. Neuron 10(2): 243–254

Matsuo M (1993) Age-related alterations in antioxidative defense. In: Yu BP (ed) Free radicals in aging. CRC Press, Boca Raton, pp 143–181

Multhaup G, Ruppert T, Schlicksupp A, Hesse L, Beher D, Masters CL, Beyreuther K (1997) Reactive oxygen species and Alzheimer's disease. Biochem Pharmacol 54(5): 533–559

Pappolla MA, Chyan YJ, Omar RA, Hsiao K, Perry G, Smith MA, Bozner P (1998) Evidence of oxidative stress and in vivo neurotoxicity of beta-amyloid in a transgenic mouse model of Alzheimer's disease: a chronic oxidative paradigm for testing anti-oxidant therapies in vivo. Am J Pathol 152(4): 871–877

Patterson DH (1987) The cause of Down Syndrome. Sci Am 257: 42–49

Pueschel S (1990) Clinical aspects of Down's syndrome from infancy to adulthood. Am J Med Genet 7: 52–56

Ross MH, Galaburda AM, Kemper TL (1984) Down's syndrome: is there a decreased population of neurons? Neurology 34(7): 909–916

Rothstein JD, Dykes-Hoberg M, Pardo Ca, Bristol LA, Jin L, Kuncl RW, Kanai Y, Hediger MA, Wang Y, Schielke JP, Welty DF (1996) Knockout of glutamate transporters reveals a major role for astroglial transport in excitotoxicity and clearance of glutamate. Neuron 16(3): 675–686

Schuchmann S, Muller W, Heinemann U (1998) Altered Ca2+ signaling and mitochondrial deficiencies in hippocampal neurons of trisomy 16 mice: a model of Down's syndrome. J Neurosci 18(18): 7216–7231

Scott BS, Petit TL, Becker LE, Edwards BA (1981) Abnormal electric membrane properties of Down's syndrome DRG neurons in cell culture. Brain Res 254(2): 257–270

Silei V, Fabrizi C, Venturini G, Salmona M, Bugiani O, Tagliavini F, Lauro GM (1999) Activation of microglial cells by PrP and beta-amyloid framenhts raises intracellular calcium through L-type voltage sensitive calcium channels. Brain Res 818(1): 168–170

Sinet PM, Michelson AM, Bazin A, Lejeune J, Jerome H (1975) Increase in glutathione peroxidase activity in erythrocytes from trisomy 21 subjects. Biochem Biophys Res Comm 67: 910–915

Sumarsono SH, Wilson TJ, Tymms MJ, Venter DJ, Corrick CM, Kola R, Lahoud MH, Papas TS, Seth A, Kola I (1996) Down's syndrome-like skeletal abnormalities in Ets2 transgenic mice. Nature 379(6565): 534–537

Tanzi RE, Kovacs DM, Kim TW, Moir RD, Guenette SY, Wasco W (1996) The gene defects responsible for familial Alzheimer's disease. Neurobiol Dis 3(3): 159–168

Trotti D, Rolfs A, Danbolt NC, Brown RH, Hediger MA (1999) SOD1 mutants linked to amyotrophic lateral sclerosis inactivate a glial glutamate transporter. Nat Neurosci 2(5): 427–433

van Leeuwen FW, de Kleijn DP, van den Hurk HH, Neubauer A, Sonnemans MA, Sluijs JA, Koycu S, Ramdjielal RDJ, Salehi A, Martens GJM, Grosveld FG, Peter J, Burbach H, Hol EM (1998) Frameshift mutants of beta amyloid precursor protein and ubiquitin-B in Alzheimer's and Down patients. Science 279(5348): 242–247

Wermeth B (1982) Enzymology of carbonyl metabolism. AR Liss, New York, pp 261–274

Winyard PG, Blake DR (1997) Antioxidants, redox-regulated transcription factors, and inflammation. Adv Pharmacol 38: 403–421

Yan SD, Yan SF, Chen X, Fu J, Chen M, Kuppusamy P, Smith MA, Perry G, Godman GC, Nawroth P, et al (1995) Non-enzymatically glycated tau in Alzheimer's disease induces neuronal oxidant stress resulting in cytokine gene expression and release of amyloid beta-peptide. Nat Med 1(7): 693–699

Yan SD, Chen X, Fu J, Chen M, Zhu H, Roher A, Slattery T, Zhao L, Nagashima M, Morser J, Migheli A, Nawroth P, Stern D, Schmidt AM (1996) RAGE and amyloid-beta peptide neurotoxicity in Alzheimer's disease. Nature 382(6593): 685–691

Yarom R, Sagher U, Havivi Y, Peled IJ, Wexler MR (1986) Myofibers in tongues of Down's syndrome. J Neurol Sci 73(3): 279–287

Yarom R, Sherman Y, Sagher U, Peled IJ, Wexler MR, Gorodetsky R, Elevated (1987) concentrations of elements and abnormalities of neuromuscular junctions in tongue muscles of Down's syndrome. J Neurol Sci 79(3): 315–326

Yarom R, Sapoznikov D, Havivi Y, Avraham KB, Schickler M, Groner Y (1988) Premature aging changes in neuromuscular junctions of transgenic mice with an extra human CuZnSOD gene: a model for tongue pathology in Down's syndrome. J Neurol Sci 88(1–3): 41–53

Yim MB, Chock PB, Stadtman ER (1993) Enzyme function of copper, zinc superoxide as a free radical generator. J Biol Chem 268: 4099–4105

Yu BP (1994) Cellular defenses against damage from reactive oxygen species. Physiol Rev 74: 139–162

Authors' address: Prof. I. Kola, Centre for Functional Genomics and Human Disease, Institute of Reproduction and Development, Monash Medical Centre, 246 Clayton Road, Clayton, VIC 3168, Australia, e-mail: ismailko@silas.cc.monash.edu.au

J Neural Transm (1999) [Suppl] 57: 269–281

Expression of the transcription factor ETS2 in brain of patients with Down Syndrome—evidence against the overexpression-gene dosage hypothesis

S. Greber-Platzer[1], **D. Schatzmann-Turhani**[1], **N. Cairns**[2], **B. Balcz**[1], and **G. Lubec**[1]

[1] Department of Pediatrics, Division of Cardiology, University of Vienna, Vienna, Austria
[2] Institute of Psychiatry, Brain Bank, London, United kingdom

Summary.   Overexpression of the transcription factor ETS2 and other genes localized in the socalled critical Down Syndrome region of chromosome 21 due to a gene dosage effect, is an attractive hypothesis for the explanation of the Down Syndrome phenotype. The overexpression of ETS2, however, has never been demonstrated in a human organ. We therefore challenged this hypothesis determining ETS2 levels in several brain regions of patients with Down Syndrome as compared to controls.

We used a highly sensitive and quantitative RT-PCR method for the determination of ETS2 mRNA steady state levels in frontal, parietal, temporal, occipital lobe and cerebellum of 9 adult Down Syndrome patients and 9 adult controls.

Significantly decreased ETS2 mRNA steady state levels (16,9 ± 26,7 attogram mRNA ETS2/10 ng total RNA versus 87,7 ± 92,9 in controls) in frontal lobe of Down Syndrome brain and decreased ETS2 mRNA steady state levels (6,99 ± 6,4 attogram mRNA ETS2/100 pg beta-actin versus 19,8 ± 15,7 in controls) in temporal lobe of Down Syndrome brain were found. In the other brain regions no statistically significant difference was detected.

Our data provide evidence against the overexpression hypothesis for the development of the Down Syndrome phenotype. Decreased ETS2 transcripts found in temporal and frontal lobe of patients with Down Syndrome, however, may be involved in the pathogenesis of Down Syndrome including specific neurodegenerative processes and deteriorated plasticity of the brain taking place in Down Syndrome brain, as the concerted action of transcription factors may be seriously impaired.

Abbreviations

DS Down Syndrome, *C* Controls, *FL* frontal lobe, *PL* parietal lobe, *TL* temporal lobe, *OL* occipital lobe, *CB* Cerebellum.

Introduction

Trisomy 21, the chromosome abnormality found in more than 95% of Down Syndrome (DS) patients, is the most common trisomy with an overall incidence of one in 700 live births and the most common genetic cause of mental retardation (Shen, 1995). Mental retardation is seen in nearly 100% of cases but other traits, including the classic facial features, have variable penetrance and expressivity within the DS population (Epstein, 1991; Korenberg, 1994). Molecular analysis of rare patients with DS and partial chromosome 21 duplications and monosomies has led to the association of certain chromosomal regions with specific DS phenotypes and it is now widely accepted that genes in a single region of chromosome 21 are largely responsible for a given phenotype with the penetrance when the expressivity of the trait are the same in individuals with full trisomy 21 and in individuals with duplications of the single region (Huret, 1987, 1995). Therefore, the region D21S55 is of main interest due to ist association with several phenotypic features and mental retardation (Sinet, 1994).

The human ETS2 and ERG genes are members of the ETS gene family and are located in the critical region D21S55 to MX1 on chromosome 21. The ETS family of transcription factors (TF) is defined by the presence of a highly conserved domain of 85 amino acids, termed the ETS domain, which mediated specific binding to DNA purine rich sequences characterized by an invariant GGA/TA core element. ETS family members play a critical role in the transcriptional regulation of many genes and the ETS-DNA binding motif has been found in the regulatory control regions of numerous genes including growth factors, protooncogenes, viruses, TFs, genes important for immune functions and others (Wasylyk, 1993). ETS2 is ubiquitously expressed in human tissues including brain (Bhat, 1987). Studies in mice have shown that during organogenesis ETS2 expression is found in high concentrations in the cortex, epithelial cells and mesenchyme and it is suggested that the ETS2 TF may play an important role in CNS development and in the maintenance of certain neuronal functions (Maraoulakou, 1994).

Overexpression of ETS2 in DS is an attractive hypothesis for the explanation for the DS phenotype: ETS2, a TF which is widely distributed, would explain dysmorphic features, immunological and brain symptoms present in DS as the concerted action of TFs is necessary to enable normal brain development. Triplication of the ETS2 gene in DS could interfere with organogenesis of different tissues but so far only one report was published showing ETS2 overexpression in a single DS fibroblast cell line by using northern blot hybridization and RNAse protection assay (Kola, 1995) and there is no information on overexpression in DS organs.

The availability of a highly sensitive method for the determination of transcripts at the attogram level (Schatzmann-Turhani, in press) and of several DS brain regions from a brain bank made us challenging the overdosage hypothesis of this gene in DS brain.

Material and methods

Patients

5 different brain regions including the frontal lobe, parietal lobe, temporal lobe, occipital lobe and the cerebellum of 9 adult Down Syndrome individuals (n = 9; 3 females, 6 males; 56,1 ± 7,1 years) and 9 adult control persons without any known neurological and psychiatric disorder (n = 9; 5 females, 4 males; 72,6 ± 9,6 years) were analysed (see Table 1).

Reagents

— RNAzol B: Molecular Research Center, Inc., Cincinnati, OH,
— pSP64 Poly(A) Vector: Promega Corporation, Madison, WI, USA,
— TA Cloning Kit: INFaF' from Invitrogen Corporation, San Diego, CA, USA,
— restriction enzymes and Rapid DNA Ligation Kit: Boehringer Mannheim GmbH, Mannheim, Germany,
— Oligo(dt) Cellulose Columns, DNA and RNA molecular weight standards, yeast tRNA and cRNA SP6 transcription kit — SP6 Polymerase, DDT, buffer from Life Technologies, Gibco BRL, Maryland, USA,
— cDNA transcription reagents — RNase Inhibitor, Oligo $d(T)_{16}$, dNTPs, MuLV Reverse Transcriptase, buffer and PCR reagents — dNTPs, primers, $MgCl_2$, AmpliTaq DNA Polymerase, buffer: PE Applied Biosystems, Foster City, CA, USA and Roche Molecular System, Inc, Branchburg, NeW Jersey, USA, — GELase TM Agarose Gel-Digestion Preparation Epicentre Technologies, Madison, WI, USA, PCR purification kit: Qiagen Inc, Santa Clarita, CA, USA,
— oligosynthesis: Vienna Biocenter, Vienna Austria and Applied Biosystems GmbH, Weiterstadt, Germany,
— Chemicals as phenol, chloroform, isopropanol, different salts, agarose were purchased from Sigma Chemical Co,
— Reagents as Formamide, POP4, capillaries, 10xbuffer with EDTA and TAMRA 500 for the ABI Prism 310 Genetic Analyzer: PE Applied Biosystems, Foster City, CA, USA

Instruments

PCR was performed on a Perkin — Elmer GeneAmp PCR System 9600.

The nested PCR products labeled with FAM were analysed on the ABI-310 Genetic Analyzer from Applied Biosystem containing a 47 cm × 50 μm long capillary with Performance Optimized Polymer 4 (POP4) and as matrix standard TAMRA 500.

Synthesis of RNA Internal Standard for the protooncogene ets-2 and the house keeping gene beta-actin

We performed a synthetic RNA internal standard using a Competitor DNA — cloning strategy. Primer sequences were selected with the OLIGO Program from Rychlik and Rhoads (1989) for ets-2 and with the primer express program (Primer Express, PE Applied Biosystems, Foster City, CA, USA) for beta-actin.

Firstly we performed the internal standard for the protooncogene ets-2 cDNA. Prof. R. H. Scheuermann from the Southwestern Medical Center, Dallas was so kind to send us

Table 1. Human ets-2 primerpairs

Primerpair		Primersequence	cDNA position of the primer	Length of the amplified segment (bp)
primerpair A	forward	5'-GCAGCGGCAGGATGAATGAT-3'	281 bp	596 bp
	reverse	5'-CTCTGTGCCAAAACCTAATGT AGGAACGGAGGTGAGGTGTGAA-3'	834 bp	
primerpair B	forward	5'-TTCACACCTCACCTCCGTTCCT ACATTAGGTTTTGGCACAGAG-3'	874 bp	420 bp
	reverse	5'-GGCTTATTGAGGCAGAGAGAC-3'	1,251 bp	
primerpair 1 linked with BAM HI and Hind III restriction enzyme sites				1,006 bp
	forward	5'-GGAAGCTTGCAGCGGCAGGATGAATGAT-3'	281 bp	
	reverse	5'-GGGATCCGGCTTATTGAGGCAGAGAGAC-3'	1,251 bp	
primerpair 2 for nested PCR				wildtype 665 bp / internal standard 647 bp
	forward	5'-TGGAGTGAGCAACAGGTATG-3'	607 bp	
	reverse	5'-GGCTTATTGAGGCAGAGAGAC-3'	1,251 bp	
primerpair 3 — FAM labeled				wildtype 284 bp / internal standard 266 bp
	forward	5'-CTGGAGCTGGCACCTGACTT-3'	733 bp	
	reverse	5'-GACTTGGGGAACATCTGAAACT-3'	995 bp	

a Bluescript vector containing the ets2-cDNA, which was linearized by the restriction enzyme EcoR 1.

As first step PCR amplification procedure we performed two DNA fragments of the ets2-cDNA (fragment A 596 bp and fragment B 420 bp), each containing a 5'linkage of 21 or 22 bp respectively producing an overlapping region of 43 bp for both fragments (primerpair A and primerpair B ets-s see Table 2). Equal concentrations of both DNA fragments (fragment A and fragment B) were used as templates in the following PCR amplification reaction producing a DNA fragment of 1,006 bp length (see Table 2 — primerpair 1 ets-2). The difference between this 1,006 bp DNA fragment and the wild-type ets2-cDNA is a lacking of 18 bp between nucleotide position 856 and 874. Because each primer was linked with a restriction enzyme site (forward primer: Hind III, reverse primer: BamHI) the 1,006 bp fragment could be ligated into the pSP64 Poly(A) Vector from Promega between the Hind III and BamHI restriction enzyme sites to perform in vitro sRNA transcription.

Following a human beta-actin internal standard was constructed which lacks a 41 bp sequence between nucleotide position 985 and 1,026 compared to the wild-type cDNA human beta-actin sequence. Therefore serial PCR amplification and purification steps were performed. The first PCR amplification reaction produced with primerpair 1 (see Table 3 — primerpair 1) and 1 µg of reverse transcribed human heart cDNA as template a 1,600 bp fragment of the house keeping gene beta-actin cDNA. After treatment with the Qiagen PCR purification kit 100 ng of the 1,006 bp fragment were used to perform in two different PCR amplification reactions a 975 bp and a 624 bp fragment, each with an overlapping sequence of 39 nucleotides by using a reverse primer with a 20 bp linkage and a forward primer with a 19 bp linkage respectively (see Table 3 — primerpair A and primerpair B). The third step PCR amplification reaction was performed with primerpair 2 (see Table 3) by 5'linkage of the SP6 promotor and poly T tail sequences to produce a 1,277 bp fragment. After gel extraction the 1,277 bp fragment was used as template for in vitro sRNA transcription.

Extraction of the DNA fragments was performed with the GELase Agarose Gel-Digesting Preparation Kit^R from Epicentre Technologies after separation on a 1% low melting agarose gel and quantification by uv absorbance spectrophotometry. Following PCR reagents and conditions were used:

10x reaction buffer II from Perkin Elmer, 2 mM MgCl$_2$, 150 µM each dNTP, 300 nM each primer and 1 U Taq Polymerase in a 50 µl reaction volume. The amplification procedure was performed in a Perkin Elmer thermocycler 9,600 starting with a 3 min. denaturing step at 94°C, followed by a 35 cycle step with denaturing step for 30 sec. at 94°C, annealing step for 30 sec. at 60°C and elongation step for 30 or 60 sec. at 72°C and finished by an elongation step for 7 min. at 72°C.

We confirmed the correct sequences of the ets-2 and beta-actin internal standards by direct sequencing with fluorescent labeled nucleotides using the ABI Prism 310 Genetic Analyzer from Perkin Elmer.

In vitro RNA-synthesis of the ets-2 standard and beta-actin standard

After transforming E.coli cells, growing in culture medium and purification with the Qiagen Plasmid Extraction System (Qiagen, Inc) the pSP64 Poly(A) Vector containing our constructed insert of the ets-2 standard was linearized by the restriction enzyme EcoRI. For the beta-actin standard the purified 1,277 bp fragment containing the SP6 promotor region and the poly A tail was used. The following protocol was performed for both standard in separate reactions.

For in vitro RNA synthesis 5x Gibco buffer (200 mM Tris-HCl, 30 mM MgCl2, 10 mM spermidine hydrochloride), 10 mM DTT, 20 µM each rNTP, 10 U human placental RNase Inhibitor and 35 U SP6 polymerase were mixed with 3 µg linearized plasmid for the ets-2 standard and with 100 ng 1,277 bp fragment for the beta-actin standard and incubated in

Table 2. Human beta-actin primerpairs

Primerpair		Primersequence	cDNA position of the primer	Length of amplified segment (bp)
primerpair 1	forward	5'-GCCAGCTCACCATGGATGAT-3'	31 bp	1,601 bp
	reverse	5'-GCACGAAGGCTCATCATTCA-3'	1,612 bp	
primerpair A	forward	5'-GCCAGCTCACCATGGATGAT-3'	31 bp	974 bp
	reverse	5'-CGCTCAGGAGGAGCAATGAT TCTGCATCCTGTCGGCAAT-3'	966 bp	
primerpair B	forward	5'-ATTGCCGACAGGATGCAGA ATCATTGCTCCTCCTGAGCG-3'	1,026 bp	625 bp
	reverse	5'-GCACGAAGGCTCATCATTCA-3'	1,612 bp	
primerpair 2	forward	5'linked SP6 promotor 5'-GTATCATACACACGATTTA GGTGACACTATAGAACCATGTA CGTTGCTATCCAGGC-3'	433 bp	wildtype 1,277 bp / internal standard 1,236 bp
	reverse	5'linked poly T 5'-GAATTCGGTTTTTTTTTTT TTTTTTTTTTTTTTGGGAGGC ACGAAGGCTCATCATTC-3'	1,612 bp	
pimerpair 3 — FAM labeled				
	forward	5'-CCATCATGAAGTGTGACGTG-3'	883 bp	wildtype 224 bp / internal standard 183 bp
	reverse	5'-ACATCTGCTGGAAGGTGGAC-3'	1,087 bp	

Table 3. Human beta-actin primerpairs

Primerpair		Primersequence	cDNA position of the primer	Length of amplified segment (bp)
primerpair 1	forward	5'-GCCAGCTCACCATGGATGAT-3'	31 bp	1,601 bp
	reverse	5'-GCACGAAGGCTCATCATTCA-3'	1,612 bp	
primerpair A	forward	5'-GCCAGCTCACCATGGATGAT-3'	31 bp	974 bp
	reverse	5'-CGCTCAGGAGGAGCAATGAT TCTGCATCCTGTCGGCAAT-3'	966 bp	
primerpair B	forward	5'-ATTGCCGACAGGATGCAGA ATCATTGCTCCTCCTGAGCG-3'	1,026 bp	625 bp
	reverse	5'-GCACGAAGGCTCATCATTCA-3'	1,612 bp	
primerpair 2	forward	5'linked SP6 promotor 5'-GTATCATACACATACGATTTA GGTGACACTATAGAACCATGTA CGTTGCTATCCAGGC-3'	433 bp	wildtype 1,277 bp / internal standard 1,236 bp
	reverse	5'linked poly T 5'-GAATTCGGTTTTTTTTTTTTTT TTTTTTTTTTTTTGGGAGGC ACGAAGGCTCATCATTC-3'	1,612 bp	
primerpair 3 — FAM labeled				
	forward	5'-CCATCATGAAGTGTGACGTG-3'	883 bp	wildtype 224 bp / internal standard 183 bp
	reverse	5'-ACATCTGCTGGAAGGTGGAC-3'	1,087 bp	

a 25 μl volume at 40°C for 1 hour. After treatment with RNase-free DNase I (Gibco BRL) for 15 min the in vitro synthesized RNA fragments were recovered by ethanol precipitation, resolved in DNase and RNase free water and aliquots were visualized on an 1% agarose gel electrophoresis by ethidiumbromide staining (fragment size was confirmed by comparison with a bp RNA ladder). Thereafter the synthetic RNA fragments were purified by oligo(dt) affinity chromatography using Oligo(dt) Cellulose Columns (Gibco BRL). Quantity and purity of the synthetic RNAs were determined by OD 260/280 absorbance spectrophotometry.

Reverse transcription PCR (RT-PCR) amplification from the synthetic RNAs were confirmed by PCR amplification reaction with an RNAse-treated sample used as a negative control to ensure no contamination of any DNA template.

Several synthetic RNA aliquots ranging from 48 femtogram/μl to 0,016 attogram/μl for the ets-2 internal standard and from 500 to 5 picogram/μl for the beta-actin internal standard were diluted in yeast tRNA (1 mg/ml) in silanised tubes (silanization solution SERVA) to prevent RNA adhesion to plasticware. Several PCR reactions performed with yeast tRNA as template were negative and therefore interferences between the synthetic RNA and yeast tRNA during transcription and amplification could be excluded. Storage was performed at −80°C for 3 months maximum.

Competitive RT-PCR

Serial dilutions of the synthetic RNAs for the ets-2 (48 femtogram/μl to 0,016 attogram/μl) and beta-actin internal standard (500 picogram/ml to 5 picogram/μl) were added to equal concentrations of total RNA (10 ng) extracted from 5 different brain regions of DS patients and control persons in series of RT-PCR reaction mixtures. We could confirm linearity for all samples (example: see Fig. 3). Reverse transcription was performed in a 20 μl volume containing 4 mM MgCl 2, PCR Buffer II from Perkin Elmer, 1 mM dGTP, 1 mM dATP, 1 mM dTTP, 1 mM dCTP, 1 U/μL RNase Inhibitor, 2.5 μM Oligo d(T) 16 2.5 U/μL Moloney murine leukemia virus reverse transcriptase (all components from Perkin-Elmer). The temperature profile started at room temperature for 10 minutes, followed by 15 minutes at 42°C, heat inactivation of the enzyme at 99°C for 5 minutes and cooled to 5°C for 5 minutes. Immediately 5 μl of the produced cDNA were used for nested or direct labeled PCR reaction. 15 μl were stored at −80°C.

Nested PCR amplification reaction was performed for ets-2 quantification in a 50 ml volume. The first step PCR- fragment produced a 665 bp fragment for wild-type ets2 and a 647 bp fragment for the synthetic cDNA. PCR mixture contained buffer II from Perkin Elmer, 2 mM MgCl2, 150 μM each dNTP, 400 nM each primer (see Table 2 — primerpair 2: forward 5′TGG AGT GAG CAA CAG GTA TG3′ and reverse 5′GGC TTA TTG AGG CAG AGA GAC3′ — see Table 3) and 2 U Taq Polymerase. For amplification the temperature profile started with denaturing for 3 min. at 94°C, followed by 20, 25 or 30 cycles with a denaturing step for 30 sec. at 94°C, annealing step for 30 sec. at 60°C and elongation step for 1 min. at 72°C and finished by an elongation step for 7 min. at 72°C.

The second PCR amplification profile was equal to the first PCR procedure except the elongation time was shorten to 30 sec. instead of 1 minute. The reation mixture contained a 5′FAM-labeled primer, PCR buffer from Perkin Elmer, 2 mM MgCl2, 150 μM each dNTP, 400 nM each primer (see Table 2 — primerpair 3: forward FAM-labeled 5′CTG GAG CTG GCA CCT GAC TT 3′and reverse 5′GAC TTG GGG AAC ATC TGA AAC T 3′) and 1 U AmpliTaq DNA polymerase. It produced a 284 bp fragment for the wild-type ets2-cDNA and a 266 bp fragment for the synthetic cDNA standard.

For the beta-actin PCR amplification reaction the mixture contained buffer II from Perkin Elmer, 2 mM MgCl$_2$, 150 μM each dNTP and 600 nM each primer (see Table 3 — primerpair 3: forward primer 5′FAM-labeled 5′CCATCATGAAGTGTGACGTG

3′and reverse primer 5′ ACATCTGCTGGAAGGTGGAC 3′) and 1 U AmpliTaq DNA polymerase and produced a 225 bp fragment for the wild-type beta-actin cDNA and a 184 bp fragment for the synthetic cDNA standard. Temperature profile consisted of 25 cycles and followed the above mentioned protocol for the nested PCR amplification procedure of ets-2.

1 μl of each nested PCR product was solved in 20 μl formamide containing 0,5 μl TAMRA standard solution from Perkin Elmer for quantification on the ABI PRISM 310 genetic analyzer of Perkin Elmer.

Results

We have determined ETS2 and beta-actin mRNA levels in the frontal lobe, parietal lobe, temporal lobe, occipital lobe and cerebellum of 9 adult DS patients and 9 control persons. Mean ETS2 mRNA levels (SD) are shown in Fig. 1 as values in the attogram range in 10 ng total RNA and in Fig. 2 as calculated values versus 100 pg beta-actin, a house keeping gene.

We could find very low ETS2 mRNA levels in parietal lobe and the highest values in frontal lobe in both groups, the DS patients and the controls.

No significant differences of the ETS2 mRNA levels were shown in parietal and occipital lobe and cerebellum at the P < 0,05 level, when either related to 10 ng total RNA or to 100 pg beta-actin.

In temporal lobe a significant decrease of the ETS2 mRNA/100 pg beta-actin (P = 0,0124) was shown in the DS group.

In frontal lobe a significant decrease of the ETS2 mRNA/ 10n total RNA (P = 0,0326) was found in the DS group.

Discussion

The attractive gene dosage hypothesis for the explanation of abnormal organogenesis by ETS2 overexpression was supported by a study in transgenic

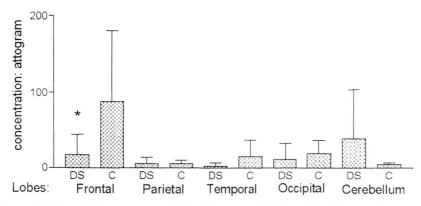

Fig. 1. Mean ETS2 mRNA level ± SD in 10 ng total RNA of 5 different brain regions in DS patients and the control group. The Students t test has shown no significant differences in 4 brain regions. Only the frontal lobe (p = 0,0326) shows a significant difference between both groups (*)

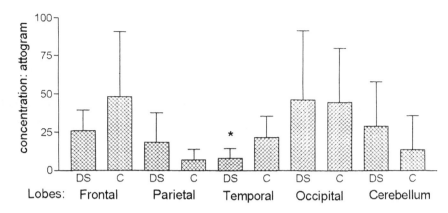

Fig. 2. Mean ETS2 mRNA level ± SD in 100 pg beta-actin of 5 different brain regions in DS patients and the control group. The Students t test has shown no significant differences in 4 brain regions. Only the temporal lobe (p = 0,0124) shows a significant difference between both groups (*)

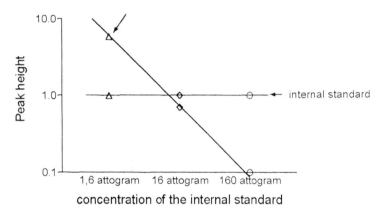

Fig. 3. Linearity of the quantitative, competitive RT-PCR method shown as example: a linear increase of ETS2 internal standard (1,6–16–160 attogram) decreases the peak height of the unknown concentration of ETS2 mRNA in 10 ng total RNA of the same sample — frontal lobe DS 8

mice overexpressing ETS2: the animals developed skeletal abnormalities similar to those found in DS patients (Sumarsano, 1996).

This observation however was the only scientifique in vivo support which was published so far and in a critical review Shapiro was already challenging this hypothesis, discussing whether traits in DS may be the direct consequence of gene products but rather suggests that a major portion of the DS phenotype is due to the disruption of developmental and physiological homeostasis i.e. aneuploidy per se (Shapiro, 1994). Performing a quantitative study on ETS2 in heart of patients with DS we did not find overexpression (Greber-Platzer, in press) and as shown in the Results and Fig. 1 and 2, no upregulation of ETS2 could be found in any of the areas or brain regions investigated.

Instead of overexpression, downregulation of ETS2 was found in two individual brain regions of patients with DS. In temporal lobe mRNA steady state levels of ETS2 were significantly reduced when normalized versus the housekeeping gene beta-actin. Normalization versus the housekeeping gene beta actin not only rules out degraded RNA but also compensates the potential influence of neuronal (or simply cell-) loss in the brain, which has been demonstrated in patients with DS. DS brain shows cell loss anatomically and microscopically: DS brain shows an abnormally rounded contour of the brain, narrow superior temporal gyrus and hypoplasia of the inferior frontal gyrus (Kumar, 1992), histological changes are those found in Alzheimer's disease (AD) and, indeed, close to 100% of DS patients develop AD neuropathology in the brain (Lott, 1992). All our patients with DS examined show the development of AD neuropathology in all the areas examined (Seidl, 1997). This fact has to be taken into account for the evaluation of the relevance of the finding of reduced ETS2 in temporal lobe and therefore the decrease of ETS2 in temporal lobe may reflect AD lesions rather than DS pathology in the brain.

The observation of significantly reduced frontal lobe ETS2 mRNA steady stage level may well be due to cell loss as normalization versus beta-actin resulted in levels comparable to controls. Methodologically, the construction and use of an internal standard (ETS2 minus 18 bp) allowed the evaluation of absolute ETS2 levels, i.e. levels expressed in attograms ETS2/10 ng total RNA.

The use of postmortem tissue for the evaluation of mRNAs has been justified by own observations of patent beta-actin and ETS2 mRNA and was shown to be appropriate in literature (Burke, 1991). The influence of postmortem interval and storage on gene expression in the brain has been studied in our laboratory and was not a confounding factor (unpublished results).

Our data show that in adult DS brain no evidence for the ETS2 — overexpression hypothesis can be obtained. It may well be that ETS2 during brain development in the fetus may be upregulated as postulated by the gene dosage hypothesis, on the other hand, but there is no reason that increased brain ETS2 levels should normalize with age as the trisomic state naturally persists.

We are in the process of collecting and examining fetal brain samples to bypass the development of secondary changes of AD pathology in DS brain.

Acknowledgements

The authors wish to thank R. H. Scheuermann from the Southwestern Medical Center, Dallas for the supply of Oncoquant, to the Red Bull Company and the "Österreichische Nationalbank-Fond" for generous financial support of the study.

References

Bhat NK, Fisher RJ, Fujiwara S, Ascione R, Papas TS (1987) Temporal and tissue-specific expression of mouse ets genes. Proc Natl Acad Sci USA 84: 3161–3165

Burke WJ, O'Malleay KL, Chung HD, Harmon SK, Miller JP, Berg L (1991) Effect of pre- and postmortem variables on specific mRNA levels in human brain. Brain Res Mol Brain Res 11(1): 37–41

Epstein CJ, Korenberg JR, Anneren G, Antonarakis SE, Ayme S, Courchesne E, Epstein LB, Fowler A, Groner Y, Huret JL, Kemper TI, Lott IT, Lubin BH, Magenis E, Opitz JM, Patterson D, Priest JH, Pueschel SM, Rapoport SI, Sinet PM, Tanzi RE, de la Cruz F (1991) Protocols to establisch genotype-phenotype correlations in Down syndrome. Am J Hum Genet 49: 207–235

Greber-Platzer S, Schatzmann-Turhani D, Wollenek G, Lubec G (1999) Evidence against the current hypothesis of "gene dosage effects" of trisomy 21: ets-2, encoded on chromosome 21 is not overexpressed in hearts of patients with Down Syndrome. Biochem Biophys Res Comm (in press)

Huret JL, Delabar JM, Marlhens F, Aurias A, Nicole A, Berthier M, Tanzer J, Sinet PM (1987) Down syndrome with duplication of a region of chromosome 21 containing the CuZn superoxide dismutase gene without detectable karyotypic abnormality. Hum Genet 75: 251–257

Huret JL, Leonard C, Chery M, Phillipe C, Shafei-Benaissa E, Lefaure G, Labrune B, Gilgenkranzt S (1995) Monosomy 21q: two cases of del (21q) and review of the literaure. Clin Genet 48: 140–147

Kola I, Christiano F, de Haan JB, Thomas R, Sumarsono S, Corrick CM, Tymms M (1995) Genes, embryogenesis and Down Syndrome. In: Moeloek FA, Affandi B, Trounson AO (eds) Advances in human reproduction. Casterton-Parthenon, NY, pp 309–320

Korenberg JR, Chen XN, Schipper R, Sun Z, Gonsky R, Gerwehr S, Carpenter N, Daumer C, Dignan P, Disteche C, Graham JM, Hugdins L, McGillivray B, Miyazaki K, Ogasawara N, Park JP, Pagon R, Pueschel S, Sack G, Say B, Schuffenhauer S, Soukup S, Yamanaka T (1994) Down syndrome phenotypes: the consequences of chromosomal imbalance. Proc Natl Acad Sci USA 91: 4997–5001

Kumar AJ, Naidich TP, Stetten G, et al (1992) Chromosomal disorders: background and neuroradiology. AJNR Am J Neuroradiol 13: 577–593

Lott IT (1992) The neurology of Alzheimer disease in Down Syndrome. Prog Clin Biol Res 379: 1–14

Maraoulakou IG, Papas TS, Green JE (1994) Differential expression of ets-1 and ets-2 proto-oncogenes during murine embryogenesis. Oncogene 9: 1551–1565

Rychlik W, Rhoads RE (1989) A computer program for choosing optimal oligonucletide primers for filter hybridization, sequencing and in vitro amplification. Nucl Acids Res 17: 8543–8551

Schatzmann-Turhani D, Greber-Platzer S, Cairns N, Lubec G (1999) Determination of the protooncogene ets-2 gene transcript in human brain at the attogram-level by the use of competitive RT/PCR. Amino Acids (in press)

Seidl R, Gerber S, Schuller E, Bernert G, Cairns N, Lubec G (1997) Evidence against increased oxidative DNA-damabe in Down syndrome. Neurosci Lett 235: 137–140

Shapiro BL (1994) The environmental basis of the Down Syndrome phenotype. Dev Med Child Neurol 36(1): 84–90

Shen JJ, Williams BJ, Zipursky A, Doyle J, Sherman SL, Jabobs PA, Shugar AL, Soukoup SW, Hassold TJ (1995) Cytogenetic and molecular studies of Down syndrome individuals with leukemia. Am J Hum Genet 56: 915–9254

Sinet PM, Theophile D, Rahmani Z, Chettouh Z, Blouin JL, Prieur M, Noel B, Delabar JM (1994) Mapping of the Down Syndrome phenotype on chormosome 21 at the molecular level. Biomed Pharmacother 48: 247–252

Sumarsono SH, Wilson TJ, Tymms MF, Venter DJ, Corrick CM (1996) Down's syndrome-like skeletal abnormalities in Ets 2 transgenic mice. Nature 379: 534–537

Wasylyk B, Hahn SL, Giovane A (1993) The Ets family transcription factors. Eur J Biochem 211(1–2): 7–18

Authors' address: Prof. Dr. G. Lubec, University of Vienna, Department of Pediatrics, Währinger Gürtel 18–20, A-1090 Vienna, Austria, e-mail: gert.lubec@akh-wien.ac.at

J Neural Transm (1999) [Suppl] 57: 283–291
© Springer-Verlag 1999

Neuronal apoptosis inhibitory protein (NAIP)-like immunoreactivity in brains of adult patients with Down syndrome

R. Seidl[1], M. Bajo[2], K. Böhm[1], E. C. LaCasse[3], A. E. MacKenzie[3], N. Cairns[4], and G. Lubec[1]

[1] Department of Pediatrics, University of Vienna, Austria
[2] Institute of Neuroimmunology, SAS, Bratislava, Slovakia
[3] Apoptogen Inc., CHEO Research Institute, Ottawa, Ontario, Canada
[4] Brain Bank, Institute of Psychiatry, University of London, London, United Kingdom

Summary. In Down syndrome (DS), enhanced apoptosis (programmed cell death) may play a role in the pathogenesis of characteristic early mental retardation and precocious neurodegeneration of Alzheimer-type.

The human IAP (inhibitor of apoptosis proteins) genes (NAIP, c-IAP-2/ HIAP-1, c-IAP-1/Hiap-2, XIAP, survivin) are an evolutionary conserved family of proteins which prevent cell death across species, implying that they act at a central, highly conserved point in the cell death cascade. Evidence for downregulation of NAIP-mRNA in fetal DS (23rd week of gestation), as found by subtractive hybridization technique challenged studies at the protein level in adult DS brain specimen.

NAIP-like immunoreactivity was determined in four different regions of cerebral cortex and cerebellum in 9 adult DS patients with Alzheimer-like neuropathologic lesions, 9 Alzheimer disease (AD) patients as compared to 9 controls. For the first time, NAIP-IR could be demonstrated in different cortical regions of the human brain. Compared to control subjects, western blotting demonstrated significantly decreased levels in parietal and occipital cortex in DS and in frontal and occipital cortex in AD. While the mode of NAIP action is unknown, inhibition of certain caspases has already been demonstrated for other IAP-family members (c-IAP1, c-IAP2 and XIAP). Although decreased NAIP-IR of certain brain regions in DS and AD awaits further confirmation, the results suggest that alterations of apoptosis regulatory (inhibitory) proteins may be another feature of neurodegeneration in DS and AD.

Introduction

Down syndrome (DS) is a genetic disease with developmental brain abnormalities (Wisniewski and Kida, 1994) resulting in early mental retardation and precocious, age dependent Alzheimer-type neurodegeneration (Wisniewski et al., 1985; Nadel and Epstein, 1992; Epstein, 1995). So far, the pathogenesis of neurodegeneration is not clear yet. Besides the role of enhanced apoptosis in Alzheimer disease (AD) and other neurologic diseases (see Bredesen, 1995; DelaMonte, 1997; Nagy and Esiri, 1997), there is accumulating evidence that apoptosis may be involved in disease-related neuronal loss in DS (Busciglio and Yankner, 1995; DelaMonte et al., 1998; Nagy et al., 1997). In vitro cultured neurons from fetal DS brain exhibited an increased occurrence of apoptosis (Busciglio and Yankner, 1995). In autopsy brain samples, technical limitations of the DNA end labeling (TUNEL assay) method for apoptosis detection have prompted analyses of protein pathways involved in the initiation of apoptotic neuronal death (Sawa et al., 1997; delaMonte et al., 1998; Nagy and Esiri, 1997). The proteins involved in apoptosis can be classified into two broad groups, cell surface receptor proteins and intracellular proteins. Recently, we were able to demonstrate higher frontal, temporal lobe and cerebellar protein levels of p53 in adult patients with DS (Seidl et al., 1999).

Performing gene hunting by subtractive hybridization technique in fetal DS brain (23rd week of gestation), we observed downregulated NAIP-mRNA. The human IAP (inhibitor of apoptosis proteins) genes (NAIP, c-IAP-2/HIAP-1, c-IAP-1/Hiap-2, XIAP, survivin) are an evolutionary conserved family of proteins which prevent cell death across species, implying that they act at a central, highly conserved point in the cell death cascade (Roy et al., 1997). No data exist on involvement of IAPs such as NAIP in neuronal loss in DS. This challenged studies at the protein level and led us to assess the expression of NAIP in adult DS, AD and control brain specimen.

Materials and methods

The levels of NAIP-like immunoreactivity (NAIP-IR) were quantified by Western blotting analyses in 5 different brain regions of 9 adult DS individuals with AD-like neuropathologic lesions (3 females, 6 males; mean age of 55.7 ± 7.5 years), 9 patients with Alzheimer disease (4 females, 5 males; mean age 58.7 ± 5.7 years) and 9 controls (3 females, 6 males; mean age 54.3 ± 10.2 years). Post-mortem brain samples (gray matter — frontal, temporal, parietal, occipital and cerebellar cortex) were obtained from the Medical Research Council's London Brain Bank for Neurodegenerative Diseases, Institute of Psychiatry, London, UK. Corresponding brain regions of the contralateral hemisphere were histologically/neuropathologically checked for cellular damage and no obvious differences (no morphometry performed) were detected. In all DS brains there was evidence of abundant β-amyloid plaques and neurofibrillary tangles. AD patients fulfilled the National Institute of Neurological and Communicative Disorders and Stroke and the Alzheimer Disease and Related Disorders Association (NINCDS/ADRDA) criteria for probable AD (Tierney et al., 1988). The histological diagnosis of AD was established consistent with the CERAD criteria (Mirra et al., 1991) for a "definite" diagnosis of AD. The controls were brains from individuals with no history of neurological or psychiatric illness. The major cause of death was bronchopneumonia in DS and AD patients and heart disease in the controls. Post-mortem interval of brain dissection in DS, AD and controls was 32.8 ± 29.2, 29.2 ± 19.8 and 38.1 ± 22.7 hours, respectively. Tissue samples were stored at $-70°\,C$ and freezing chain was never interrupted. Frozen brain samples (about 100–150 mg) were thawed on ice in the presence of protease inhibitor Pantinol® (Gerot, Austria) 500 KIE/ml homogenization solution (0.25 M sucrose, 1 mM EDTA, 3 mM imidazole, 0.1% ethanol, pH 7.2 and then homogenized for 30 sec (6 strokes) at 440 rev/min in a Potter-Elvehjem homogenizer on ice. The homogenate was centrifuged for 5 min at 3,000 g and 4°C. The supernatant was used for Western blotting and determination of total protein. The protein content of each sample was determined by the method of Lowry (Lowry et al., 1951). For Western blot analysis, equal amounts of brain protein (25 μg) were separated by electrophoresis on a 8–18% gradient sodium dodecyl sulfate-polyacrylamide gel electrophoresis (SDS-PAGE) on a Multiphor II (Pharmacia Biotech, Uppsala, Sweden), transferred to polyvinylidene fluoride (PVDF) membranes (Millipore, Bedford, USA) following the manufacturer's guidelines. An affinity-purified polyclonal rabbit antibody, raised against n-terminal 80 kDa portion of human NAIP was kindly provided by Apoptogen Inc. (Ottawa, Ontario, Canada). After transfer membranes were washed 3 × 5 min in TBS and blocked for 2 hours in a blocking/wash buffer containing Tris-HCl (TBS-buffer: 10 mM Tris-HCl, 150 mM NaCl, 1 mM MgCl adjusted to pH 7.5), 1.5% non-fat dried milk (Bio-Rad, Richmond, USA) and 0.1% Tween 20. After blocking, the membranes were immunoblotted with polyclonal rabbit anti-NAIP (1:5,000, v/v, TBST) overnight at 4°C. After 3 × 15 min washing, the bound antibodies were visualized by HRP-conjugated goat anti-rabbit IgG (# 4050-05; Southern Biotechnology Inc., Birmingham, USA) secondary antibody (1:3,500, v/v, TBST), and after further 3 × 15 min washing, by a chemiluminiscence detection method (NENTM, Life Science Products, Inc., Boston, USA). Exposed and developed films (Kodak Blue XB-1, Rochester, USA) were scanned and internal densities of NAIP-like IR signaling bands were calculated by RFLPScan™ 2.1 software program (Scanalytics, Billerica, USA).

Statistics

Results were expressed as means ± standard deviation (SD). Between-group differences were investigated by non-parametric Mann-Whitney U test. Within-group correlations were done with non-parametric Spearman rank-correlation coefficient procedure. The level of significance was set at $P < 0.05$. All analyses were run with the SPSS-PC V 8.0 statistical software package.

Results

The used anti-NAIP antibody, produced in Dr. Alex MacKenzie's laboratory, has been tested in human neuroblastoma cells (sh-sy-5y), where a consistent 90 kDa band and to a lesser extent a 150 kDa band has been demonstrated (Alex MacKenzie, personal communication, 1999). The Western blot pattern of NAIP-like immunoreactive material in human brain samples indicates a band at about 90 kDa (Fig. 1), a further band at 150 kDa was not detected. NAIP-like IR level of DS samples was significantly decreased in parietal (Fig. 1) and occipital cortex, insignificantly decreased in frontal cortex, but was comparable to controls in temporal cortex and cerebellum (Table 1). In AD subjects NAIP-like IR levels were significantly lower in frontal and occipital cortex, insignificantly decreased in parietal (Fig. 1) and temporal cortex, and comparable to controls in cerebellar cortex (Table 1). Age, post-mortem interval, sex and agonal state in each group revealed no influence on NAIP-like IR (data not shown).

Fig. 1. Results of parietal lobe comprising 8 AD (lane 1–8), 8 DS (lane 9–16) and 8 control samples (lane 17–24). Western blot indicating a 90 kDa band (marker not shown) of NAIP-like immunoreactivity. 25 μg of protein per lane were separated by reducing SDS-PAGE, NAIP-IR detected by a polyclonal rabbit anti-NAIP (1:5,000) followed by a HRP-conjugated goat anti-rabbit antibody (1:3,500) and chemiluminiscence detection. Double band in one DS sample (lane 10) and an additional band in another DS sample (lane 11) are suspected to reflect isoforms (lane 10), further processed or cleaved NAIP-IR (lane 11)

Table 1. NAIP-like immunoreactivity in 5 different brain regions of DS, AD and control subjects

	Frontal cortex	Temporal cortex	Parietal cortex	Occipital cortex	Cerebellum
Control	4.18 ± 2.22	2.41 ± 1.19	3.60 ± 2.02	1.55 ± 0.81	0.91 ± 0.65
	n = 8	n = 7	n = 8	n = 8	n = 7
DS	2.85 ± 1.30	1.83 ± 1.19	1.55 ± 0.53	0.71 ± 0.37	1.20 ± 0.47
	n = 8	n = 7	n = 8	n = 8	n = 7
	(P = 0.27)	(P = 0.41)	(P = 0.015)	(P = 0.014)	(P = 0.14)
AD	1.09 ± 1.86	1.68 ± 0.89	2.10 ± 2.02	0.55 ± 0.46	0.89 ± 0.59
	n = 8	n = 7	n = 8	n = 8	n = 8
	(P = 0.009)	(P = 0.23)	(P = 0.115)	(P = 0.012)	(P = 0.87)

Optical density, arbitrary units in mean ± SD; Statistics: Mann Whitney U-test, DS and AD subjects compared to normal controls

Discussion

This is the first study providing data on NAIP-like immunoreactivity in the human brain. Compared to matched controls, NAIP-IR was significantly decreased in parietal and occipital cortex of adult DS individuals with Alzheimer-like neuropathology, whereas it was decreased in frontal and occipital cortex in AD brain samples.

The human IAP (inhibitor of apoptosis proteins) genes (NAIP, cIAP2/HIAP1, cIAP1/Hiap-2, XIAP, survivin) encode an evolutionary conserved family of proteins which prevent cell death across species, implying that they act at a central, highly conserved point in the cell death cascade (Roy et al., 1997; LaCasse et al., 1998). They block a higher spectrum of apoptotic triggers than does Bcl-2, possibly reflecting a site of activity further downstream than that of the Bcl-2 family. Following apoptotic triggers have been reported to be blocked by different IAPs: TNF-α, Fas ligand and transducers of the TNF receptor superfamily (e.g. RIP, FADD, v-Rel, proapoptotic members of the

Bcl-2 family — BIK, BAK, BAX -, cytochrome C, chemotherapeutic drugs, ionizing or UV radiation, oxidative stress, ischemia, potassium withdrawal, viral infection or viral proteins, growth factor withdrawal, caspases and several other triggers (for review see LaCasse et al., 1998; Simons et al., 1999). In addition, they have a distinct but overlapping pattern of expression in fetal and adult tissues (Liston et al., 1996). Neuronal apoptosis inhibitory protein (NAIP), which maps to the spinal muscular atrophy (SMA)-region on chromosome 5q13.1 (Roy et al., 1995), is 1,403 amino acids in length, with a size of 156 kDa (Chen et al., 1998). Deletion mutations of NAIP were found in the majority of infants with the most severe form of the neurodegenerative disorder spinal muscular atrophy (SMA) (Roy et al., 1995). While the SMN-gene has been found to be causative (Lefebvre et al., 1995), expression of NAIP is suggested to modify SMA phenotype. Motor neurons missing NAIP withstand the low cellular levels of SMN protein poorly (MacKenzie, 1998) possibly dying by pathological continuation or reactivation of developmental neuronal apoptosis (Liston et al., 1997).

Expression of NAIP has been shown in the human spinal cord by immunofluorescence largely restricted to the cytoplasm of anterior horn cells and intermediolateral neurons (Liston et al., 1997). In the present study for the first time, a 90 kDa NAIP-IR band was detected throughout all cortical brain regions of the human brain, including cerebellum. In control brains, highest levels of NAIP-IR were present in frontal, followed by parietal cortex, although comparing levels of different regions on separate membranes might be too artificial for drawing conclusions on different expression patterns. However, the fact that NAIP-IR was found in all investigated cortical regions, speaks against an expression solely in motor neurons as stated by Liston et al. (1996). We have no explanation why we were not able to detect full-length NAIP protein at about 160 kDa. Studies in human neuroblastoma cells using the same antibody demonstrated also a consistent 90 kDa band and only to a lesser extent a 150 kDa band (Alex MacKenzie, personal communication, 1999).

In the rat, NAIP-like immunoreactivity (NAIP-LI) was found to be widely distributed in neurons of central nervous system (CNS) including neocortex and cerebellum (Xu et al., 1997a), but not in non-neuronal cell types like those associated with vessels or glia. Neuronal populations with the highest levels of anti-NAIP immunoreactivity correspond to neuronal populations reported to degenerate in SMA. With respect to cortical neurons, low to moderate levels of NAIP-LI (compared to facial nerve neurons), were detected in neurons of cortical regions with the most prominent labeling observed in the apical dentrites of layer V pyramidal neurons (Liston et al., 1996). Although widely distribution in the CNS, NAIP seems to be particularly expressed in motor neurons, probably acting as a negative regulator of motor-neuron apoptosis (Liston et al., 1996; Xu et al., 1997b). Attenuated apoptotic resistance due to decreased NAIP-expression could disturb functional and temporal longevity of neurons. In the cerebellum, high NAIP-LI was observed in cortical Purkinje cells and deep cerebellar nuclei (Liston et al., 1996). Consistent with these findings, NAIP mRNA was barely and NAIP immunoreactivity not detectable in cerebellar granule neurons in vitro (Simons et al., 1999). Progressive neuronal loss with subsequent gliosis is a hallmark of neuropathology in DS and AD (Nadel and Epstein, 1992). As NAIP is only expressed in neuronal cells (Xu et al., 1997), one must be aware that lower levels in DS and AD brain specimen must be discussed with respect to advanced pathology of postmortem brain samples. One possibility to cope with this potential pitfall could be to calculate a ratio of NAIP to relevant cellular, and in particular neuronal versus glial protein markers (now in progress). On the other hand, NAIP-IR in temporal lobe and cerebellum was comparable in DS, AD and controls, a fact which might speak against a decrease of NAIP-IR only due to decreased neuronal density. Further clarification with NAIP-immunohistochemistry is necessary. Another elegant approach would be to investigate early stages of disease, before neuronal loss has reached a substantial degree. In DS, one further aspect is of crucial relevance. In contrast to sporadic AD, DS pathology starts early during pre- and postnatal development (Becker, 1991; Schmidt-Sidor et al., 1990; Wisniewski and Kida, 1994) which might dipose neurons to exaggerated apoptosis and AD-like neurodegeneration during early adult life (Busciglio and Yankner, 1995). Consistent with this view might be the observed NAIP-mRNA downregulation in fetal DS brain (Seidl et al., this volume). However, further studies on protein levels at different ages are necessary.

NAIP has been shown in vitro to block apoptosis of several different cell types induced by a variety of non-related triggers, including serum deprivation, treatment with a free radical inducer menadione, TNF-α, potassium withdrawal, and chemotherapeutic agents etoposide and adriamycin (Liston et al., 1996; LaCasse et al., 1998; Simons et al., 1999). NAIP overexpression in vivo reduces transient ischemic forebrain damage in the rat hippocampus suggesting that NAIP may also play a role in preventing ischemic damage (anti-apoptotic activity) (Xu et al., 1997b). Furthermore, transient forebrain ischemia selectively elevated NAIP-levels in ischemia-resistant rat hippocampal neurons, indicating that endogenous NAIP has a functional role as anti-apoptotic protein (Xu et al., 1997b). Although cIAP1, cIAP2 and XIAP were found to bind to and inhibit two members of the caspase protease family, caspase-3 and -7, the mode of NAIP action is unknown (Roy et al., 1997). Theoretically, NAIP could also be a potent inhibitor of certain caspases (Roy et al., 1997). Virally-mediated expression of NAIP in cultured cerebellar granule neurons inhibited N-acetyl-Asp-Glu-Val-Asp (DEVD)-specific caspase but failed to directly inhibit caspase-3 and 7, suggesting an "upstream" level of action on the apoptotic pathway (Simons et al., 1999). Inhibition of DEVD-specific caspase could be either due to inhibition of active caspases and/or prevention of the processing of procaspases to subunits that form active heterodimers (Simons et al., 1999). Identification of caspase inhibitors as regulators of complex proteolytic cell-death systems seems not surprising and the IAP protein family is until now the only one to have mammalian members (Thornberry and Lazebnik, 1998).

Members of the IAP family were identified as gene targets of NF-κB transcriptional activity (Wang et al., 1998), a fact that could explain NF-κB antiapoptotic properties. NF-κB activation induces TRAF1, TRAF2, cellular IAPs (c-IAP1, c-IAP2) along with caspase 8 suppression, leading to protection against TNF-induced apoptosis (Wang et al., 1998). In addition, cIAP1 and cIAP2 have also been shown to activate NF-κB possibly forming a positive feed-back loop (Chu et al., 1997). So, cellular inhibitors of apoptosis (IAPs) like NAIP, are targets and inducers of NF-κB transcriptional activity (Wang et al., 1998; Chu et al., 1997) and might well play an antiapoptotic role in neurodegenerative conditions that involve oxidative stress, a condition discussed in DS neuronal (apoptotic) death (Busciglio and Yankner, 1995). In parallel to SMA, one hypothesis would be that NAIP downregulation leads to a defective antiapoptotic reserve, with consecutive reduced cellular function and cell longevity (Liston et al., 1997).

In conclusion, downregulation of NAIP-mRNA in fetal DS (23rd week of gestation), as found by subtractive hybridization technique challenged studies at the protein level in adult DS brain specimen. NAIP-IR was demonstrated in all cortical brain regions investigated. Although decreased NAIP-IR of certain brain regions in DS and AD awaits further confirmation, the results suggest that alteration of apoptosis regulatory proteins may be another feature of neurodegeneration in DS and AD.

Acknowledgement

We (G. L.) are highly indebted to the Red Bull Company, Salzburg, Austria, for generous support of the study.

References

Becker LE, Mito T, Takashima S, Onodera K (1991) Growth and development of the brain in Down syndrome. Prog Clin Biol Res 373: 133–152

Bredesen DE (1995) Neural apoptosis. Ann Neurol 38: 839–851

Busciglio J, Yankner BA (1995) Apoptosis and increased generation of reactive oxygen species in Down's syndrome neurons in vitro. Nature 378: 776–779

Chen QF, Baird SD, Mahadevan M, BesnerJohnston A, Farahani R, Xuan JY, Lefebvre C, Ikeda JE, Korneluk RG, MacKenzie AE (1998) Sequence of a 131-kb region of 5q13.1 containing the spinal muscular atrophy candidate genes SMN and NAIP. Genomics 48: 121–127

Chu ZL, McKinsey TA, Liu L, Gentry JJ, Malim MH, Ballard DW (1997) Suppression of tumor necrosis factor-induced cell death by inhibitor of apoptosis c-IAP2 is under NF kappa B control. Proc Natl Acad Sci USA 94: 10057–10062

DelaMonte SM, Sohn YK, Wands JR (1997) Correlates of p53- and Fas (CD95)-mediated apoptosis in Alzheimer's disease. J Neurol Sci 152: 73–83

DelaMonte SM, Sohn YK, Ganju N, Wands JR (1998) p53- and CD95-associated apoptosis in neurodegenerative diseases. Lab Invest 78: 401–411

Epstein CJ (1995) Down Syndrome (Trisomy 21). In: Scriver SR, Beaudet AL, Sly WS, Valle D (eds) The metabolic and molecular bases of inherited disease. McGraw-Hill, New York, pp 749–794

LaCasse EC, Baird S, Korneluk RG, MacKenzie AE (1998) The inhibitors of apoptosis (IAPs) and their emerging role in cancer. ONCOGENE 17: 3247–3259

Lefebvre S, Burglen L, Reboullet S, Clermont O, Burlet P, Viollet L, Benichou B, Cruaud C, Millasseau P, Zeviani M, Le Paslier D, Frezal J, Cohen D, Weissenbach J, Munnich A, Melki J (1995) Identification and characterization of a spinal muscular atrophy-determining gene. Cell 80: 155–165

Liston P, Roy N, Tamai K, Lefebvre C, Baird S, Cherton-Horvat G, Farahani R, McLean M, Ikeda JE, MacKenzie A, Korneluk RG (1996) Suppression of apoptosis in mammalian cells by NAIP and a related family of IAP genes. Nature 379: 349–353

Liston P, Young SS, MacKenzie AE, Korneluk RG (1997) Life and death decisions: the role of the IAPs in the modulating programmed cell death. Apoptosis 2: 423–441

Lowry OH, Rosebrough NJ, Farr AL, Randall RJ (1951) Protein measurement with the folin phenol reagent. J Biol Chem 193: 265–275

MacKenzie AE (1998) Reply to Burghes [letter]. Am J Hum Genet 62: 485–488

Mirra SS, Heyman A, McKeel D, Sumi SM, Crain BJ (1991) The consortium to establish a registry for Alzheimer's disease (CERAD). II. Standardisation of the neuropathological assessment of Alzheimer's Disease. Neurol 4: 479–486

Nadel L, Epstein CJ (1992) Down Syndrome and Alzheimer disease. Wiley-Liss, New York (Prog Clin Biol Res 379)

Nagy Z, Esiri MM, Smith AD (1997a) Expression of cell division markers in the hippocampus in Alzheimer's disease and other neurodegenerative conditions. Acta Neuropathol Berl 93: 294–300

Nagy Z, Esiri MM (1997b) Apoptosis-related protein expression in the hippocampus in Alzheimer's disease. Neurobiol Aging 18: 565–571

Roy N, Deveraux QL, Takahashi R, Salvesen GS, Reed JC (1997) The c-IAP-1 and c-IAP-2 proteins are direct inhibitors of specific caspases. Embo J 16: 6914–6925

Roy N, Mahadevan MS, McLean M, Shutler G, Yaraghi Z, Farahani R, Baird S, Besner-Johnston A, Lefebvre C, Kang X, Salih M, Aubry H, Tamai K, Guan X, Ioannou P, Crawford TO, de Jong PJ, Surh L, Ikeda J-E, Korneluk RG, MacKenzie AE (1995) The gene for neuronal apoptosis inhibitory protein is partially deleted in individuals with spinal muscular atrophy. Cell 80: 167–178

Sawa A, Oyama F, Cairns NJ, Amano N, Matsushita M (1997) Aberrant expression of bcl-2 gene family in Down's syndrome brains. Mol Brain Res 48: 53–59

Schmidt-Sidor B, Wisniewski KE, Shepard TH, Sersen EA (1990) Brain growth in Down syndrome subjects 15 to 22 weeks of gestational age and birth to 60 months. Clin Neuropathol 9: 181–190

Seidl R, Fang-Kircher S, Bidmon B, Cairns N, Lubec G (1999) Apoptosis-associated proteins p53 and APO-1/Fas (CD95) in brains of adult patients with Down syndrome. Neurosci Lett 260: 9–12

Simons M, Beinroth S, Gleichmann M, Liston P, Korneluk, MacKenzie AE, Bahr M, Klockgether T, Robertson GS, Weller M, Schulz JB (1999) Adenovirus-mediated gene transfer of inhibitors of apoptosis proteins delays apoptosis in cerebellar granule neurons. J Neurochem 72: 292–301

Thornberry NA, Lazebnik Y (1998) Caspases: enemies within. Science 281: 1312–1316

Tierney MC, Fisher RH, Lewis AJ, Torzitto ML, Snow WG, Reid DW, Nieuwstraten P, Van Rooijen LAA, Derks HJGM, Van Wijk R, Bischop A (1988) The NINCDA-ADRDA work group criteria for the clinical diagnosis of probable Alzheimer's disease. Neurol 38: 359–364

Wang CY, Mayo MW, Korneluk RG, Goeddel DV, Baldwin AS (1998) NF-kappa B antiapoptosis: induction of TRAF1 and TRAF2 and c-IAP1 and c-IAP2 to suppress caspase-8 activation. Science 281: 1680–1683

Wisniewski KE, Kida E (1994) Abnormal neurogenesis and synaptogenesis in Down syndrome brain. Dev Brain Dysfunct 7: 289–301

Wisneiwski KE, Wisniewski HM, Wen GY (1985) Occurrence of neuropathological changes and dementia of Alzheimer's disease in Down syndrome. Ann Neurol 17: 278–282

Xu DG, Korneluk RG, Tamai K, Wigle N, Hakim A, MacKenzie A (1997a) Distribution of neuronal apoptosis inhibitory protein-like immunoreactivity in the rat central nervous system. J Comp Neurol 382: 247–259

Xu DG, Crocker SJ, Doucet JP, St. Jean M, Tamai K, Hakim AM (1997b) Elevation of neuronal expression of NAIP reduces ischemic damage in the rat hippocampus. Nat Med 3: 997–1004

Authors' address: Prof. Dr. G. Lubec, Department of Pediatrics, University of Vienna, Währinger Gürtel 18-20, A-1090 Vienna, Austria, e-mail: gert.lubec@akh-wien.ac.at

J Neural Transm (1999) [Suppl] 57: 293–303
© Springer-Verlag 1999

The "gene dosage effect" hypothesis versus the "amplified developmental instability" hypothesis in Down syndrome

M. A. Pritchard and **I. Kola**

Centre for Functional Genomics and Human Disease, Institute of Reproduction and
Development, Monash University, Clayton, Victoria, Australia

Summary.   Two hypotheses exist to explain the Down syndrome (DS) phenotype. The "gene dosage effect" hypothesis states that the phenotype is a direct result of the cumulative effect of the imbalance of the individual genes located on the triplicated chromosome or chromosome region. In a nut shell, the phenotype results directly from the overexpression of specific chromosome 21 genes. The "amplified developmental instability" hypothesis contends that most manifestations of DS may be interpreted as the results of a non-specific disturbance of chromosome balance, resulting in a disruption of homeostasis. This hypothesis was proposed in an attempt to explain the similarities between the phenotypes of different aneuploid states and the observation that all of the phenotypic traits in DS are also seen in the general population but at lower frequency, with less severity and usually only present as a single trait. Herein, we review recent data and present evidence to support the theory that the phenotypic traits of aneuploid syndromes, and DS in particular, result from the increased dosage of genes encoded on the triplicated chromosome.

Introduction

Down syndrome (DS) or trisomy 21 is the most common chromosomal aneuploidy that comes to term in humans. With an incidence of 1 in 700 live births DS is the most frequent genetic cause of mental retardation (Hassold and Jacobs, 1984). DS affects every major organ system in the body and includes abnormalities, such as an increased risk of childhood leukaemias, congenital heart and gastrointestinal tract defects, and characteristic craniofacial anomalies. DS individuals also develop Alzheimer's-like amyloid plaques and neurofibrillary tangles by approximately the third decade of life, with a significant number of these individuals developing presenile dementia (Epstein, 1986).

Two models exist to explain the DS phenotype:1) the phenotype is a direct result of the cumulative effect of the imbalance of the individual gene, or groups of genes, located on the triplicated chromosome or chromosome region, or perhaps put more succinctly, the phenotype results directly from the overexpression of specific chromosome 21 genes. And 2), most manifestations of DS may be interpreted as the results of a non-specific disturbance of chromosome balance, resulting in a disruption of homeostasis. This so called, "amplified developmental instability" hypothesis was first proposed by Hall (1965) and later expanded on by Shapiro (1970, 1975, 1983, 1989, 1994). Several authors have explained their findings in terms of amplified developmental instability (Witkop et al., 1988; Blum-Hoffmann et al., 1988; Barden, 1980; Langenbeck et al., 1984; Greber-Platzer et al., 1998).

The "amplified developmental instability" hypothesis was proposed in an attempt to explain two observations: 1) the commonalties in the phenotypes of different aneuploid states and 2) all of the phenotypic traits in DS are also seen in the general population, albeit at lower frequency, with less severity and usually only present as a single trait. The hypothesis is based on the idea that environmental variance is a measure of the degree of homeostasis, and that those traits which are highly variable in the normal population (due to a greater environmental influence) are the traits most frequently and severely disturbed in aneuploidy. Proponents of the "amplified developmental instability" hypothesis have measured various physical parameters such as palate length, palmer atd angle, tooth size, tooth eruption rates and stature (reviewed in Shapiro, 1983) and found these to be more variable in DS compared to the normal population, and concluded that this is due to a greater disruption of homeostasis.

Viewed simply, the major disagreement between the proponents of either the "amplified developmental instability" hypothesis or the "gene dosage effect" hypothesis is based on the consideration of whether the phenotypic similarities observed between aneuploid syndromes are more important than the differences. It is our contention that it is the differences that count, and as time goes on, more and more differences are being revealed. We will review recent data and present evidence to support the theory that the phenotypic traits of aneuploid syndromes, and DS in particular, result from the increased dosage of genes encoded on the triplicated chromosome.

Conceptual considerations

Shapiro argues that a substantial body of physiognomic and autopsy data has shown that no trisomy is identifiable by any single clinical diagnostic trait. We have no problem with this -a syndrome, by definition, is always a collection of traits or stigmata which when taken together, distinguishes the syndrome from others. We also have no problem with the notion that similarities exist between syndromes and we agree with Optiz (1982) that "human organs are evidently capable of responding to a huge number of diverse dysmorphic causes with production of only a limited repertoire of malformations". Phenotype is quite a gross measurement -aberrations in more than one pathway may produce the same phenotype. For instance, cancer was the global term used to describe what later emerged as very specific types of lesion caused by mutations in a number of specific genes.

A particular part of the phenotype might also be caused by the disturbance of a single pathway. Generally, the components of any pathway are gene products encoded on a variety of chromosomes, so the disturbance may be caused by the aberrant actions of any one of these products, all leading to the same phenotypic manifestation. Shapiro (1989) used the observation that taurodontism was found in DS, X chromosome aneuploidy and in the deletion syndromes 4p-, 5p- and 18q- to support the concept that chromosomal abnormalities disrupt the developmental homeostasis of tooth form and that taurodontism is not the result of the actions of specific genes on the chromosome involved. We would argue that a common or related pathway(s) in tooth development has been disrupted, directly or indirectly, by aberrations in genes in the pathway which are located on the chromosomes involved.

We still cannot answer the question of how three copies of normal genes contribute to the abnormal phenotype of DS, and to date, there is no convincing functional link in the human between any of the various traits that constitute DS and the overexpression of any one gene on human chromosome 21 (HSA21). However, this link is being convincingly established in the mouse (as discussed later). Furthermore, as Epstein (1989) pointed out, this is not surprising as we still do not know the functions of every gene on HSA21 (indeed, we still have not identified all the genes) and we know very little about the regulation of development and morphogenesis. This situation is rapidly changing -genes are being discovered at a rapid rate and the dissection of pathways and biological processes is gathering pace.

Having stated that we are comfortable with the idea that similarities between syndromes exist, we would challenge this notion on the basis that as we learn more, look harder and our tools of investigation become more sophisticated, we are revealing the subtle differences between states previously lumped together under the heading of "mental retardation", for example. Animal models have been particularly instructive in this regard. For instance, the Ts65Dn and Ts1Cje partial trisomy 16 mice have the same chromosomal region triplicated but in the case of Ts65Dn, the region is slightly larger, and this is reflected in subtle differences in their behavioural and cognitive functions (see later).

Aneuploidies

We do not believe that aneuploidy per se can explain the specificity of phenotypes observed in the various aneuploid syndromes. Although trisomies 13, 18 and 21 are all associated with mental and growth retardation and cardiac malformations, there are very obvious differences in the manifestations of these aneuploid states. Trisomy 13 is characterised by cleft palate and lip, by renal abnormalities that include, polycystic kidney, hydronephrosis, horseshoe kidney and duplicated ureters (Jones, 1988). These features are extremely rare in DS. Indeed, they occur at the same rate as in the general population, so these features cannot be a non-specific effect of aneuploidy. Dunlap and colleagues (1986) studied the neuromuscular phenotype of the forelimb in human trisomies 13, 18 and 21. They found that there was specificity of the phenotypes with features unique to each aneuploidy state. This is perhaps an example of looking harder to reveal differences.

The DS phenotype is also very distinctive from the phenotypes associated with other aneupliodies. Females with Turner's syndrome (46,X), often have no overt phenotype but are short in stature and infertile due to ovarian failure. Klinefelter syndrome males (47,XXY), have mild learning difficulties, gynaecomastia and are sterile. DS individuals are short and the males are sterile, but their phenotype on the whole encompasses every major organ system and their mental disability is much more severe than that seen in Klinefelter's. Females with a 47,XXX genotype represent about 0.1% of the population and have no physical nor mental abnormalities (Mueller and Young, 1998). Males with 47,XYY are emotionally immature, display impulsive behaviour, and in some cases display antisocial and criminal behaviour. DS individuals possess good social skills and have pleasant personalities, indicating a quite distinct behavioural phenotype.

In DS there are very specific and characteristic cognitive deficits which distinguish it from other forms of mental retardation. Trauner and co-workers (1989) compared the neurological features of Williams and Down syndromes. Williams syndrome (WS) is the consequence of a deletion of chromosome 7q11. The neurological testing indicated that WS patients had greater difficulty with gross and fine motor co-ordination, oromotor skills and cerebellar function than did those with DS, however, WS patients demonstrated more advanced linguistic skills than their DS counterparts. There is no indication that the mental retardation observed in any aneuploid state is non-specific. The neurological profiles of other aneuploid states are also quite distinctive (e.g. DiGeorge syndrome, Karayiorgou et al., 1995).

Shapiro (1994) concedes that because there are differences among the aneuploid phenotypes, genes on the triplicated chromosome must contribute, but he suggests that these differences may be attributable to the quantity of genetic material triplicated, and not to the genes themselves. It is true that human chromosomes 13 and 18 are larger than chromosome 21 and that the anomalies of trisomies 13 and 18 are much more severe, resulting in the death of these individuals shortly after birth, but other examples discredit this argument. Cases of double aneuploidy have been reported, and include males with 48,XXY,+21 or 48,XYY,+21 and females with 48,XXX,+21 or 46,X,+21. These individuals have usually shown typical DS. In addition, those males with 48,XXY,+21 had testes compatible with those found in Klinefelter syndrome. Some females with 46,X,+21 had features of Turner's syndrome but their appearance was mostly that of DS (Breg, 1977). If the characteristics and severity of the phenotype were a reflection of the amount of extra chromosomal material, then these double trisomy phenotypes should have been non-specific and much more severe. Conversely, there are patients with DS without any obvious chromosomal abnormality (Lee and Jackson,

1973; Ahlbom et al., 1996). It is conceivable that these patients have a small triplicated region of HSA21 which has escaped detection or that abnormalities of a few genes on HSA21 cause the phenotype. Nonetheless, if we assume a small duplication, then the tiny amount of extra chromosomal material that these people carry is enough to cause the relatively severe DS phenotype.

Mouse models

In 1970, trisomies were generated for each of the 19 mouse autosomes (reviewed in Gearhart et al., 1986), and based solely on pathological characteristics (at this time nothing was known about human:mouse genetic homology), it was predicted that the Ts16 mouse could serve as an animal model for DS.

Since it was noted that DS results from the triplication of the smallest human chromosome, investigators examined trisomy of the smallest mouse chromosome, MMU19, in the belief that the amount of chromosomal material, not the gene dosage effects, gives rise to the phenotypic characteristics associated with autosomal trisomy. Again, there was no evidence for non-specific mechanisms, the phenotypes were distinct (Gearhart et al., 1986; Oster-Granite and Lacey Casem, 1995). Ts19 mice survive whereas Ts16 mice die in utero and display many gross abnormalities, affecting development and morphogenesis of organs such as the brain, heart and skeleton not observed in Ts19. Many of the features observed in Ts16 mice are similar to those described in humans with DS, again pointing to specific genes affecting the phenotype rather than the amount of extra genetic material.

Changes in electrical membrane properties of dorsal root ganglion nerves were specific to Ts16 and not a general effect of trisomy as the same effects did not occur in Ts19 (Caviedes et al., 1990). Also, a comparison of the synaptic neurochemistry in Ts16 and Ts19 mice revealed numerous differences (reviewed in Oster-Granite and Lacey Casem, 1995).

Two mouse models (Ts65Dn and Ts1Cje) exist that contain partial trisomy 16, which corresponds to HSA21 from marker APP to MX1 in Ts65Dn and from SOD1 to MX1 (a slightly smaller region) in Ts1Cje. Ts65Dn mice (Davisson et al., 1993) display a variety of phenotypic abnormalities similar to human DS, including early developmental delay, reduced birth weight, muscular trembling, male sterility, abnormal facies and impairment in both non-spatial and spatial learning (Reeves et al., 1995; Escorihuela et al., 1995; Holtzman et al., 1996; Coussons-Read and Crnic, 1996). In Ts1Cje, both sexes are fertile and, although their phenotype overlaps with Ts65Dn as would be expected since they contain the same chromosomal region in triplicate, they exhibit subtle differences from Ts65Dn. The learning deficits of Ts1Cje are less severe than those of Ts65Dn in that they have no defect in non-spatial learning tasks and a less severe deficit in spatial learning. Also, the degeneration of basal forebrain cholinergic neurones observed in Ts65Dn is absent from Ts1Cje (Sago et al., 1998). We can infer from such studies that overexpression of specific genes located in the region App to Sod1 are required for the impairment in the non-spatial learning tasks, as well as for the neuronal atrophy to develop in the Ts65Dn mouse.

The generation of YAC transgenic mice also lends credence to our contention that the overexpression of specific genes lead to specific phenotypes. Strains of mice carrying different HSA21 YACs displayed different characteristics. Mice carrying YACs 152F2 and 230E8 both showed specific but different defects in learning and memory compared to transgenic mice containing YACs 141G6 and 285E6 which showed no abnormalities. The 230E8 mice had a significantly increased neuronal density in the cerebral cortex compared to non-transgenics and mice transgenic for the other YACs (Smith et al., 1997). Thus the differences must be due to the expression of genes contained within the YACs, and not due to non-specific defects resulting simply from the presence of large segments of exogenous DNA.

We stated earlier that no single gene has been shown to cause any phenotypic trait in the human condition of DS, but the use of animal models is yielding powerful insights into the biological functions of HSA21 genes and the pathologies they may cause. The idea behind the generation of single gene transgenics was to try to dissect the DS phenotype and to determine the contributions, if any, of single genes, in producing any part of the syndrome. DS is considered to be a complex disease, resulting from the interactions of numerous aberrantly expressed genes and the disruption of numerous pathways. So, the notion that overexpressing one gene could produce any of the syndrome's features was beyond the expectations of many investigators. However, the results have been spectacular.

Transgenic mice harbouring candidate genes for DS have been generated and most model, at least part, of the DS phenotype (reviewed in Kola and Herzhog, 1997, 1998; Kola and Pritchard, 1999). Transgenic mice overexpressing SOD1, an enzyme involved in the metabolism of oxygen free radicals exhibit some abnormalities characteristic of DS patients. They display thymic hypoplasia, thickened tongues, neuromuscular junction abnormalities, increased levels of serotonin in the plasma and are deficient in hippocampal-dependent learning. Mice overexpressing the liver-type phosphofructokinase gene (PFKL) have altered liver glycolysis, as do individuals with DS. Mice transgenic for HMG14, a protein implicated in chromatin structure modulation, develop the thymic cysts also present in DS and mice overexpressing S100β, a neurotrophic factor released by astroglial cells, show defects in spatial memory and learning, abnormal dendritic development and astrocytosis, all features of the DS phenotype. Moreover, overexpression of both SOD1 and S100β has been implicated in accelerated aging, another DS trait. Mice that overexpress the Ets2 transcription factor develop skeletal/bone defects akin to those seen in DS, particularly those affecting the craniofacial region, such as altered head-shape and brachycephaly. Based on this substantial body of evidence we are convinced that parts of the DS phenotype are caused by the overexpression of specific HSA21 genes and that the phenotype in its entirety is caused by the cumulative effects of specific gene interactions.

Specific abnormalities of DS

It is true that no physical nor behavioural findings occur uniquely in DS, but that many of the pathologies also occur at lower frequencies in the general population. Some of these pathologies are so characteristic of DS that it is difficult to argue that they can be caused by mechanisms other than the aberration of HSA21 genes. For instance, 70% of all endocardial cushion defects (ECDs) are associated with DS (Ferencz et al., 1989) and atrioventricular septal defects (AVSD), the predominant form of ECD in DS, occurs very rarely in the non-DS population (2.8% of isolated cardiac lesions) (Wilson et al., 1993). Ninety-six per cent of Tri16 mice have a heart defect with 56% of these having an AVSD (Miyabdra et al., 1982). Increased adhesion of fibroblasts leads to the failure of cushion fusion in trisomy 21 (Kurnit et al., 1985) and intriguingly, two cell adhesion molecules have been mapped to HSA21 (Paoloni-Giacobino et al., 1997; Yamakawa et al., 1998). A twofold increase in cell surface expression of such molecules can increase the adhesion properties of cells more than thirty-fold (Paoloni-Giacobino et al., 1997).

Duodenal stenosis occurs in the general population at a frequency of 1 in 10,000–40,000 live births but in 4% to 7% of infants with DS. Further, 35% of all congenital duodenal stenosis is associated with DS (Korenberg et al., 1992). Indeed, a search of the literature failed to find duodenal stenosis as a characteristic trait associated with any other syndrome.

One type of leukaemia (transient leukaemia) occurs almost exclusively in DS and the other form, acute nonlymphocytic leukaemia subtype M7 (ANLL-M7) accounts for 50% of DS leukaemia and is extremely rare in the non-DS population (Shen et al., 1997).

DS individuals are more susceptible to infection. The incidence of infectious hepatitis in institutionalised subjects with DS was 28% compared to 3% in other mentally retarded individuals (Shapiro, 1983). The basis for this increased susceptibility to infection and leukaemia is unknown but it has been proposed that, at least for leukaemia, an increase in interferon sensitivity could be responsible (Zihni, 1994). Chromosome 21 encodes a number of genes involved in interferon/immune responses, including those for various receptor molecules. It is conceivable that the overexpression of these receptors results in increased sensitivity to interferons, leading to an abnormal immune response in DS.

Alzheimer's-like pathology is not seen in other aneuploid states so it was reasonable to assume that a gene involved in Alzheimer's disease (AD) mapped to HSA21. This is indeed the case and it is interesting to note, in support of our argument that specific genes cause specific phenotypes, a report of a 78 year old woman with partial trisomy 21 and DS but without evidence of AD. Molecular analysis revealed that she had only 2 copies of the APP gene (Prasher et al., 1998).

Cancer of the stomach is a common neoplasm in the general population. A study of the genotypes of gastric tumours led to the suggestion that at least two tumour suppressor genes map to HSA21 (Sataka et al., 1997). Satge and colleagues (1998a) postulated that if this was the case, then stomach cancer should be a rare event in DS due to the gene dosage effect of the tumour suppressor genes. This is precisely the case -stomach cancer rarely occurs in association with DS.

Other cancers are also unusually underrepresented in DS. Breast cancers have rarely been reported, and in children, nephroblastomas and neuroblastomas are extremely rare (reviewed in Satge et al., 1998a). A European survey of 6,724 infants and children with neuroblastoma, detected no cases of DS (Satge et al., 1998b) and it was suggested that S100β may be responsible for this protection. S100β is the product of a gene on HSA21 which is over-expressed in DS and was found to be abundant in those neuroblastomas which were classified as having a good prognosis.

DS individuals seem to age prematurely. Many of the processes associated with normal ageing, eg the occurrence of AD-like pathology in the brain, senile dementia, perceptive hearing loss, skin dryness and reduced elasticity, cataracts and thymic involution are accelerated in DS. Proponents of the "amplified developmental instability" hypothesis argue that because all these pathologies are evident in the general population, then in DS they must be the result of disturbed homeostasis due to a greater intolerance to environmental traumas in these individuals, and therefore, non-specific. We would argue that the pathologies of ageing are due to the cumulative effects of somatic mutations or genetic aberrations (presumably caused by environmental stresses) that accumulate with age and disrupt normal cellular processes and pathways due to alterations in specific gene products. In support of this is the observation that cells from AD patients exhibited an increased frequency of aneuploidy, and in particular trisomy 21, compared to age matched normal individuals (Li et al., 1997).

Concluding remarks

Shapiro (1989, 1994) states that two of the major defects in the notion of specificity of phenotypes in aneuploid syndromes and their total explanation in terms of the triplicated genes are: 1) the complete denial of the role of the environment in the production of these traits, and 2) the failure to accept that certain developmental pathways are more vulnerable than others to disruptive genetic and environmental factors. Neither of these is disregarded by the "gene dosage effect" hypothesis. No geneticist would dispute the idea that epigenetic/epistatic/environmental influences play a role in the expressivity/ penetrance of a particular trait. This is best illustrated by monozygotic twins who clearly do show some differences. Indeed, it is these influences that explain the variability in both the penetrance and expressivity of traits seen in DS. Proponents of the "gene dosage effect" hypothesis would also agree that some developmental pathways may be particularly vulnerable and therefore less able to tolerate disturbances in the doses of gene products.

In conclusion, as Shapiro (1989) notes, the points of disagreement between the hypotheses of "amplified instability" and "gene dosage effect" may be due to differences in perspective. Much of the early data could be used to support either argument, depending on your point of view. Still there is no definitive proof of either, but we believe that the weight of evidence is heavily on the side of the "gene dosage effect" hypothesis. We hope, with our review of recent data, and their presentation in the context of the "gene dosage effect" hypothesis, that we have convinced our readers that increased dosage of specific HSA21 genes causes the DS phenotype.

References

Ahlbom BE, Goetz P, Korenberg JR, Pettersson U, Seemanova E, Wadelius C, Zech L, Anneren G (1996) Molecular analysis of chromosome 21 in a patient with a phenotype of Down syndrome and apparently normal karyotype. Am J Med Genet 63: 566–572

Barden HS (1980) Fluctuating dental asymmetry: a measure of developmental instability in Down syndrome. Am J Phys Anthropol 52: 169–173

Blum-Hoffmann E, Rehder H, Langenbeck U (1988) Skeletal anomalies in trisomy 21 as an example of amplified developmental instability in chromosome disorders: a histological study of the feet of 21 mid-trimester fetuses with trisomy 21. Am J Med Genet 29: 155–160

Breg WR (1977) Down syndrome: a review of recent progress in research. Pathobiol Annu 7: 257–303

Caviedes P, Ault B, Rapoport SI (1990) Electrical membrane properties of cultured dorsal root ganglion neurons from trisomy 19 mouse fetuses: a comparison with the trisomy 16 mouse fetus, a model for Down syndrome. Brain Res 511: 169–172

Coussons-Read ME, Crnic LS (1996) Behavioral assessment of the Ts65Dn mouse, a model for Down syndrome: altered behavior in the elevated plus maze and open field. Behav Genet 26: 7–13

Davisson MT, Schmidt C, Reeves RH, Irving NG, Akeson EC, Harris BS, Bronson RT (1993) Segmental trisomy as a mouse model for DS. In: Epstein CJ (ed) The phenotypic mapping of Down syndrome and other aneuploid conditions. John Wiley, New York, pp 117–133

Dunlap SS, Aziz MA, Rosenbaum KN (1986) Comparative anatomical analysis of human trisomies 13, 18 and 21: the forelimb. Teratology 33: 159–186

Epstein CJ (1986) The consequences of chromosome imbalance: principles, mechanisms and models. Cambridge University Press, New York

Epstein CJ (1988) Specificity versus nonspecificity in the pathogenesis of aneuploid phenotypes. Am J Med Genet 29: 161–165

Escorihuela RM, Fernández-Teruel A, Vallina IF, Baamonde C, Lumbreras MA, Dierssen M, Tobeña A, Flórez J (1995) Behavioral assessment of Ts65Dn mice: a putative DS model. Neurosci Lett 199: 143–146

Ferencz C, Neill CA, Boughman JA, Rubin JD, Brenner JI, Perry LW (1989) Congenital cardiovascular malformations associated with chromosome abnormalities: an epidemiologic study. J Pediatr 114: 79–86

Gearhart JD, Davisson MT, Oster-Granite ML (1986) Autosomal aneuploidy in mice: generation and developmental consequences. Brain Res Bull 16: 789–801

Greber-Platzer S, Schatzmann-Turhani D, Wollenek G, Lubec G (1999) Evidence against the current hypothesis of "gene dosage effects" of trisomy 21: Ets-2, encoded on chromosome 21 is not overexpressed in hearts of patients with Down Syndrome. Biochem Biophys Res Commun 254: 395–399

Hassold TJ, Jacobs PA (1984) Trisomy in man. Annu Rev Genet 18: 69–97

Holtzman DM, Santucci D, Kilbridge J, Chua-Couzens J, Fontana DJ, Daniels SE, Johnson RM, Chen K, Sun Y, Carlson E, Alleva E, Epstein CJ, Mobley WC (1996) Developmental abnormalities and age-related neurodegeneration in a mouse model of DS. Proc Natl Acad Sci USA 93: 13333–13338

Jones KL (ed) (1988) Smith's recognizable patterns of human malformation, 4th ed. WB Saunders, Philadelphia

Karayiorgou M, Morris MA, Morrow B, Shprintzen RJ, Goldberg R, Borrow J, Gos A, Nestadt G, Wolyniec PS, Lasseter VK, Eisen H, Childs B, Kazanian HH, Kucherlapati R, Atonarakis SE, Pulver AE, Housman (1995) Schizophrenia susceptibility associated with interstitial deletions of chromosome 22q11. Proc Natl Acad Sci USA 92: 7612–7616

Kola I, Hertzog P (1997) Animal models in the study of the biological function of genes on HSA21 and their role in the pathophysiology of DS. Hum Mol Genet 6: 1713–1727

Kola I, Herzog PJ (1998) Down Syndrome in mouse models. Curr Opin Genet Dev 8: 316–321

Kola I, Pritchard M (1999) Animal models of Down syndrome. Mol Med Today 5: 276–277

Korenberg JR, Bradley C, Disteche CM (1992) Down syndrome: molecular mapping of the congenital heart disease and duodenal stenosis. Am J Hum Genet 50: 294–302

Kurnit DM, Aldridge JF, Matsuoka R, Matthysse S (1985) Increased adhesiveness of trisomy 21 cells and atrioventricular canal malformations in Down syndrome: a stochastic model. Am J Med Genet 30: 385–399

Langenbeck U, Blum E, Wilkert-Walter C, Hansmann (1984) Developmental pathogenesis of chromosome disorders: report on two newly recognized signs of Down syndrome. Am J Med Genet 18: 223–230

Lee LG, Jackson JF (1972) Diagnosis of Down's syndrome: clinical versus laboratory. Clin Pediatr 11: 353–356

Li J, Xu M, Zhou H, Ma J, Potter H (1997) Alzheimer presenilins in the nuclear membrane, interphase kinetochores, and centrosomes suggest a role in chromosome segregation. Cell 90: 917–927

Mueller RF, Young ID (1998) Chromosome disorders. In: Emery's Elements of Medical Genetics. Churchill Livinstone, pp 245–264

Opitz J (1982) The developmental field concept in clinical genetics. J Pediatr 101: 805–809

Oster-Granite ML, Lacey-Casem ML (1995) Neurotransmitter alterations in the trisomy 16 mouse: a genetic model system for studies of Down syndrome. MRDD Res Rev 1: 227–236

Paoloni-Giacobino A, Chen H, Antonarakis SE (1997) Cloning of a novel human neural cell adhesion molecule gene (NCAM2) that maps to chromosome region 21q21 and is potentially involved in Down syndrome. Genomics 43: 43–51

Prasher VP, Farrer MJ, Kessling AM, Fisher EM, West RJ, Barber PC, Butler AC (1998) Molecular mapping of Alzheimer-type dementia in Down's syndrome. Ann Neurol 43: 380–383

Reeves RH, Irving NG, Moran TH, Wohn A, Kitt C, Sisodia SS, Schmid C, Bronson RT, Davisson M (1995) A mouse model for DS exhibits learning and behaviour deficits. Nat Genet 11: 177–184

Sago H, Carlson EJ, Smith D, Kilbridge J, Rubin EM, Mobley WC, Epstein CJ, Huang TT (1998) Ts1Cje, a partial trisomy 16 mouse model for DS, exhibits learning and behavioral abnormalities. Proc Natl Acad Sci USA 95: 6256–6261

Sakata K, Tamura G, Nishizuka S, Maesawa C, Suzuki Y, Iwaya T, Terashima M, Saito K, Satodate R (1997) Commonly deleted regions on the long arm of chromosome 21 in differentiated adenocarcinoma of the stomach. Genes Chromosomes Cancer 18: 318–321

Satge D, Sasco AJ, Geneix A, Malet P (1998a) Another reason to look for tumor suppressor genes on chromosome 21. Genes Chromosomes Cancer 21: 1

Satge D, Sasco AJ, Carlsen NL, Stiller CA, Rubie H, Hero B, de Bernardi B, de Kraker J, Coze C, Kogner P, Langmark F, Hakvoort-Cammel FG, Beck D, von der Weid N,

Parkes S, Hartmann O, Lippens RJ, Kamps WA, Sommelet D (1998b) A lack of neuroblastoma in Down syndrome: a study from 11 European countries. Cancer Res 58: 448–452

Shapiro BL (1975) Amplified developmental instability in Down's syndrome. Ann Hum Genet 38: 429–437

Shapiro BL (1983) Down syndrome-A disruption of homeostasis. Am J Med Genet 14: 241–296

Shapiro BL (1989) The pathogenesis of aneuploid phenotypes: the fallacy of explanatory reductionism. Am J Med Genet 33: 146–150

Shapiro BL (1994) The environmental basis of the Down syndrome phenotype. Dev Med Child Neurol 36: 84–90

Shen JJ, Williams BJ, Zipursky A, Doyle J, Sherman SL, Jacobs PA, Shugar AL, Soukup SW, Hassold TJ (1995) Cytogenetic and molecular studies of Down syndrome individuals with leukemia. Am J Hum Genet 56: 915–925

Smith DJ, Stevens ME, Sudanagunta SP, Bronson RT, Makhinson M, Watabe AM, O'Dell TJ, Fung J, Weier HU, Cheng JF, Rubin EM (1997) Functional screening of 2 Mb of human chromosome 21q22.2 in transgenic mice implicates minibrain in learning defects associated with Down syndrome. Nat Genet 16: 28–36

Trauner DA, Bellugi U, Chase C (1989) Neurologic features of Williams and Down syndromes. Pediatr Neurol 5: 166–168

Wilson L, Curtis A, Korenberg JR, Schipper RD, Allan L, Chenevix-Trench G, Stephenson A, Goodship J, Burn J (1993) A large, dominant pedigree of atrioventricular septal defect (AVSD): exclusion from the Down syndrome critical region on chromosome 21. Am J Hum Genet 53: 1262–1268

Witkop CJ Jr, Keenan KM, Cervenka J, Jaspers MT (1988) Taurodontism: an anomaly of teeth reflecting disruptive developmental homeostasis. Am J Med Genet [Suppl]4: 85–97

Yamakawa K, Huot YK, Haendelt MA, Hubert R, Chen XN, Lyons GE, Korenberg JR (1998) DSCAM: a novel member of the immunoglobulin superfamily maps in a Down syndrome region and is involved in the development of the nervous system. Hum Mol Genet 7: 227–237

Zihni L (1994) Down's syndrome, interferon sensitivity and the development of leukaemia. Leuk Res 18: 1–6

Authors' address: Prof. I. Kola, Centre for Functional Genomics and Human Disease, Institute of Reproduction and Development, Monash Medical Centre, 27-31 Wright St., Clayton, Victoria, 3168 Australia, e-mail: ismailko@silas.cc.monash.edu.au

J Neural Transm (1999) [Suppl] 57: 305–314
© Springer-Verlag 1999

Downregulation of the transcription factor scleraxis in brain of patients with Down Syndrome

K. Yeghiazaryan[1*], **D. Turhani-Schatzmann**[1], **O. Labudova**[2], **E. Schuller**[1], **E. N. Olson**[3], **N. Cairns**[4], and **G. Lubec**[1]

[1] Department of Pediatrics, University of Vienna, Austria
[2] Department of Radiobiology, University of Bonn, Federal Republic of Germany
[3] Southwestern Medical Center, University of Texas, Dallas, Tx, USA
[4] Brain Bank, Institute of Psychiatry, London, United Kingdom

Summary. Performing gene hunting in fetal Down Syndrome (DS) brain, we found a downregulated sequence with 100% homology to the basic — helix — loop — helix transcription factor (TF) scleraxis (Scl).

It was the aim of the study to evaluate Scl — mRNA steady state levels in adult DS brain with Alzheimer's disease (AD) neuropathological changes, brain of patients with AD, and controls in order to find out whether Scl — downregulation is linked to DS per se or simply to neurodegeneration, common to both disorders.

Determination of Scl — mRNA steady state levels was carried out by a blotting method in frontal, parietal, temporal, occipital lobe and cerebellum.

We found significantly decreased Scl — transcripts in brain of DS and AD, both, when normalized versus the house-keeping gene beta actin or total RNA.

We demonstrate the significant decrease of Scl — mRNA steady state levels in the pathogenesis of DS and AD suggesting a tentative role for this transcription factor in the development of the neurodegenerative processes known to occur in both disorders. More specifically, the biological meaning of the downregulation of Scl may be the involvement in the pathogenesis of impaired neuronal plasticity and wiring observed in DS and AD, phenomena regulated by the concerted action of the many transcription factors expressed in human brain.

Introduction

The novel transcription factor scleraxis (Scl) was cloned recently and was shown to be a representative of the basic helix — loop — helix (bHLH) family. Members of the bHLH family of transcription factors (TF) have been shown to regulate growth and differentiation of numerous cell types. Cell-type specific bHLH proteins typically form heterodimers with ubiquitous

---

* K. Y. is sponsored by a fellowship of the International Society for Amino Acid Research

bHLH proteins, and bind a DNA consensus sequence known as an E-box and this was found to be the case also for Scl. The expression pattern in the mouse, DNA-binding properties and transcriptional activity of Scl suggest that it is a regulator of gene expression within mesenchymal cell lineages. It is expressed in the sclerotome, in mesenchymal precursors of bone, cartilage and in connective tissue (Cserjesi, 1995). Scl was recently shown to be also expressed in osteoblasts where it was modulated by type-beta transforming growth factor (Liu, 1996). Liu and coworkers demonstrated the expression of Scl-mRNA in myoblasts and provided evidence for the regulation by a morphogenetic protein (Liu, 1997a). Information on the role of Scl is limited and the one report on a tentative function describes that this TF enhances aggrecan gene expression: overexpression of Scl induced aggrecan gene expression via binding to the E-box containing aggrecan gene with subsequent induction of of matrix proteins (Liu, 1997b).

Performing gene hunting in fetal Down Syndrome brain we found a sequence highly homologous to Scl and this gene was never reported before in mammalian or human brain. Moreover, subtractive hybridization revealed that in fetal DS brain Scl mRNA was reduced, and we therefore decided to study mRNA steady state levels in control, Down Syndrome (DS) and Alzheimer's disease (AD) brain as TFs in general are essential determinants for the normal development, wiring and functional maintenance of the brain. The availability of adult DS with AD neuropathology and control brain on the one, and the availability of AD brain on the other hand allowed the comparison not only between DS and controls but in addition between neurodegeneration in DS and in AD.

Methods

Substractive hybridization protocol (Labudova 1998a,b)

Fetal brain samples of two fetuses with Down Syndrome and two age and sex matched control, 23rd week of gestation, were obtained from the Brain Bank of the Institute of Psychiatry (Denmark Hill, London, UK). Hippocampus (gyrus parahippocampalis) was taken into liquid nitrogen and ground for the isolation of mRNA.

Isolation of mRNA was performed using the Quick Prep Micro mRNA purification kit (Pharmacia Biotech Inc., Uppsala, Sweden, cat.27 92 55 01). 1 microgram of mRNA from each (of the two) preparation was quality — checked by cDNA cloning kit (Gibco, Life Technologies, Eggenstein, Germany, cat. 18248-013) using the incorporation of [alpha — ^{32}P] dATP (Amersham, Buckinghamshire, UK, cat AA0004)) with subsequent electrophoresis on 1% agarose followed by autoradiography. The Reflection film (Dupont NEF 496) was exposed to the gel for a period of two hours at room temperature.

Construction of the substractive library: 10 micrograms each of mRNA from brain of DS and control were biotinilated by UV irradiation at 360 nm according to the instructions supplied in the subtractor kit (Invitrogen, Leek, Netherlands, cat K4320-01). 1 microgram of mRNA — pools each from the DS brain sample was subject to reverse transcriptase reaction (subtractor kit, Invitrogen) and the cDNA — pools were hybridized with the corresponding biotinilated mRNAs from controls.

The substractive hybridization mixture was incubated with streptavidin according to the subtractor kit given above and thus the biotinilated molecules (non-induced biotinilated mRNAs and the hybrid [biotinilated mRNAs / cDNAs]) complexed. The streptavidine complexes were removed by repeated phenol-chloroform extraction and subtracted cDNAs were separated from the aquous phase by alcohol precipitation (subtractor kit).

In order to amplify and clone subtracted cDNAs, they were ligated with Not I — linkers followed by Not I — digestion. These Not I linked cDNAs were ligated to Not I site of sPORT 1 cloning vector (cDNA cloning kit, Gibco).

To enable visualization of subtracted cDNAs the cloned cDNAs were amplified using universal primers

I 5'-GTAAAACGACGGCCAGT-3'
II 5'-ACAGCTATGACCATG-3'

from multiple cloning site of the sPORT-1 vector (cDNA cloning kit, Gibco). Amplified cDNAs were analysed on 1% agarose electrophoresis.

Cloning of subtracted Not I — linked cDNAs

Not I linked cDNAs ligated with sPORT 1 vector were used for the transformation of highly competent INFalpha F' E. coli cells (Invitrogen, Leek, Netherlands, cat C2020-03) and plated clones were analysed by plasmid isolation kit (Quiagen, Hilden, Germany, cat 12245) and digestion with Eco RI / Hind III.

Recombinant clones were sequenced by K.Granderath, MWG — Biotech (Ebersberg, Germany).

Homologies were determined by computer assisted comparison of data from the genebank sequence library: fastA@ebi.ac.uk (GBALL, EMBL Gen Bank Heidelberg, Germany).

Subtractive hybridization was performed cross-wise i.e. DS sample mRNA subtraction from control and vice versa at the 1:3 level (DSmRNA:control mRNA).

Northern and dot blot determination and quantification of Scl — transcript

Northern and dot blot analyses were carried out on postmortem brain samples from adult patients with DS, AD and controls.

The brain regions temporal, frontal, occipital, parietal cortex and cerebellum of patients with DS (n = 9; 3 females, 6 males; 56.1 ± 7.1 years old), AD (n = 9; 6 females, 3 males; 72.3 ± 7.6 years old) and controls (n = 9; 5 females, 4 males; 72.6 ± 9.6 years old) were mainly characterized in a previous publication (Seidl, 1997) and used for the studies at the transcriptional level. Briefly, post mortem brain samples were obtained from the MRC London Brain Bank for Neurodegenerative Diseases, Institute of Psychiatry). In all DS brains there was evidence of abundant beta A plaques and neurofibrillary tangles. The AD patients fulfilled the National Institute of neurological and Communicative Disorders and Stroke and Alzheimer Disease and Related Disorders Association (NINCDS/ADRDA) criteria for probable AD (Mirra, 1991). The histological diagnosis of AD was established and was consistent with the CERAD criteria (Tierney, 1988) for a "definite" diagnosis of AD. The controls were brains from individuals with no history of neurological or psychiatric illness. The major cause of death was bronchopneumonia in DS and AD patients and heart disease in controls. Post mortem interval of brain dissection in AD, DS and controls was 34.1 ± 13.7, 30.6 ± 17.5 and 34.8 ± 15.0hrs. Tissue samples were stored at −70°C and the freezing chain was never interrupted.

Biopsies from frontal, temporal, parietal, occipital cortex and cerebellum were obtained at autopsy and taken immediately into liquid nitrogen.

Northern and slot blots were performed according to the principle given previously (Hardmeier, 1997): Frozen samples were ground and mRNA extraction was performed using the TotalRNA Purification kit (Pharmacia). Subsequently, RNA was applied onto 1.4% agarose gel after denaturation with glyoxal and dimethylsulfoxide according to the method by McMaster and Carmichael (1977) and electrophoresed at 3–4 V/cm for 2.5 hrs in circulating 0.01 M phosphate buffer pH 7.0. RNA was then transfered to a positively charged nylon membrane (Hybond N+, Dupont, NEF 986) by capillary blotting (Southern, 1975) and fixed with 0.05 N sodium hydroxide for 5 min at room temperature and finally equilibrated at pH 7.0 with 3 washes in 2XSSC.

The probe for human beta-actin (ATCC 9800) was bought and the cDNA for human Scl was obtained from the scleraxis-plasmid pBSIISK, plasmid size 3.90 kb, subcloned into the Sal I / NotI sites of pBSIISK and linearized with Sal I and kindly supplied by E.N.Olson, Dallas, TX).

The probe was denatured prior to labelling by boiling for 5 min and subsequent cooling on ice, and labeled with fluorescein-12-dUTP using the Renaissance Random Primer Fluorescein-12 d-UTP labeling kit (Dupont, NEL 203).

After fixation of bound RNA, the nylon membrane was incubated in pre-hybridization solution (0.25 M phosphate buffer pH 7.2, containing 5% SDS w/v, 1 mM EDTA and 0.5% blocking reagent [from Dupont NEL 203]) for 12 hrs at 65°C in a hybridization oven. The blots were hybridized overnight at 65°C with the labelled probes each (50 ng/ml of pre-hybridization buffer).

After hybridization, nonspecifically bound material was removed by post-hybridization washes with 0.5X and 0.1X pre-hybridization buffer 2 × 10 min each at 65°C. The 0.5X and 0.1X pre-hybridization buffer was brought up to 65°C prior to use and the second wash was performed at room temperature.

Hybridized blots were blocked with 0.5% blocking reagent in 0.1 M Tris HCl pH 7.2 and 0.15 M NaCl for 1 hr at room temperature. Membranes were then incubated with antifluorescein HRP antibody (Dupont NEL 203) at a 1:1000 dilution in the solution given above for 1 hr under constant shaking.

Membranes were washed 4X for 5 min each in the solution given above.

The Nucleic Acid Chemiluminescence Reagent (Dupont NEL 201) was added to the membranes and incubated for 1 min. Excess detection reagent was removed by the use of filter paper, the membrane was placed in Sarawrap paper and exposed to autoradiography Reflection films (Dupont NEF 496) for 15 min at room temperature.

Slot blots were performed according to the method of White and Bancroft (1982). This procedure consisted of placing 2 ug of total RNA dissolved in 10 ul double distilled water mixed with 500 ul of 100% formamide, 162 ul of 37% formaldehyde, 100 ul of 10X MOPS(4-morpholinepropanesulfonic acid) buffer. This mixture was incubated for 10 min at 65°C and cooled down subsequently on ice. Samples were placed onto the membrane by the Manifolds filtration equipment (slot blot apparatus Bio slot TM, BIORAD) and the hybridization was performed as described above.

Denistometry of films was performed using the Hirschmann elscript 400 densitometer and calibration performed as given in the instrument's software program (Germany).

Statistical calculations were performed using the ANOVA followed by Kruskal-Wallis or Wilcoxon test, when appropriate. $P < 0.05$ was considered as statistical significant and is indicated as an asterisks in Fig. 2.

Results

Subtractive hybridization

The corresponding sequence with a fragment length of 339 bp found by subtractive hybridization from clone 89 is given in Fig. 1 and reveals 100%

```
Scleraxis, 89


89                                           ¹GGACGTGCCCAAGTTCGCCAGCTGG²⁵
                                             :::::::::::::::::::::::::::
hsSC  CTGACCCGCCTGTTCAGCGAGCCCGGCCTTCTCTCGGACGTGCCCAAGTTCGCCAGCTGG¹²⁰
        L  T  R  L  F  S  E  P  G  L  L  S  D  V  P  K  F  A  S  W


89    GGCGACGGCGAAGACGACGAGCCGAGGAGCGACAAGGGCGACGCGCCGCCACCGCCACCG⁸⁵
      ::::::::::::::::::::::::::::::::::::::::::::::::::::::::::::::
hsSC  GGCGACGGCGAAGACGACGAGCCGAGGAGCGACAAGGGCGACGCGCCGCCACCGCCACCG¹⁸⁰
        G  D  G  E  D  D  E  P  R  S  D  K  G  D  A  P  P  P  P  P


89    CCTGCGCCCGGGCCAGGGGCTCCGGGGCCAGCCCGGGCGGCCAAGCCAGTCCCTCTCCGT¹⁴⁵
      ::::::::::::::::::::::::::::::::::::::::::::::::::::::::::::::
hsSC  CCTGCGCCCGGGCCAGGGGCTCCGGGGCCAGCCCGGGCGGCCAAGCCAGTCCCTCTCCGT²⁴⁰
        P  A  P  G  P  G  A  P  G  P  A  R  A  A  K  P  V  P  L  R


89    GGAGAAGAGGGGACGGAGGCCACGTTGGCCGAGGTCAAGGAGGAAGGCGAGCTGGGGGGA²⁰⁵
      ::::::::::::::::::::::::::::::::::::::::::::::::::::::::::::::
hsSC  GGAGAAGAGGGGACGGAGGCCACGTTGGCCGAGGTCAAGGAGGAAGGCGAGCTGGGGGGA³⁰⁰
        G  E  E  G  T  E  A  T  L  A  E  V  K  E  E  G  E  L  G  G


89    GAGGAGGAGGAGGAAGAGGAGGAGGAAGAAGGACTGGACGAGGCGGAGGGCGAGCGGCCC²⁶⁵
      ::::::::::::::::::::::::::::::::::::::::::::::::::::::::::::::
hsSC  GAGGAGGAGGAGGAAGAGGAGGAGGAAGAAGGACTGGACGAGGCGGAGGGCGAGCGGCCC³⁶⁰
        E  E  E  E  E  E  E  E  E  G  L  D  E  A  E  G  E  R  P


89    AAGAAGCGCGGGCCCAAGAAGCGCAAGATGACCAAGGCGCGCTTGGAGCGCTCCAAGCTT³²⁵
      ::::::::::::::::::::::::::::::::::::::::::::::::::::::::::::::
hsSC  AAGAAGCGCGGGCCCAAGAAGCGCAAGATGACCAAGGCGCGCTTGGAGCGCTCCAAGCTT⁴²⁰
        K  K  R  G  P  K  K  R  K  M  T  K  A  R  L  E  R  S  K  L


89    CGGCGGCAGAAGGC³³⁹
      :::::::::::::
hsSC  CGGCGGCAGAAGGCAACGCGCGGGAGCGCAACCGCATGCACGACCTGAACGCAGCCCTG⁴⁸⁰
        R  R  Q  K  A
```

Fig. 1. Alignment of the nucleic acid sequence found by subtractive hybridization as clone 89 with the Scl sequence hsSC obtained from the EMBL gene bank Heidelberg, U 58681. 100% of homology was found

homology with human Scl accession number U58681 according to data from EMBL NETWORK (File Server <netserv@ebi.ac.uk>).

Northern blots and slot blots

Northern blots revealed a single band indicating the specificity of the hybridization process.

Densitometry of slots revealed the following:

As shown in Fig. 2a, Scl was manifold and significantly downregulated in frontal, temporal, parietal, occipital lobe and cerebellum of patients with DS and with AD, when expressed as mRNA DNAse I normalized versus the housekeeping gene beta actin in comparison with controls.

As shown in Fig. 2b, Scl was manifold and significantly downregulated when normalized versus 50 micrograms of total RNA in frontal, temporal, parietal, occipital lobe and cerebellum of patients with DS and AD in comparison to the control group.

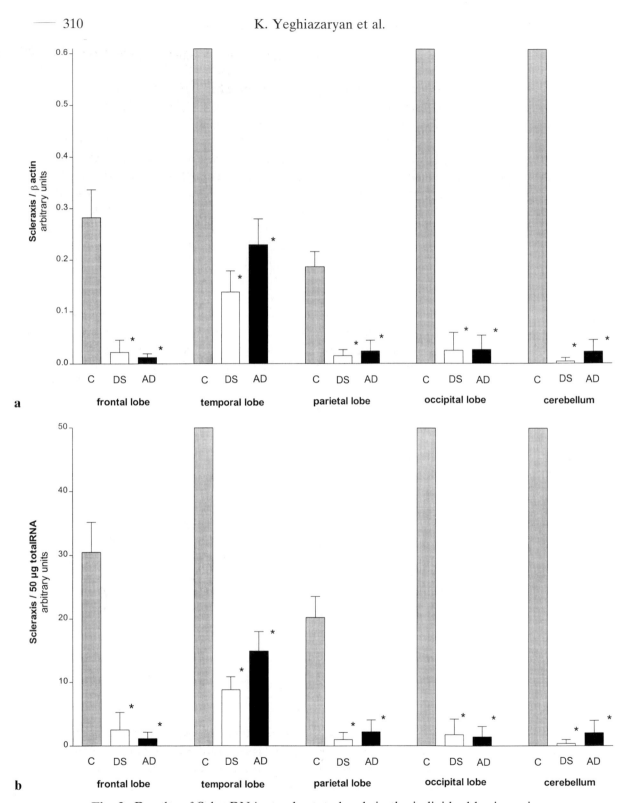

Fig. 2. Results of Scl-mRNA steady state levels in the individual brain regions

Discussion

As shown in the Results we detected a human Scl transcript by subtractive hybridization in fetal human brain. This transcript was downregulated at least threefold as the mRNA level set at mRNA subtraction in controls was three times higher. This information from gene hunting performed in fetal brain when secondary pathological changes could be ruled out, was tested in brain from normal adults, adults with DS and AD — pathology and in adults with AD.

Using this experimental design we could show that downregulation of Scl was found in both neurodegenerative disorders and was not specific for DS pathology.

It is well-documented and widely accepted that nearly 100% of patients with DS develop AD pathology in brain from the fourth decade (Burger, 1973; Wisniewski, 1985). Information drawn from the two fetal brain samples with DS would suggest that decreased Scl-mRNA is an event taking place early in (prenatal) life. And indeed, dysregulation of a transcription factor in fetal brain would help to explain DS brain pathology found already early in life (Epstein, 1992): The large number of TFs, their diverse sequence-specific interactions with DNA sites and with other TFs, and their ability to be modified in response to a variety of environmental cues and intracellular signals provide combinatorial codes for highly complex and yet highly organized patterns of gene expression likely to underlie the determination of diversity of neuronal phenotypes (He, 1991). For the deeper understanding of the complex system of TF in brain development, regulation of gene expression, differentiation, growth and migration the reader must be refered to the excellent review by Arenander and de Vellis (1994).

In adult brain pathologies downregulation of Scl was found when Scl — mRNA steady state levels were expressed per RNA or when normalized versus the housekeeping gene beta-actin. This housekeeping gene was determined to show the integrity of RNA and to compensate cell loss in DS and AD, which may have been a confounding factor. The biological meaning of decreased Scl mRNA steady state levels in adult brain with neurodegeneration is not known so far but could be seen in the context of deteriorated TFs already observed in DS and AD: we have recently shown the decrease of junD in several brain regions of the identical cohort of DS but not in brain of patients with AD (Labudova, 1998c). In a comparable panel of patients with DS we found overexpression of the ERCC2 and ERCC3 genes, identical to the p89 subunit of a general transcription factor (Ma, 1994) and associated with a class II transcription factor (Schaeffer, 1994), in several brain regions (Fang-Kircher, 1999). — The possible biological relevance may be that both genes are involved in "transcription coupled repair" and TAF — DBP (Adams, 1991) and EF1A (Adams, 1992) showed patterns paralleling mRNA-distribution and levels of repair genes. This again could reflect the coordinated activation of repair, transcription and translation in DS brain.

Finally, the ERCC2/3 — related transcription factor TFIIH was recently proposed to be involved in a p53 — mediated apoptotic pathway (Wang, 1996; Warbrick, 1996), a cascade, however speculative it may be, that could be activated in DS, where apoptosis is a hallmark of the disease (Busciglio, 1995; Nagy, 1997; Kitamura, 1998; Kim, 1997). In addition, we demonstrated that ETS2, a protooncogene and TF encoded on chromosome 21 and already proposed to be responsible for the development of the DS-phenotype (Kola, 1995), was decreased in individual brain regions of patients with DS (own unpublished results, in preparation).

The deterioration of TFs including Scl found in adult DS brain with AD neuropathological changes and in AD brains may therefore point to an involvement in apoptotic processes, repair mechanisms, development of AD neuropathological changes and impaired plasticity (Morris, 1997) of both neurodegenerative disorders.

Acknowledgement

This study was supported generously by the Red Bull Company, Salzburg, Austria.

References

Adams MD, Kelley JM, Gocayne JD, Dubnick M, Polymeropoulos MH, Xiao H, Merril CR, Wu A, Olde BO, Moreno RF, Kerlavage AR, McCombie WR, Venter JC (1991) Complementary DNA sequencing: expressed sequence tags and human genome project. Science 252: 1651–1656

Adams MD, Dubnick M, Kerlavage AR, Moreno R, Kelley JM, Utterback RT, Nagle JW, Fields C, Venter JC (1992) Sequence identification of 2,375 human brain genes. Nature 355: 632–634

Arenander AT, De Vellis J (1994) Development of the nervous system. In: Siegel GJ, Agranoff BW, Albers RW, Molinoff PB (eds) Basic neurochemistry, 5th edn. Raven Press, New York, pp 573–606

Burger PC, Vogel FS (1973) The development of pathologic changes of Alzheimer's disease and senile dementia in patients with Down's syndrome. Am J Pathol 73: 457–476

Busciglio J, Yankner BA (1995) Apoptosis and increased generation of reactive oxygen species in Down's syndrome neurons in vitro. Nature 378: 776–779

Cserjesi P, Brown D, Ligon KL, Lyons GE, Copeland NG, Gilbert DJ, Jenkins NA, Olson EN (1995) Scleraxis: a basic helix-loop-helix protgein that prefigures skeletal formation during mouse embryogenesis. Development 121: 1099–1110

Epstein CJ (1992) Down Syndrome (Trisomy 21) In: Scriver CR, Beaudet AL, Sly WS, Valle D (eds) The metabolic and molecular basis of inherited disease. McGraw Hill, New York, pp 749–794

Fang-Kircher S, Labudova O, Kitzmueller E, Rink H, Cairns N, Lubec G (1999) Increased steady state mRNA levels of DNA repair genes XRCC1, ERCC2 and ERCC3 in brain of patients with Down Syndrome. Life Sci (in press)

Hardmeier R, Hoeger H, Fang-Kircher S, Khoshsorur A, Lubec G (1997) Transcription and activity of superoxide dismutase, catalase and glutathione peroxidase following irradiation in radiation resistant and radiation sensitive mice Proc. Natl Acad Sci (USA) 94: 7572–7576

He X, Rosenfeld MG (1991) Mechanisms of complex transcriptional regulation: implications for brain development. Neuron 7: 183–196

Kim TW, Pettingell WH, Jung YK, Kovacs DM, Tanzi RE (1997) Alternative cleavage of Alzheimer-associated presenilins during apoptosis by a caspase — 3 family protease. Science 277: 373–376

Kitamura Y, Shimohama S, Kamoshima W, Ota T, Matsuoka Y, Nomura Y, Smith MA, Perry G, Whithouse PJ, Taniguchi T (1998) Alteration of proteins regulating apoptosis, bcl-2, bcl-x, bax, bak, bad, ICH-1 and CPP32 in Alzheimer's disease. Brain Res 780: 260–269

Kola I, Christiano F, de Haan JB, Thomas R, Sumarsono S, Corrick CM, Tymms M (1995) Genes, embryogenesis and Down Syndrome. In: Moeloek FA, Affandi B, Trounson AO (eds) Advances in human reproduction. Casterton-Parthenon, NY, pp 309–320

Labudova O, Lubec G (1998) cAMP upregulates the transposable element mys-1: a possible link between signaling and mobile DNA. Life Sci 62: 431–437

Labudova O, Fang-Kircher S, Cairns N, Lubec G (1998) Upregulation of vasopressin in brain of patients with Down Syndrome. Brain Res 806: 55–59

Labudova O, Krapfenbauer K, Moenkemann H, Rink H, Kitzmueller E, Cairns N, Lubec G (1998) Decreased transcription factor junD in brain of patients with Down Syndrome. Neurosci Lett 252: 159–162

Liu Y, Cserjesi P, Nifuji A, Olson EN, Noda M (1996) Sclerotome-related helix-loop-helix type transcription factor (scleraxis) mRNA is expressed in osteoblasts and its level is enhanced by type-beta transforming growth factor. J Endocrinol 151: 491–499

Liu Y, Nifuji A, Tamura M, Wozney JM, Olson EN, Noda M (1997) Scleraxis messenger ribonucleic acid is expressed in C2C12 myoblasts and its level is down-regulated by bone morphogenetic protein — 2 (BMP2). J Cell Biochem 67: 66–74

Liu Y, Watanabe H, Nifuji A, Yamada Y, Olson EN, Noda M (1997) Overexpression of a single hclix-loop-helix type transcription factor, scleraxis, enhances aggrecan gene expression in osteoblastic osteosarcoma ROS17/2.8 cells. J Biol Chem 272: 29880–29885

Ma L, Westbroek A, Jochemsen AG, Weeda G, Bosch A, Bootsma D, Hoeijmakers JHJ, van der Erb AJ (1994) Mutational analysis of ERCC3, which is involved in DNA repair and transcription initiation: identification of domains essential for the DNA repair function. Mol Cell Biol 14: 4126–4134

McMaster GK, Carmichael GG (1977) Analysis by single and double stranded nucleic acids on polyacrylamide and agarose gels by using glyoxal and acridine orange. Proc Natl Acad Sci (USA) 74: 4835–4841

Mirra SS, Heyman A, McKeel D, Sumi S, Crain BJ (1991) The consortium to establish a registry for Alzheimer disease (CERAD). II. Standardisation of the neuropathological assessment of Alzheimer's disease. Neurology 41: 479–486

Morris BJ (1997) Neuronal plasticity. In: Davies RW, Morris BJ (eds) Molecular biology of the neuron. Bios Scientifique Publishers, Oxford, pp 323–335

Nagy ZS, Esiri MM (1997) Apoptosis — related protein expression in the hippocampus in Alzheimer's disease. Neurobiol Aging 18: 565–571

Schaeffer L, Moncollin V, Roy R, Staub A, Mezzina M, Sarasin A, Weeda G, Hoeijmakers JHJ, Egly JM (1994) The ERCC2 / DNA repair protein is associated with the class II BTF2 / TFIIH transcription factor. EMBO J 13: 2388–2392

Seidl R, Greber S, Schuller E, Bernert G, Cairns N, Lubec G (1997) Evidence against increased oxidative DNA damage in Down Syndrome. Neurosci Lett 235: 137–140

Southern EM (1975) Detection of specific sequences among DNA fragments separated by gel electrophoresis. J Mol Biol 98: 503–511

Tierney MC, Fisher RH, Lewis AJ, Torzitto ML, Snow WG, Reid DW, Nieuwstraaten P, Van Rooijen LAA, Derks HJGM, Van Wijk R, Bischop A (1988) The NINCDA-

ADRDA work group criteria for the clinical diagnosis of probable Alzheimer's disease. Neurology 38: 359–364

Wang XW, Vermeulen W, Coursen JD, Gibson M, Lupold SE, Forrester K, Xu G, Elmore L, Yeh H, Hoeijmakers JHJ, Harris CC (1996) The XPB and XPD DNA helicases are components of the p53 mediated apoptosis pathway. Genes Dev 10: 1219–1232

Warbrick E (1996) Apoptosis: a new twist to the tale? Curr Biol 6: 1057–1059

White BA, Bancroft FC (1982) Cytoplasmic dot hybridization. Simple analysis of mRNA levels in multiple small cell or tissue samples. J Biol Chem 257: 8569–8574

Wisniewski KE, Wisniewski HM, Wen GY (1985) Occurrence of neuropathological changes and dementia of Alzheimer's disease in Down's syndrome. Ann Neurol 17: 278–282

Authors' address: Prof. Dr. G. Lubec, Department of Pediatrics, University of Vienna, Währinger Gürtel 18, A-1090 Vienna, Austria, e-mail: gert.lubec@akh-wien.ac.at

J Neural Transm (1999) [Suppl] 57: 315–322

Heat-shock protein 70 levels in brain of patients with Down Syndrome and Alzheimer's disease

B. C. Yoo[1]**, R. Seidl**[1]**, N. Cairns**[2]**, and G. Lubec**[1]

[1]Department of Pediatrics, University of Vienna, Austria
[2]Institute of Psychiatry, MRC Brain Bank, University of London, United Kingdom

Summary. Heat-shock proteins are proteins serving as molecular chaperones, involved in the protection of cells from various forms of stress. Since the expression of these proteins is closely related to that of amyloid precursor protein (APP), heat-shock protein has been studied in brain of patients with Alzheimer's disease (AD) and furthermore, brain Hsp70 mRNA levels were related to the agonal state. The aim of our study was to demonstrate the presence of Hsp70 — immunoreactive protein in brain of controls, patients with AD and Down Syndrome (DS) in individual brain regions. The rationale for the study was to test the hypothesis that expression of Hsp70, a protein involved in apoptosis would be altered in brain of these patients with neurodegenerative disorders where (neuronal) apoptosis is a hallmark of the disease. Brain immunoreactive-Hsp70 — protein (Hsp70) was determined by Western blotting using specific monoclonal antibody in five different brain regions (frontal, parietal, occipital, temporal cortex and cerebellum) from controls, DS and AD patients. Hsp70 expression was significantly increased in temporal cortex of patients with AD (arbitrary units: means ± SD; 0.35 ± 0.49 for controls, 0.97 ± 0.70 for DS patients, 1.16 ± 0.56 for AD patients). In frontal and parietal cortex from DS patients, there was a strong correlation between Hsp70 levels and the length of post-mortem interval (r = 0.95, P < 0.01 and r = 0.82, P < 0.021).

Introduction

The Hsp70 family of heat-shock proteins is one of the major proteins rapidly synthesized in response to various stresses such as heat, anoxia, ischaemia, ethanol, heavy metals and etc (Morimoto et al., 1994). The fundamental role of the Hsp70 family has been shown to be protection against subsequent, more severe stress, cross-tolerance to different stress forms, and stress-induced apoptosis (Mosser et al., 1997). Hsp70 and cognate Hsp70 (Hsc70) participate in many biological processes, in which protein folding is involved as e.g. protein translocation, protein translation, protein assembly and disassembly, and protein degradation (Hartl, 1996).

Stress-induced apoptosis could be prevented by Hsp70 which interferes with stress-activated protein kinase/c-Jun N-terminal kinase (SAPK/JNK) signaling pathway, and blocks the caspase proteolytic cascade (Mosser et al., 1997). Activation of SAPK/JNK signaling pathway is induced by heat stress, which can increase intracellular ceramide levels via activation of spingomyelinase, and exogenous addition of ceramide. The caspase cascade can be blocked in Hsp70-expressing cells. Interestingly, the SAPK/JNK signaling pathway could be also blocked by inducible overexpression of Hsp70 only, but not by permanent overexpression. Heat-shock proteins might protect by regulating apoptosis–inducing proteins: Apoptosis is basically is regulated by interactions between death-promoting (e.g. Bax, Bad, and Bcl-xs), and inhibiting proteins (e.g. Bcl-2, Bcl-xL and Mcl-1) (Nunez et al., 1994; Oltvai et al., 1994; Reed, 1994; Werner, 1996) but Strasser et al. (1995) proposed that heat-shock proteins act by a mechanism independent of Bcl-2 since a thermotolerance-inducing pretreatment can enhance resistance to apoptosis in Bcl-2 expressing cells. As apoptosis is a hallmark of neurodegenerative disorders DS and AD (de la Monte et al., 1997, 1998; Sawa et al., 1997; Busciglio et al., 1995) and brain Hsp70 protein levels were not available in literature, the determination of Hsp70 was of highest interest.

Harrison et al. (1993) demonstrated that Hsp70 mRNA was concentrated in pia matter and glia, and also present in neurons of frontal cortex from normal and demented patients. In AD the authors found increased Hsp70 mRNA in frontal cortex white matter and this made them suggest a role for Hsp70 in the pathophysiology of AD. We were able to confirm increased expression of Hsp70 in AD at the protein level, although in a different cortical brain region.

Material and methods

Human brain samples

Five different brain regions (gray matter of frontal, temporal, parietal, occipital cortex and cerebellum) were obtained from the Medical Research Council's London Brain Bank for Neurodegenerative Diseases, Department of Neuropathology, Institute of Psychiatry, London, UK.

The major cause of death in ten DS patients (three females, seven males) with a mean age of 54.4 ± 7.9 years, and in ten AD patients (four females, six males), aged 58.7 ± 5.7 years was bronchopneumonia. Ten controls (four females, six males), 54.4 ± 6.8 years old were individuals with no history of neurological or psychiatric illness and the major cause of death was heart disease. In all DS brains there was evidence of abundant beta-amyloid plaques and neurofibrillary tangles. AD patients satisfied the National Institute of Neurological and Communicative Disorders and Stroke and the Alzheimer Disease and Related Disorders Association (NINCDS/ADRDA) criteria for probable AD (Tierney et al., 1988). The diagnosis of AD was established using the CERAD criteria (Mirra et al., 1991) for a "definite" diagnosis of AD. Post-mortem interval of brain dissection in DS, AD, and controls were 32.0 ± 29.2, 29.2 ± 19.8 and 38.1 ± 22.7 hours respectively. All samples were stored at $-70°C$ until use.

Brain sample preparation for Western blot

Frozen brains were thawed and homogenized in TBST buffer containing 10 mM Tris-HCl (pH 7.5), 150 mM NaCl, 0.05% (V/V) Tween 20 in a Potter-Elvehjem homogenizer at 4°C. Protein concentrations of brain homogenates were estimated by the dye binding method (Bradford, 1976) using bovine serum albumin as the standard.

Western blot method

For Western blotting, 4,000 g supernatants of brain homogenates containing equivalent amounts of protein (20 microgram) were subjected to sodium dodecyl sulfate-polyacrylamide gel electrophoresis using the principle of Laemmli (1970). Electrophoresis was carried out using an electrophoresis apparatus (Hoefer, San Francisco, USA) at 200 volt with 12% polyacrylamide for the separating and 4% for the stacking gel. Following electrophoresis, proteins were transferred in a blotting chamber (Hoefer, San Francisco, USA) to polyvinylidene fluoride (PVDF) membranes (Millipore, Bedford, USA) at 120 volts for 3 hours. For blocking membranes were incubated in 5% non-fat dry milk (Bio-Rad, Richmond, USA) in TBST for 2 hours at room temperature on a rocking platform. The membranes were washed 3×5 min with TBST and put into a glass vessel containing 2,000 times diluted (0.5 microgram per 1 ml TBST) mouse anti-Hsp70 monoclonal antibody (StressGen, Victoria, Canada). After 2 hours of incubation with the Hsp70-antibody at room temperature, the membranes were washed 4×5 min with TBST and incubated with 5,000 times diluted (0.2 microgram antibody per 1 ml TBST) HRP-conjugated goat anti-mouse immunoglobulin G1 (Southern Biotechnology Associates, Inc., Birmingham, USA) for 1 hour at room temperature. The membranes were washed 6×5 min with TBST followed by NEN chemiluminescence reagent (NEN™ Life Science Products, Inc., Boston, USA) for 1 min and exposed to films (Kodak Blue XB-1, Rochester, USA) for 5 seconds. Developed films were scanned and densities of Hsp70-immunoreactive bands on the films were calculated by RFLP Scan 2.1 software program (Scanalytics, Billerica, USA).

All results were expressed as the mean ± standard deviation (SD). Between-group differences were calculated by non-parametric Mann-Whitney U-test and within-group-correlations were done using the Spearman rank-coefficient. The significance was set at the $P < 0.05$ level and all statistical analyses were performed by the SPSS-PC V 6.0.1 software.

Results

As shown in Table 1 and the scattergram in Fig. 1 Hsp70 was significantly increased in temporal cortex of patients with AD at the $P < 0.05$ level. Sex and age in each group had no effect on Hsp70 levels (Data not shown).

Linear regression analysis revealed a strong correlation between Hsp70 levels in frontal and parietal cortex from DS patients and postmortem interval ($r = 0.95$, $P < 0.01$ and $r = 0.82$, $P < 0.021$) as given in Fig. 2.

In all three groups, the highest amount of Hsp70 was in the cerebellum. The differences compared to cerebellum were significant in temporal cortex ($P < 0.05$) in the control group; in frontal cortex ($P < 0.05$), temporal cortex ($P < 0.05$), parietal cortex ($P < 0.05$), occipital cortex ($P < 0.05$) in DS; and

Table 1. The Hsp70 levels in five different brain regions

	Control	DS	AD
Frontal cortex	0.69 ± 0.28 (n = 8)	1.28 ± 0.71 (n = 7)	1.23 ± 1.49 (n = 9)
Temporal cortex	0.35 ± 0.49 (n = 7)	0.97 ± 0.70 (n = 9)	1.16 ± 0.56 * (n = 9)
Parietal cortex	0.52 ± 0.25 (n = 8)	0.74 ± 0.50 (n = 7)	0.68 ± 0.52 (n = 8)
Occipital cortex	1.54 ± 0.91 (n = 7)	1.43 ± 0.64 (n = 6)	1.45 ± 0.92 (n = 8)
Cerebellum	1.63 ± 1.09 (n = 6)	2.39 ± 0.53 (n = 7)	2.07 ± 0.88 (n = 7)

Values are the mean ± SD of optical density. *Significant difference ($P < 0.05$) compared with control

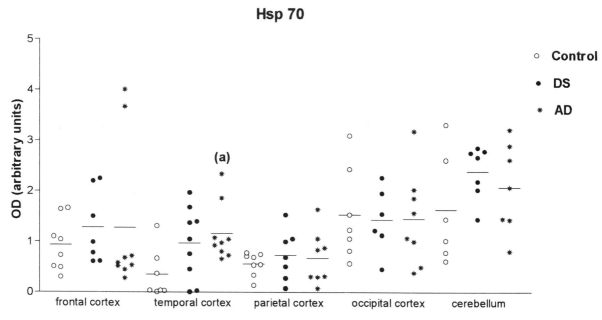

Fig. 1. The Hsp70 levels in five different brain regions: a) indicates significant difference compared with control

in temporal cortex ($P < 0.05$), parietal cortex ($P < 0.05$) and occipital cortex ($P < 0.05$) in AD.

As demonstrated in Fig. 3 immunoreactive Hsp70 revealed a single band at 70 kD on Western blots.

Discussion

We have recently shown that apoptosis–related markers were affected in brain of patients with DS and AD which were also used in the present study (Seidl et al., in press). We are thus confirming a previous report on overexpression of mRNA Hsp70 (Harrison et al., 1993) at the protein level. Although mean Hsp70 levels in DS temporal cortex were comparable to those

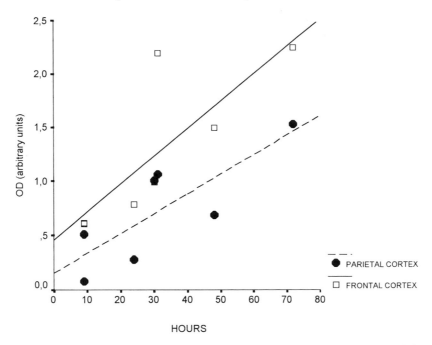

Fig. 2. Linear regression analysis revealing the correlation between Hsp70 levels with post-mortem intervals in frontal and parietal cortex of DS brain

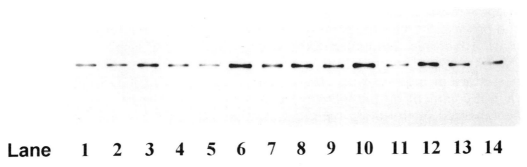

Fig. 3. The Western blot pattern of brain Hsp70 protein in temporal cortex demonstrating a single individual band at 70 kD (lanes 1 to 5 — Control; lanes 6 to 9 — DS, and lanes 10 to 14 — AD)

in AD, due to the wider range levels did not reach statistical significance (Fig. 1). The biological meaning of increased immunoreactive Hsp70 may be interpreted as the involvement of Hsp70 in the apoptotic process as well as in amyloid deposition, regularly taking place in DS and AD brain (de la Monte et al., 1996): The APP gene promoter contains a heat shock element and an abnormal APP heat shock response can increase accumulation of A beta, a key APP metabolite found in AD amyloid plaques (Zhong et al., 1996; Johnston et al., 1997). Heat-shock induction of Hsp70 mRNA leads to increased APP mRNA suggesting a role of Hsp70 in the accumulation of APP (Abe et al., 1991). In brain of AD patients, expression of small Hsp27 is highly induced (Renkawek et al., 1994) and it was proposed to inhibit in vitro amyloid formation by the Alzheimer's A beta (1–42) polypeptide (Kudva et al., 1997). Heat shock proteins may play either a protecting or promoting role in the formation of A 68 and/or the amyloidogenic C-terminal fragment of APP (Johnson et al., 1993). The mechanism of Hsp70 overexpression in AD may be linked to apoptosis per se but in patients with DS overexpression, although not having reached significance, increased Hsp70 may be assigned to a tentative gene dosage effect: Hsp70 was shown to be encoded on chromosome 21 (Harrison et al., 1987) but Gabriele et al. (1996) reported that in chinese hamster ovary–human hybrid cell lines containing human chromosome 21 Hsp70 was not expressed. The Hsp70 gene was shown to be located on human chromosomes 6 and 14 (Leung et al., 1991; Wakutani et al., 1995). A complex derangement of chromosomal mRNAs was found recently and this chromosomal imbalance per se could have been leading to deteriorated (although increase was statisticaly not significant) Hsp70 levels in the DS panel. There was an uneven distribution of Hsp70 in the individual brain areas with highest levels in cerebellum > occipital cortex > frontal cortex > parietal cortex and lowest levels in temporal cortex. The biological meaning of this distributional pattern cannot be interpreted but forms the basis for brain studies on Hsp70. The influence of the agonal state on brain Hsp70 mRNA levels has been studied by Morrison-Bogorad et al. (1995) who showed that Hsp70 was higher (up to 33 fold) in cerebellum of patients with AD and recorded fevel of > or = 39.2°C. However impressive these transcriptional data are, the fate and transcripts in the agonal state remains wholly unclear and it is more than questionable whether the transcripts are being translated and processed under the conditions of the agonal state and terminal metabolism. Anaerobic metabolism is a strong stimulus for Hsp70 transcription (Myrmel et al., 1994) and this may well explain the correlation of length of post-mortem interval with Hsp70 expression in brain of DS patients. We cannot comment on the influence of fever on brain Hsp70 as fever was not documented in the brain bank data available but the absolute majority of patients with DS and AD died on bronchopneumonia whereas heart failure was the cause of death in controls. The question whether infection per se has been responsible for increased brain Hsp70 remains open but is unlikely as one would expect overexpression in all areas.

This report is the first to show Hsp70 protein levels in individual areas of adult human brain and its uneven distribution. We suggest a role in the apoptotic mechanisms taking place in neurodegenerative disorders AD and DS and indicate to take into account the agonal state and post-mortem interval when designing studies on Hsp70 and heat-shock related systems in the brain. Derangement of brain Hsp70 levels may also be used to relativate or normalize expression of other brain proteins.

Acknowledgements

We are highly indebted to the Red Bull Company, Salzburg, for generous financial assistance.

References

Abe K, St. George Hyslop PH, Tanzi RE, Kogure K (1991) Induction of amyloid precursor protein mRNA after heat shock in cultured human lymphoblastoid cells. Neurosci Lett 125: 169–171

Bradford M (1976) A rapid and sensitive method for the quantitation of microgram quantities of protein utilizing the principle of protein-dye binding. Anal Biochem 72: 248–254

Busciglio J, Yankner BA (1995) Apoptosis and increased generation of reactive oxygen species in Down's syndrome neurons in vitro. Nature 378: 776–779

de la Monte SM, Xu YY, Wands JR (1996) Neuronal thread protein gene modulation with sprouting: relevevance to Alzheimer's disease. J Neurol Sci 138: 26–35

de la Monte SM, Sohn YK, Wands JR (1997) Correlates of p53- and Fas (CD95)-mediated apoptosis in Alzheimer's disease. J Neurol Sci 152: 73–83

de la Monte SM, Sohn YK, Ganju N, Wands JR (1998) p53- and CD95-associated apoptosis in neurodegenerative diseases. Lab Invest 78: 401–411

Gabriele T, Tavaria M, Kola I, Anderson RL (1996) Analysis of heat shock protein 70 in human chromosome 21 containing hybrids. Int J Biochem Cell Biol 28: 905–910

Harrison GS, Drabkin HA, Kao FT, Hart IM, Chu EH, Wu BJ, Morimoto RI (1987) Chromosomal location of human genes encoding major heat-shock protein HSP70. Somat Cell Mol Genet 13: 119–130

Harrison PJ, Procter AW, Exworthy T, Roberts GW, Najilerahim A, Barton AJ, Pearson RC (1993) Heat shock protein (hsx70) mRNA expression in human brain: effects of neurodegenerative disease and agonal state. Neuropathol Appl Neurobiol 19: 10–21

Hartl FU (1996) Molecular chaperones in cellular protein folding. Nature 381: 571–578

Johnson G, Refolo LM, Wallace W (1993) Heat-shocked neuronal PC12 cells reveal Alzheimer's disease-associated alterations in amyloid precursor protein and tau. Ann NY Acad Sci 685: 194–197

Johnston JA, Lannfelt L, Wiehanger B, Oneill C, Cowburn RF (1997) Amyloid precursor protein heat shock response in lymphoblastoid cell lines bearing presenilin-1 mutation. Biochem Biophys Acta Molecular Basis of Disease 1362: 182–192

Kudva YC, Hiddinga HJ, Butler PC, Mueske CS, Eberhart NL (1997) Small heat shock proteins inhibit in vitro A beta (1–42) amyloidogenesis. FEBS Lett 416: 117–121

Laemmli UK (1970) Cleavage of structural proteins during the assembly of the head of bacteriophage T4. Nature 277: 680–685

Leung PS, Gershwin ME (1991) The immunobiology of heat shock proteins. J Investg Allergol Clin Immunol 1: 23–30

Mirra SS, Heyman A, McKeel D, Sumi S, Crain BJ (1991) The consortium to establish a registry for Alzheimer disease (CERAD). II. Standardization of the neuropathological assessment of Alzheimer's disease. Neurology 41: 479–486

Morimoto RI, Tissieres A, Georgopoulos (1994) The biology of heat shock proteins and molecular chaperones. Cold Spring Harbor, New York, pp 1–593

Morrison-Bogorad M, Zimmerman AL, Pardue S (1995) Heat-shock 70 messenger RNA levels in human brain: correlation with agonal fever. J Neurochem 64: 235–246

Mosser DD, Caron AW, Bourget L, Denis-Larose C, Massie B (1997) Role of the human heat shock protein hsp70 in protection against stress-induced apoptosis. Mol Cell Biol 17: 5317–5327

Myrmel T, McCully JD, Malikin L, Krukenkamp IB, Levitsky S (1994) Heat-shock protein 70 mRNA is induced by anaerobic metabolism in rat hearts. Circulation 90: II299–305

Nunez G, Clarke MF (1994) The Bcl-2 family of proteins: regulators of cell death and survival. Trend Cell Biol 4: 399–403

Oltvai ZN, Korsmeyer SJ (1994) Checkpoints of dueling dimers foil death wishes. Cell 79: 189–192

322 B. C. Yoo et al.: Hsp 70 levels in Down Syndrome brain

Reed JC (1994) Bcl-2 and the regulation of programmed cell death. J Cell Biol 124: 1–6
Renkawek K, Bosman GJ, de-Jong WW (1994) Expression of small heat-shock protein hsp27 in reactive gliosis in Alzheimer disease and other types of dementia. Acta Neuropathol Berl 87: 511–519
Sawa A, Oyama F, Cairns NJ, Amano N, Matsushita M (1997) Aberrant expression of bcl-2 gene family in Down's syndrome brains. Mol Brain Res 48: 53–59
Seidl R, Fang-Kircher S, Cairns N, Lubec G (1999) Apoptosis-associated proteins p53 and APO-1/Fas (CD95) in brains of adult patients with Down syndrome. Neurosci Lett 260: 9–12
Strasser A, Anderson RL (1995) Bcl-2 and thermotolerance cooperate in cell survival. Cell Growth Differ 6: 799–805
Tierney MC, Fisher RH, Lewis AJ, Torzitto ML, Snow WG, Reid DW, Nieuwstraaten P, Van Rooijen LAA, Derks HJGM, Van Wiijk R, Bischop A (1988) The NINCDA-ADRDA work group criteria for the clinic diagnosis of probable Alzheimer's disease. Neurology 38: 359–364
Wakutani Y, Urakami K, Shimomura T, Takahasi K (1995) Heat shock protein 70 mRNA levels in mononuclear blood cells from patients with dementia of the Alzheimer type. Dementia 6: 301–305
Werner MH (1996) Stopping death cold. Structure 4: 879–883
Zhong S, Wu K, Black IB, Schaar DG (1996) Characterization of the genomic structure of the mouse APLP1 gene. Genetics 32: 159–162

Authors' address: Prof. Dr. G. Lubec, Department of Pediatrics, University of Vienna, Währinger Gürtel 18-20, A-1090 Vienna, Austria, e-mail: gert.lubec@akh-wien.ac.at

J Neural Transm (1999) [Suppl] 57: 323–335
© Springer-Verlag 1999

Increased levels of 14-3-3 gamma and epsilon proteins in brain of patients with Alzheimer's disease and Down Syndrome

M. Fountoulakis[1], **N. Cairns**[2] and **G. Lubec**[3]

[1] F. Hoffmann-La Roche Ltd., Basel, Switzerland
[2] Brain Bank, Institute of Psychiatry, University of London, United Kingdom
[3] Department of Pediatrics, University of Vienna, Austria

Summary. The 14-3-3 family consists of homo- and heterodimeric proteins representing a novel type of "adaptor proteins" modulating the interaction between components of signal transduction pathways. 14-3-3 isoforms interact with phosphoserine motifs on many proteins as kinases, phosphatases, apoptosis related proteins etc. Performing protein mapping by 2D electrophoresis in human brain we identified two isoforms, 14-3-3 gamma and epsilon and decided to determine these two multifunctional proteins in several brain regions of aged patients with Alzheimer's disease (AD) and Down Syndrome (DS) with AD neuropathology in comparison with control brains.

14-3-3 gamma and 14-3-3 epsilon proteins were increased in several brain regions of AD and DS patients.

These changes may contribute to the complex pathomechanisms of AD and AD in DS, evolving inevitably from the fourth decade of life. Deranged 14-3-3 isoforms gamma and epsilon may reflect impaired signaling and / or apoptosis in the brain as several kinases (protein kinase C, Ras, mitogen-activated kinase MEK) involved in signaling and apoptotic factors as bcl-2-related proteins BAD and BAG-1 are binding to 14-3-3 motifs.

Introduction

The 14-3-3 family consists of homo-and heterodimeric proteins representing a novel type of "adaptor proteins" modulating the interaction between components of signal transduction pathways. Phosphorylation of both, the binding partner and the 14-3-3 proteins, may regulate these interactions. 14-3-3 isoforms interact with novel phosphoserine motifs on many proteins (Dubois, 1997; Muslin, 1996; Yaffe, 1997).

14-3-3 proteins bind to more than 40 brain phosphoproteins of a brain extract (Ichimura, 1997), protein kinase KSR (kinase suppressor of Ras; Xing, 1997), the serine/threonine protein kinase c-Raf-1, a key mediator of mitogenesis and differentiation (Rommel, 1996; Braselmann, 1995; Fantl, 1994; Freed, 1994), protein kinase C (Aitken, 1995; Tanji, 1994; Toker, 1992;

Toker, 1990), human Cdc25C, a dual specificity protein phosphatase controlling entry into mitosis (Peng, 1997), the phosphorylated death agonist BAD (Zha, 1996), phosphorylated keratins during cell cycle progression (Liao, 1996) and histones (Chen, 1994).

Performing protein mapping in human brain, we identified two isoforms of 14-3-3 proteins, 14-3-3 gamma and epsilon.

14-3-3 gamma (Swiss-Prot: P35214, protein kinase C inhibitor protein-1, KCIP-1) is a homodimer with a molecular weight of 28,171 kD consisting of 246 amino acids and known sequence (Watanabe, 1993). It activates tyrosine and tryptophan hydroxylase in the presence of Ca2+/calmodulin dependent protein kinase II, activates PKC and serves as a multifunctional regulator of the cell signaling processes mediated by both kinases.

14-3-3 epsilon (Swiss-Prot: P42655, protein kinase C inhibitor protein-1, KCIP-1, mitochondrial import stimulation factor L subunit) is a homodimer with a molecular weight of 29,174 kD, consisting of 255 amino acids and known sequence of human (Conklin, 1995) and sheep brain (Toker, 1990). Basic functions are as given above for the gamma isomer.

Specifically, 14-3-3 epsilon (epsilon) was shown to positively regulate Ras-mediated signaling in Drosophila (Chang, 1997), binds to the insulin receptor substrate-1 thus indicating a role for regulation of insulin sensitivity by interrupting the association between the insulin receptor and insulin receptor substrate-1 (Ogihara, 1997), the insulin-like growth factor I receptor (Craparo, 1997), binds to the zinc finger protein A20, an inhibitor of tumor necrosis factor — induced apoptosis (Vincenz, 1996) and associates with cdc25 phosphatases without affecting their activity but rather mediating the association of cdc25 with Raf-1 in vivo thus participating in the linkage between mitogenic signaling and the cell cycle machinery (Conklin, 1995; Kumagai, 1998). 14-3-3 epsilon also interacts with a specific mitogen — activated protein kinase / extracellular signal-regulated kinase kinase (MEK, MEKK) and here again, binding to the kinase does not affect activity but rather serves as scaffold for protein-protein interactions (Fanger, 1998).

When 14-3-3 gamma and epsilon isoforms from protein maps of normal brain were compared to those of brain samples of patients with Alzheimer's disease (AD) and Down Syndrome (DS), clearly different patterns could be revealed.

Material and methods

Brain samples

Post mortem brain samples were obtained from the MRC London Brain Bank for Neurodegenerative Diseases, Institute of Psychiatry. In all DS brains there was evidence of abundant beta A plaques and neurofibrillary tangles. The AD patients fulfilled the National Institute of Neurological and Communicative Disorders and Stroke and Alzheimer's Disease and Related Disorders Association (NINCDS/ADRDA) criteria for probable AD (Mirra, 1991). The histological diagnosis of AD was established and was consistent with the CERAD criteria (Tierney, 1988) for a ïdefiniteÓ diagnosis of AD. The brain regions temporal, frontal, occipital, parietal cortex and cerebellum of karyotyped patients with DS (n = 9; 3 females, 6 males; 56.1 ± 7.1 years old), AD (n = 9; 6 females, 3 males; 72.3 ± 7.6 years old) and controls (n = 9; 5 females, 4 males; 72.6 ± 9.6 years old; Seidl 1997) were used for the studies at the protein level. The controls were brains from individuals with no history of neurological or psychiatric illness. The major cause of death was bronchopneumonia in DS and AD patients and heart disease in controls. Post mortem interval of brain dissection in AD, DS and controls was 34.1 ± 13.7, 30.6 ± 17.5 and 34.8 ± 15.0 h, respectively. Tissue samples were stored at −70°C and the freezing chain was never interrupted.

Sample preparation

Brain tissue was suspended in 0.5 ml of sample buffer consisting of 40 mM Tris, 5 M urea (Merck, Darmstadt, Germany), 2 M thiourea (sigma, St.Louis, MO, USA), 4% CHAPS (Sigma), 10 mM 1,4-dithioerythritol (Merck), 1 mM EDTA and a mixture of protease inhibitors, 1 mM PMSF and 1 mg/ml of each pepstatin A, chymostatin, leupeptin and antipain. The suspension was sonicated for approximately 30 s and centrifuged at 10,000 × g for 10 min and the supernatant was centrifuged further at 150,000 × g for 45 min to sediment undissolved material. The protein content in the supernatant was determined by the Coomassie blue method (Bradford, 1976). The average protein concentration was 8 mg/ml.

Two dimensional (2D) gel electrophoresis

The 2D-gel electrophoresis was performed essentially as reported (Langen, 1997). Samples of approx. 1.5 mg were applied on immobilized pH 3–10 non-linear gradient strips (IPG, immobilized pH-gradient strips, Pharmacia Biotechnology, Uppsala, Sweden), at both, the basic and acidic ends of the strips. The proteins were focused at 300 V for 1 hr, after which the voltage was gradually increased to 3500 V within 6 h. Focusing was continued at 3500 V for 12 h and at 5000 V for 48 h. The proteins were then separated on 9–16% linear gradient polyacrylamide gels (chemicals from Serva, Heidelberg, Germany and Bio-Rad, Hercules, CA, USA). After protein fixing with 40% methanol containing 5% phosphoric acid for 12 h, the gels were stained with colloidal Coomassie blue (Novex, San Diego, CA, USA) for 48 h. The molecular mass was determined by running standard protein markers at the right side of selected gels. The size markers (Gibco, Basel, Switzerland) covered the range 10–200 kDa. pl values were used as given by the supplier of the IPG strips. The gels were destained with water and scanned in a Molecular Dynamics Personal densitometer. The images were processed using Adobe Photoshop and PowerPoint software. Protein spots were quantified using the ImageMaster 2D Elite software (Amersham Pharmacia Biotechnology).

Matrix assisted laser desorption ionization mass spectroscopy

MALDI-MS analysis was performed as described (Fountoulakis, 1997) with minor modifications. Briefly, spots were excised, destained with 50% acetonitrile in 0.1 M ammonium bicarbonate and dried in a speedvac evaporator. The dried gel pieces were reswollen with 3 ml of 3 mM Tris-HCl, pH 8.8, containing 50 ng trypsin (Promega, Madison, Wl, USA). After 15 min, 3 ml of H2O were added and left at room temperature for 12 h. Two ml of 30% acetonitrile, containing 0.1% trifluoroacetic acid were added, the content was vortexed, centrifuged for 3 min and sonicated for 5 min. One microliter was applied onto the dried matrix spot. The matrix solution consisted of 15 mg nitrocellulose (Bio-Rad) and 20 mg a-cyano 4 hydroxycinnamic acid (Sigma) dissolved in 1 ml acetone:isopropanol 1:1 (v/v). 0.5 microliters of the matrix solution were applied on the sample target. Specimen were analysed in a time-of-flight Voyager Elite mass spectrometer (PerSeptive Biosystems, Cambridge, MA, USA) equipped with a reflectron. An accelerating voltage of 20 kV was used. Calibration was internal to the samples. The peptide masses were matched with the theoretical peptide masses of all proteins from all species of the SWISS-PROT database. For protein search, monoisotopic masses were used and a mass tolerance of 0.0075% was allowed.

Results

Quantification of 14-3-3 proteins

The protein extracts from the frontal, parietal, temporal and occipital cortex regions and the cerebellum of 9 patients with DS, 9 with AD and 9 controls were separated by 2-D gel electrophoresis. The 2-D gels with proteins from the corresponding brain regions of controls, DS and AD patients were compared with each other. Several differences were observed, which could represent allelic differences among the individuals (data not shown).

Figure 1 shows the protein spots representing 14-3-3 proteins epsilon and gamma — forms in samples from the frontal cortex regions of a control (Fig. 1A), a patient with Down syndrome (Fig. 1B) and a patient with Alzheimer's disease (Fig. 1C). The proteins were unambiguously identified by MALDl-MS as 14-3-3 protein epsilon (SWISS-PROT accession number P42655) and 14-3-3 protein gamma form (SWISS-PROT accession number P35214). The latter was identified as the rat protein as the human counterpart was not included in the database. Figure 2 shows the mass spectra of the peptides generated from the tryptic digestion of the corresponding protein spots.

The 14-3-3 proteins were mainly represented by weak spots in all three groups. An increase of the intensities of the spots representing 14-3-3 epsilon and 14-3-3 gamma proteins in the patients with DS (Fig. 1B) and AD (Fig. 1C) as compared to the controls (Fig. 1A) was observed. The spots were quantified using specific software (Figs. 3 and 4). In the frontal, parietal, temporal and occipital regions of DS and AD brain, the 14-3-3 e protein levels were increased approximately 1.5 fold in comparison with the control (Fig. 3A–D). The largest increase was found in the temporal and occipital cortex regions (Fig. 3C and 3D, respectively). In the cerebellum, 14.3.3 epsilon levels were decreased (Fig. 3E). Similarly, the 14-3-3 gamma protein levels were increased in the frontal, parietal, temporal and occipital cortex regions (Fig. 4A–D), with the largest differences observed in the frontal and occipital cortex (Fig. 4A and 4D), in comparison with the control brain. In the cerebellum, the 14-3-3 gamma levels were not increased (Fig. 4E). Differences in the

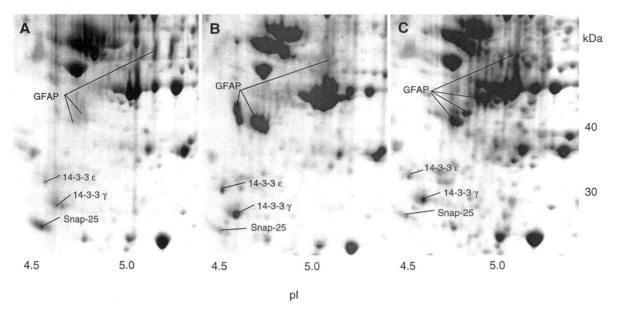

Fig. 1. Partial two-dimensional gel images of human brain proteins from a healthy (**A**), a patient with Down syndrome (**B**) and a patient with Alzheimer's disease (**C**). The proteins from the frontal lobe of cortex were extracted and separated on 3–10 nonlinear IPG strips, followed by a 9–16% SDS-polyacrylamide gels, as stated under Materials and Methods. The gels were stained with Coomassie blue. The proteins were identified by MALDI-MS. The spots corresponding to 14-3-3 ε and γ forms are indicated. The spots representing synaptosomal associated protein 25 (snap-25) and glial fibrillary acidic protein (GFAP), two proteins with different levels in the control and diseased groups, are also shown

expression level of the 14-3-3 proteins were observed for the various brain regions of the same patient. Similar has been observed for other brain proteins as well (data not shown).

A comparison of the 2-D gels (Fig. 1) shows, except of the 14-3-3 proteins, the presence of additional differences in the protein expression level between control and diseased brain. Some of the differences have been investigated, as for example increased levels of glial fibrillary acidic protein (GFAP) and decreased levels of synaptosomal associated protein 25 (snap-25) have been assigned to DS and AD (Greber, 1999). We have also observed variable levels of clathrin and tropomyosin (data not shown). Several of the other differences may be of allelic nature or may be disease-related. Many of the minor protein spots have not been identified yet. As our knowledge about expression levels and protein identification increases, we will learn more about disease-related differential protein expression.

Discussion

Brain levels of 14-3-3 protein gamma and epsilon were higher in cortical brain regions of patients with AD and DS than in controls. Information on the

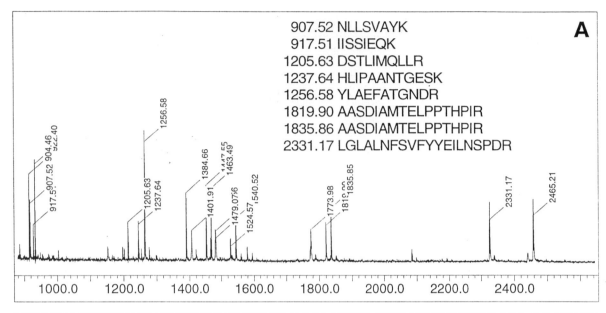

907.52 NLLSVAYK
917.51 IISSIEQK
1205.63 DSTLIMQLLR
1237.64 HLIPAANTGESK
1256.58 YLAEFATGNDR
1819.90 AASDIAMTELPPTHPIR
1835.86 AASDIAMTELPPTHPIR
2331.17 LGLALNFSVFYYEILNSPDR

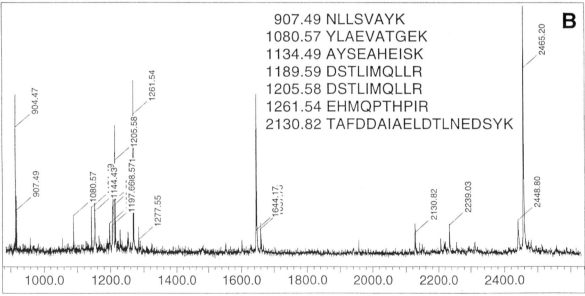

907.49 NLLSVAYK
1080.57 YLAEVATGEK
1134.49 AYSEAHEISK
1189.59 DSTLIMQLLR
1205.58 DSTLIMQLLR
1261.54 EHMQPTHPIR
2130.82 TAFDDAIAELDTLNEDSYK

m/z

Fig. 2. Mass spectrometry analysis of the 14-3-3 proteins ε (**A**) and γ (**B**). The spots representing the 14-3-3 ε and γ proteins (Fig. 1) were excised from the gels and digested wit trypsin and the peptides generated were analyzed by MALDI-MS as stated under Materials and Methods. A Search in the SWISS-PROT database resulted in the unambiguous identification of the protein as 14-3-3 ε, human form (P42655). The masses and the sequences of the 8 matching out of 20 total peptides are indicated. B Search in the SWISS-PROT database resulted in the unambiguous identification of the protein as 14-3-3 γ, rat form (P35214). The masses and the sequences of the 7 matching out of 18 total peptides are indicated. The peptides with masses 904.46 and 2465.20 represent internal standards. For protein search, corrected peptide masses and a mass tolerance window of 0.0075% were used

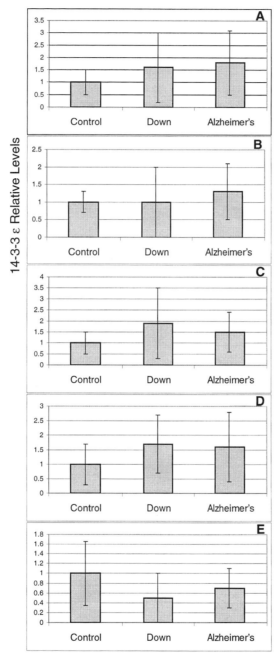

Fig. 3. Quantification of 14-3-3 ε protein levels in the brain regions of patients with DS and AD. The proteins from the frontal (**A**), parietal (**B**), temporal (**C**), occipital (**D**) cortex lobes and the cerebellum (**E**) of the patients and the control group were separated on 2-D gels and visualized following stain with colloidal Coomassie blue. In partial gel images, including 14-3-3 ε and the neighboring proteins, the intensities of the spots representing 14-3-3 ε were quantified, compared to the total proteins present. The quantification was performed using the ImageMaster 2D software. In each donor, the percentage of the volume of the spot representing 14-3-3 ε was determined. The average 14-3-3 ε values in the DS and AD groups were normalized to the mean value of the controls and the relative values are indicated. The bars indicate the standard deviation of 14-3-3 ε levels in the three groups

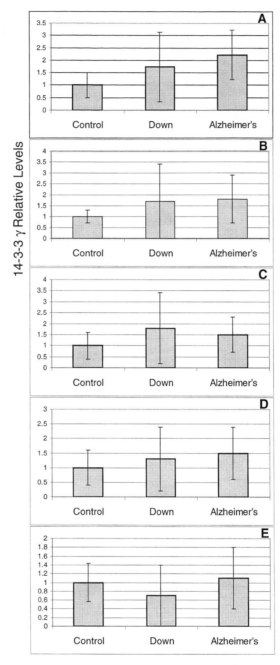

Fig. 4. Quantification of 14-3-3 γ protein levels in the brain regions of patients with DS and AD. The proteins from the frontal (**A**), parietal (**B**), temporal (**C**), occipital (**D**) cortex lobes and the cerebellum (**E**) of the patients and the control group were separated and quantified as stated under legend to Fig. 2. The bars indicate the standard deviation of 14- 3-3 γ levels in the three groups

gamma isoform is limited and not available from human brain. Watanabe and coworkers have shown that gamma is present in mammalian brain with a definite developmental pattern: they described type I neurons with high levels of expression throughout development, type II neurons showing high expression during early development with a subsequent decrease at maturing and adult stages and type III neurons showing consistently low levels of expression throughout development (Watanabe, 1993).

In DS increased gamma could be incriminated for impaired development and differentiation of the brain but we have been testing adult brain samples only, in which AD pathology has been developed (Seidl, 1997). The presence of increased gamma in both, DS and AD brains, makes the interpretation of data in context of common AD-neuropathology / neurodegeneration more likely. 14-3-3 proteins including gamma and epsilon are localized in synaptosomes and were isolated in purified synaptic membranes from the rat brain (Martin, 1994; Jones, 1995). Moreover, 14-3-3 proteins have been shown to be involved in synaptic functions as exocytosis (Chamberlain, 1995; Broadie, 1997; Roth, 1993) and 14-3-3 mutants were deficient in olfactory learning (Skoulakis, 1996). Increased gamma in brain of DS and AD patients cannot be simply explained as a consequence of synaptic pathology as synaptic loss is a hallmark of DS (Beyreuther, 1993; Becker, 1991; Petit, 1984) and we could show in identical brain regions of the identical cohort of DS and AD patients that the synaptosomal protein SNAP-25, a marker of synaptosomal density, was clearly decreased (manuscript in preparation).

An attractive concept is the possible involvement of epsilon (and possibly gamma) in the process of apoptosis, which unequivocally takes place in AD and DS with AD pathology (Dawbarn, 1995): Hsu and coworkers described the binding of 14.3.3 proteins to BAD, a Bcl-xL / Bcl-2-associated death promoter indicating that 14-3-3 proteins may be involved in modulating apoptosis (Hsu, 1997). Furthermore, Zha and coworkers found that the phosphorylated death agonist BAD bound to 14-3-3 protein (1996). Wang and coworkers provided evidence for a role of 14-3-3 proteins in apoptosis showing that the anti-apoptotic protein BAG-1, which in turn enhances the anti-apoptotic effect of bcl-2, binds to 14-3-3 proteins (1996). The increase of epsilon may be therefore representing the involvement and contribution of epsilon in the apoptotic process in brain of AD and DS with AD pathology by modulating the bcl-mediated pathomechanisms and 14-3-3 proteins have been localized to tangles and plaques in AD brain (Layfield, 1996).

Additional biological effects of increased epsilon brain levels in AD and DS with AD may include impairment of epsilon functions as regulation of Ras-mediated signaling (Chang, 1997), of cdc25 phosphatases (Conklin, 1995), modulation of glucose metabolism by insulin receptor substrate-1 and insulin like growth factor l-binding (Ogihara, 1997; Craparo, 1997) and centrosome functions (Pietromonaco, 1996). Although epsilon was described originally as kinase C binding protein-1 (KCIP-1), the function of epsilon as a PKC-inhibitor is not unequivocal and several 14-3-3 isoforms have been described to be both, stimulators and inhibitors of PKC (Toker, 1992, 1990; Chen, 1994; Robinson, 1994; Isobe, 1992). In a recent study we have shown unchanged PKC activity in brains of patients with AD and DS with AD, an observation which does not contradict our present findings but would not support a role for the regulation of PKC activity (Bernert, 1996). Deranged 14-3-3 gamma and epsilon proteins in AD and DS with AD neuropathological findings maybe involved in the complex pathomechanisms of neuro-degenerative disorders as e.g. apoptosis but this remains speculative.

Reports on increased 14-3-3 proteins in cerebrospinal fluid (CSF) of human spongiform encephalopathies (Weber, 1997; Rosenmann, 1997; Lee, 1997; Hsieh, 1996; Zerr, 1998) are challenging the investigation of 14-3-3 proteins in CSF of AD and DS patients; 14-3-3 epsilon and gamma may be candidates for a potentially promising tool for the CSF-diagnosis of both neurodegenerative diseases.

Acknowledgement

We are highly indebted to the Red Bull Company, Austria, for generous support of the study.

References

Aitken A, Howell S, Jones D, Madrazo J, Martin H, Patel Y, Robinson K (1995) Posttranslationally modified 14-3-3 isoforms and inhibition of protein kinase C. Mol Cell Biochem 149: 41–49

Becker L, Mito T, Takashima S, Onodera K (1991) Growth and development of the brain in Down Syndrome. Progr Clin Biol Res 373: 133–152

Bernert G, Nemethova M, Cairns N, Lubec G (1996) Decreased cyclin dependent kinase in brain of patients with Down Syndrome. Neurosci Lett 216: 68–70

Beyreuther K, Pollwein P, Multhaup G, Monning U, Konig G, Dyrks T, Schubert W, Masters DL (1993) Regulation and expression of the Alzheimer's beta/A4 amyloid protein precursor in health, disease, and Down's Syndrome. Ann NY Acad Sci 695: 91–102

Bradford M (1976) A rapid and sensitive method for the quantitation of microgram quantities of protein utilizing the principle of protein-dye binding. Anal Biochem 72: 248–254

Braselmann S, McCormick F (1995) Bcr and Raf form a complex in vivo via 14-3-3 proteins. EMBO J 14: 4839–4848

Broadie K, Rushton E, Skoulakis EM, Davis RL (1997) Leonardo, a Drosophila 14-3-3 protein involved in learning, regulates presynaptic function. Neuron 19: 391–402

Chamberlain LH, Roth D, Morgan A, Burgoyne RD (1995) Distinct effects of alpha-SNAP, 14-3-3 proteins, and calmodulin on priming and triggering of regulated exocytosis. J Cell Biol 130: 1063–1070

Chang HC, Rubin GM (1997) 14-3-3 epsilon positively regulates Ras — mediated signaling in Drosophila. Genes Dev 11: 1132–1139

Chen F, Wagner PD (1994) 14-3-3 proteins bind to histone and affect both histone phosphorylation and dephosphorylation. FEBS Lett 347: 128–132

Conklin DS, Galaktionov K, Beach D (1995) 14-3-3 proteins associate with cdc25 phosphatases. Proc Natl Acad Sci USA 92: 7892–7896

Craparo A, Freund R, Gustafson TA (1997) 14-3-3 (epsilon) interacts with the insulin-like growth factor receptor and insulin receptor substrate-1 in a phosphoserine — dependent manner. J Biol Chem 272: 11663–11669

Dawbarn D, Allen SJ (1995) Neurobiology of Alzheimer's disease. Bios Scientifique Publishers, Oxford

Dubois T, Howell S, Amess B, Kerai P, Learmonth M, Madrazo J, Chaudhri M, Rittinger K, Scarabel M, Soneji Y, Aitken A (1997) Structure and sites of phosphorylation of 14-3-3 protein: role in coordinating signal transductiion pathways. J Protein Chem 16: 513–522

Fanger GR, Widman C, Porter AC, Sather S, Johnson GL, Vaillancourt RR (1998) 14-3-3 proteins interact with specific MEK kinases. J Biol Chem 273: 3476–3483

Fantl WJ, Muslin AJ, Kikuchi A, Martin JA, MacNicol AM, Gross RW, Williams LT (1994) Activation of Raf-1 by 14-3-3 proteins. Nature 371: 612–614

Freed E, Symons M, Macdonald SG, McCormick F, Ruggieri R (1994) Binding of 14-3-3 proteins to the protein kinase Raf and effects on its activation. Science 265: 1713–1716

Fountoulakis M, Langen H (1997) Identification of proteins by matrix-assisted laser desorption ionization-mass spectroscopy following in-gel digestion in low-salt, non-volatile buffer and simplified peptide recovery. Anal Biochem 250: 153–156

Greber S, Lubec G, Cairns N, Fountoulakis M (1999) Decreased synaptosomal associated protein 25 levels in brain of patients with Down Syndrome and Alzheimer's disease. Electrophoresis (in press)

Hsieh G, Kenney K, Gibbs CJ, Lee KH, Harrington MG (1996) The 14-3-3 brain protein in cerebrospinal fluid as a marker for transmissible spongiform encephalopathies. N Engl J Med 335: 924–930

Hsu SY, Kaipia A, Zhu L, Hsueh AJ (1997) Interference of BAD (Bcl-xL/Bcl-2-associated death promoter)-induced apoptosis in mammalian cells by 14-3-3 isoforms and P11. Mol Encodrinol 11: 1858–1867

Ichimura T, Ito M, Itagaki C, Takahashi M, Horigome T, Omata S, Ohno S, Isobe T (1997) The 14-3-3 proteins bind its target protein with a common site located towards the C-terminus. FEBS Lett 413: 273–276

Isobe T, Hiyane Y, Ichimura T, Okuyama T, Takahashi N, Nakajo S, Nakaya K (1992) Activation of protein kinase C by the 14-3-3 proteins homologous with Exo1 protein that stimulates calcium dependent exocytosis. FEBS Lett 308: 121–124

Jones DH, Martin H, Madrazo J, Robinson KA, Nielsen P, Roseboom PH, Patel Y, Howell SA, Aitken A (1995) Expression and structural analysis of 14-3-3 proteins. J Mol Biol 245: 375–384

Kumagai A, Yakowec PS, Dunphy WG (1998) 14-3-3 proteins act as negative regulators of the mitotic inducer Cdc25 in Xenopus egg extracts. Mol Biol Cell 9: 345–354

Langen H, Roeder D, Juranville J-F, Fountoulakis M (1997) Effect of the protein application mode and the acrylamide concentration on the resolution of protein spots separated by two-dimensional gel electrophoresis. Electrophoresis 18: 2085–2090

Layfield R, Fergusson J, Aitken A, Lowe J, Landon M, Mayer RJ (1996) Neurofibrillary tangles of Alzheimer's disease brains contain 14-3-3 proteins. Neurosci Lett 209: 57–60

Lee KH, Harrington MG (1997) The assay development of a molecular marker for transmissible sponsiform encephalopathies. Electrophoresis 18: 502–506

Liao J, Omary MB (1996) 14-3-3 proteins associate with phosphorylated simple epithelial keratins during cell cycle progression and act as a solubility cofactor. J Cell Biol 133: 345–357

Martin H, Rostas J, Patel Y, Aitken A (1994) Subcellular localisation of 14-3-3 isoforms in rat brain using specific antibodies. J Neurochem 63: 2259–2265

Mirra SS, Heyman A, McKeel D, Sumi S, Crain BJ (1991) The consortium to establish a registry for Alzheimer disease (CERAD). II. Standardisation of the neuropathological assessment of Alzheimer's disease. Neurology 41: 479–486

Muslin AJ, Tanner JW, Allen PM, Shaw AS (1996) Interaction of 14-3-3 with signaling proteins is mediated by the recognition of phosphoserine. Cell 84: 889–897

Ogihara T, Isobe T, Ichimura T, Taoka M, Funaki M, Sakoda H, Onishi Y, Inukai K, Anai M, Fukushima Y, Kikuchi M, Yazaki Y, Oka Y, Asano T (1997) 14-3-3 protein binds to insulin receptor substrate-1, one of the binding sites of which is in the phosphotyrosine binding domain. J Biol Chem 272: 25267–25274

Peng CY, Graves RR, Thoma RS, Wu Z, Shaw AS, Piwnica-Worms H (1997) Mitotic and G2 checkpoint control: regulation of 14-3-3 protein binding by phosphorylation of Cdc25C on serine 216. Science 277: 1501–1505

Petit TL, LeBoutillier JC, Alfano DP, Becker LE (1984) Synaptic development in the human fetus: a morphometric analysis of normal and Down's syndrome neocortex. Exp Neurol 83: 13–23

Pietromonaco SF, Seluja GA, Aitken A, Elias L (1996) Association of 14-3-3 proteins with centrosomes. Blood Cells Mol Dis 22: 225–237

Robinson K, Jones D, Patel Y, Martin H, Madrazo J, Martin S, Howell S, Elmore M, Finnen MJ, Aitken A (1994) Mechanisms of inhibition of protein kinase C by 14-3-3 isoforms. 14-3-3 isoforms do not have phospholipase A2 activity. Biochem J 299: 853–861

Rommel C, Radziwill G, Lovric J, Noeldeke J, Heinicke T, Jones D, Aitken A, Moelling K (1996) Activated Ras displaces 14-3-3 protein from the amino terminus of c-Raf-1. Oncogene 12: 609–619

Rosenmann H, Meiner Z, Kahana E, Halimi M, Lenetsky E, Abramsky O, Gabizon R (1997) Detection of 14-3-3 protein in the CSF of genetic Creutzfeld-Jakob disease. Neurology 49: 593–595

Roth D, Morgan A, Burgoyne RD (1993) Identification of a key domain in annexin and 14-3-3 proteins that stimulate calcium dependent exocytosis in permeabilized adrenal chromaffin cells. FEBS Lett 320: 207–210

Seidl R, Greber S, Schuller E, Bernert G, Cairns N, Lubec G (1997) Evidence against increased oxidative DNA-damage in Down Syndrome. Neurosci Lett 235: 137–140

Skoulakis EM, Davis RL (1996) Olfactory learning deficits in mutants for Leonardo, a drosophila gene encoding a 14-3-3 protein. Neuron 17: 931–944

Tanji M, Horwitz R, Rosenfeld G, Waymire JC (1994) Activation of protein kinase C by purified bovine brain 14-3-3: comparison with tyrosine hydroxylase activation. J Neurochem 63: 1908–1916

Tierney MC, Fisher RH, Lewis AJ, Torzitto ML, Snow WG, Reid DW, Nieuwstraaten P, Van Rooijen LAA, Derks HJGM, Van Wijk R, Bischop A (1988) The NINCDA-ADRDA work group criteria for the clinical diagnosis of probable Alzheimer's disease. Neurology 38: 359–364

Toker A, Ellis CA, Sellers LA, Aitken A (1990) Protein kinase C inhibitor proteins. Purification from sheep brain and sequence similarity to lipocortins and 14-3-3 proteins. Eur J Biochem 191: 421–429

Toker A, Sellers LA, Amess B, Patel Y, Harris A, Aitken A (1992) Multiple isoforms of a protein kinase C inhibitor (KCIP-1/14-3-3) from sheep brain. Eur J Biochem 206: 453–461

Vincenz C, Dixit VM (1996) 14-3-3 proteins associate with A20 in an isoform-specific manner and function both as chaperone and adapter molecule. J Biol Chem 271: 20029–20034

Wang HG, Takayama S, Rapp UR, Reed JC (1996) Bcl-2 interacting protein, BAG-1, binds to and activates the kinase Raf-1. Proc Natl Acad Sci 93: 7063–7068

Watanabe M, Isobe T, Ichimura T, Kuwano R, Takahashi Y, Kondo H (1993) Molecular cloning of rat cDNAs for beta and gamma subtypes of 14-3-3 proteins and developmental changes in expression of their mRNAs in the nervous system. Brain Res Mol Brain Res 17: 135–146

Weber T, Otto M, Bodemer M, Zerr I (1997) Diagnosis of Creutzfeld — Jakob disease and related human spongiform encephalopathies. Biomed Pharmacother 51: 381–387

Xing H, Kornfeld K, Muslin AJ (1997) The protein kinase KSR interacts with 14-3-3 protein and Raf. Curr Biol 7: 294–300

Yaffe MB, Rittinger K, Volinia S, Caron PR, Aitken A, Leffers H, Gamblin SJ, Smerdon SJ, Cantley SC (1997) The structural basis for 14-3-3: phosphopeptide binding specificity. Cell 91: 961–971

Zerr I, Bodemer M, Gefeller O, Otto M, Poser S, Wiltfang J, Windl O, Kretschmar HA, Weber T (1998) Detection of 14-3-3 protein in the cerebrospinal fluid supports the diagnosis of Creutzfeld-Jakob disease. Ann Neurol 43: 32–40

Zha J, Harada H, Yang E, Jockel J, Korsmeyer SJ (1996) Serine phosphorylation of death agonist BAD in response to survival factor results in binding to 14-3-3 not BCL-X(L). Cell 87: 619–628

Authors' address: Prof. Dr. G. Lubec, Department of Pediatrics, University of Vienna, Währinger Gürtel 18, A-1090 Wien, Austria, e-mail: gert.lubec@akh-wien.ac.at

J Neural Transm (1999) [Suppl] 57: 337–352

Application of *Alu*-splice PCR on chromosome 21: *DSCR1* and *Intersectin*

J. J. Fuentes[1]**, M. Dierssen**[1]**, C. Pucharcós**[1]**, C. Fillat**[1]**, C. Casas**[1]**, X. Estivill**[1]**, and M. Pritchard**[2]

[1] Medical and Molecular Genetics Center-IRO, L'Hospitalet de Llobregat,
Barcelona, Spain
[2] IRD, Monash University, Clayton, Victoria, Australia

Summary. Down syndrome (DS) is a major cause of mental retardation and congenital heart defects, with an overall incidence of one in 700 live births. DS is caused by increases in the amounts of a number of normal gene products, the exact number and identity of which are presently unknown. Elucidating the molecular basis of DS relies on the identification of the gene products whose augmentation by 50% or more causes symptoms of the disease.

With the aim of contributing to the transcriptional map of human chromosome 21 and to identify new genes with potential involvement in DS, we developed a technique to isolate expressed sequences called *Alu*-splice PCR, which is very simple to perform and is independent of gene expression patterns. Putative exons are PCR amplified in genomic DNA by virtue of their proximity to *Alu* repeats using primers designed from splice-site consensus sequences in combination with specific *Alu* repeat primers. The *Alu* repeats, which are repetitive DNA elements found exclusively and at high frequency in the genomes of primates, impart the human specificity to the method. The splice-site consensus sequences were used to direct primers to exon boundaries.

Using the *Alu*-splice technique, we have identified at least three new genes. We trapped an exon of *DSCR1* (Down Syndrome Candidate Region 1) and two different exons of a gene called human *Intersectin* (*ITSN*). Presently, we are working with another novel trapped exon to identify the corresponding gene. The major advantage of *Alu*-splice PCR is that the technique can be readily established in any laboratory which has the basic facilities for molecular biology because no specialised materials or expertise is required.

Down syndrome (DS) is a major cause of mental retardation and congenital heart defects, with an overall incidence of one in 700 live births (Hassold and Jacobs, 1984). There is a broad spectrum of physical abnormalities associated with DS, including anomalies of the gastrointestinal tract, increased risk of leukaemia, defects of the immune and endocrine systems, early onset of Alzheimer's dementia and distinct facial and physical features (Epstein,

1986). In most cases DS is due to three full copies of human chromosome 21 (HSA21) (Jacobs et al., 1959; Lejeune et al., 1959) which arise primarily by non-disjunction at maternal meiosis I (Warren et al., 1987; Antonarakis et al., 1993), but occasionally DS is due to unbalanced chromosomal translocations, which result in the triplication of only a part of chromosome 21.

DS is distinguishable from other genetic disorders. Diseases caused by a single gene defect depend upon alteration or reduction of a gene product. DS is caused by increases in the amounts of a number of normal gene products, the exact number and identity of which are presently unknown (Epstein, 1995). Elucidating the molecular basis of DS relies on the identification of the gene products whose augmentation by 50% or more causes symptoms of the disease, deciphering the mechanism by which the candidate gene produces the symptoms and explaining the possible interrelationships between genes.

HSA21 is estimated to contain between 600–1,000 genes. International gene cloning efforts have led to the identification of numerous genes and transcriptional units and more recently, large scale sequencing (http://www-eri.uchsc.edu/chr21/eridna.html) has enabled the prediction of many more. However, even though 95 genes (GDB, Feb. 21, 1999) have been assigned to chromosome 21, the functions of most of these genes remain largely unknown, as does their contribution, if any, to the DS phenotype.

The most usual techniques employed for isolating genes are cDNA selection (Lovett et al., 1991; Korn et al., 1992), exon trapping (Duyk et al., 1990; Buckler et al., 1991; Church et al., 1994), hybridisation using genomic fragments as probes against cDNA libraries (Rommens et al., 1989; Sargent et al., 1989), CpG island identification (Estivill and Williamson, 1987; Melmer et al., 1990) and the cloning and sequencing of genomic DNA followed by computer analysis of the possible coding regions (Wilson et al., 1994). All of these techniques have limitations, among them, the large numbers of subclones requiring analysis to yield a few true positives and for cDNA selection and hybridisation using genomic fragments, their dependence on gene expression patterns due to the use of mRNA or cDNA libraries. Since none of these methods permits the isolation of all the genes in a given region, to achieve complete coverage, several methods should be used to complement one another.

With the aim of contributing to the transcriptional map of human chromosome 21 and to identify new genes with potential involvement in DS, we developed a technique to isolate expressed sequences called Alu-splice PCR, which is very simple to perform and is independent of gene expression patterns. The method is based on the same philosophy as the technique developed by Lui and coworkers (1987). Putative exons are PCR amplified in genomic DNA by virtue of their proximity to Alu repeats using primers designed from splice-site consensus sequences in combination with specific Alu repeat primers (Fuentes et al., 1997a). The Alu repeats, which are repetitive DNA elements found exclusively and at high frequency in the genomes of primates, having arisen from the retrotransposition of copies of a 7SL RNA gene derived progenitor sequence (Ullu et al., 1982), impart the human specificity to the method. The splice-site consensus sequences (Krawczak et al.,

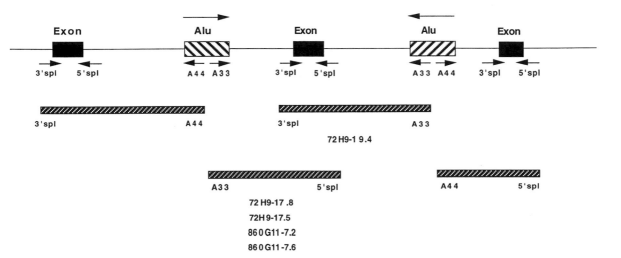

Fig. 1. Schematic representation of the *Alu*-splice PCR cloning strategy for the isolation of putative exons from a given cloned DNA region. A33 and A44 are primers derived from the consensus sequence of *Alu*-repeats, located at the 3' and 5' end, respectively. 3' spl and 5' spl are primers for consensus sequences involved in intron splicing. Combinations of splice primers (3' spl or 5' spl) with an *Alu* primer (A33 or A44) were used

1992) were used to direct primers to exon boundaries. The splice-site sequences flanking exons are recognised in pre-mRNA by the cellular splicing machinery and permit the removal of introns by a very precise cleavage and rejoining process known as splicing (Green, 1986; Padgett et al., 1986). A schematic representation of the method is presented in Fig. 1.

The sequences of the *Alu* primers were taken from (Tagle and Collins, 1992; Chumakov et al., 1992a). We added a *Sal*I restriction enzyme recognition site to facilitate subcloning plus 3 additional nucleotides to aid digestion at the 5' ends. The *Alu* primers are: *Sal*I-A33: 5'-CGCGTCGACCACTGC-ACTCCAGCCTGG GCG-3' and *Sal*I-A44: 5'-CGCGTCGACGGGGATT-ACAGGCGTGAGCCAC-3'. The splice primers were synthesised using the 5' (donor) and the 3' (acceptor) splice-site consensus sequences (Krawczak et al., 1992). The splice primers are: *Not*I-5' spl: 5'-CGCGCGGCCGCAC-WYACCW-3' and *Sac*II-3'spl: 5'-CGCCCGCGGTCNCAGG T-3', where W represents A or T, Y represents T or C and N any nucleotide. A *Not*I restriction enzyme site was added to the 5' splice primer and a *Sac*II site to the 3' splice primer and both had 3 additional nucleotides. The primers tailed with restriction enzyme sites allowed us to subclone the DNA fragments directionally. The efficient double digestion of the vector with *Sal*I / *Not*I or *Sal*I / *Sac*II should prevent the subcloning of *Alu*-*Alu* human sequences as well as splice-splice products which may arise from non-specific priming events on either the cloned human DNA or the host yeast DNA.

We have applied the *Alu*-splice method to isolate new coding sequences from human chromosome 21 using two overlapping YACs, 72H9 and 860G11 (Chumakov et al., 1992b; Fuentes et al., 1997a). To account for all the possible primer combinations we performed four PCRs for each template: A33 vs. 5′ spl, A33 vs. 3′ spl, A44 vs. 5′ spl and A44 vs. 3′ spl. The template used was total yeast DNA containing the YAC. The two YACs (72H9 and 860G11) were amplified independently. The final volume of the *Alu*-splice PCR reactions was 25 µl containing the following components: 40 mM KCl, 8 mM Tris-HCl (pH 8.3), 3 mM MgCl₂, 0.01% gelatin, 0.15 mM of each deoxynucleotide, 20 pmol of splice primer and 10 pmol of Alu primer, 300 ng of template DNA and 1U Taq polymerase (Perkin Elmer). The YAC DNA preparation was as described by (Overhauser and Radic, 1987). The PCR conditions were 94°C for 20 s, 55°C for 40 s and 74°C for 2 min, for 30 cycles. After amplification the products were digested with the appropriate enzymes and subcloned in the vector pBluescript SK⁺ (Stratagene). We picked 60 colonies from each YAC for further analysis.

Using the *Alu*-splice technique, we have identified at least three new genes. We trapped an exon of *DSCR1* (Down Syndrome Candidate Region 1) (Fig. 2), (Fuentes et al., 1995) and two different exons of a gene called human *Intersectin* (*ITSN*) (Guipponi et al., 1998; Pucharcòs et al., in press). Presently, we are working with another novel trapped exon to identify the corresponding gene. From a relatively low number of starting colonies, we have identified at least three novel chromosome 21 genes, demonstrating the efficiency of the *Alu*-splice procedure. The major advantage of *Alu*-splice PCR is that the technique can be readily established in any laboratory which has the basic

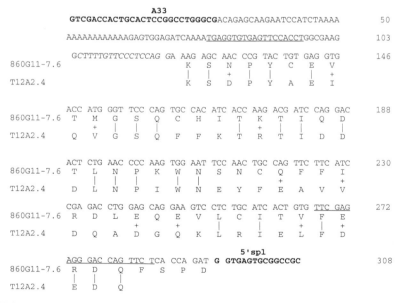

Fig. 2. DNA sequence of the trapped *DSCR1* exon. The sequences corresponding to the primers used to generate this DNA fragment are shown in bold; a putative 3′ splice sequence is italicised; the predicted amino acid sequence is shown below each codon. The search using the BLASTX program revealed 42% identity with the *C. elegans* gene product T12A2.4 (gb: R57124). Residues that are identical are indicated by a vertical line. The conservative changes are indicated by + sign. Numbers to the right of the figure indicate the nucleotide positions

facilities for molecular biology because no specialised materials or expertise is required.

From gene to function: *DSCR1* and *Intersectin* (*ITSN*)

The identification of HSA21 genes is only the first step. The next step is to determine the functions of the genes and assess their potential involvement in DS by correlating their overexpression, singly or in combination, with the presence of the DS phenotype. Various approaches have been used and to investigate the potential roles of individual genes in the phenotypic expression of DS, including the analysis of protein function, the study of the expression patterns and the generation of animal models. Mice provide a reproducible experimental system for modelling human disorders and for testing therapies. Moreover, orthologous genes are frequently linked in conserved chromosomal segments in the mouse and human genomes, and therefore, aneuploidy for regions of the mouse genome can serve to study the concerted effect of overexpression of several genes. A mouse model for DS exists, Ts65Dn (Davisson et al., 1993). These mice are trisomic for the distal segment of mouse chromosome 16 which has conserved synteny with HSA21. However, the triplicated region encompasses a large region of the chromosome and does not allow the analysis of the contribution of individual genes to the DS phenotype. The appropriate model for addressing the contribution of individual genes to complex physiological processes such as development or morphogenesis, is the generation of transgenic mice harbouring candidate genes. The overexpression of a single gene would not be expected to fully model DS, but should permit the study of its contribution to the phenotype.

Based on their expression patterns during embryogenesis, their overexpression in DS foetal brains, clues about function revealed by the presence of certain motifs in the protein sequences, and on studies of the functional roles of the orthologous genes in other species, we have postulated that overexpression of *DSCR1* and *Intersectin* may contribute in a dosage-dependent manner to certain aspects of the DS phenotype.

Human *Intersectin*

Human *Intersectin* is a HSA21 gene overexpressed in DS and exists with a high degree of similarity in flies, frogs and mammals, suggesting a conserved role in higher eukaryotes. Human *Intersectin* was isolated by two laboratories by using different approches: exon trapping (Guipponi et al., 1998) and *Alu*-splice PCR (Pucharcós et al., 1999).

The gene has the potential to code for at least two different isoforms by alternative splicing: a long transcript (*ITSN-L*) with an open reading frame of 1721 amino acids, and a short one (*ITSN-S*) that it would encode the first 1,220 amino acids of Intersectin. *ITSN* consists of at least 41 exons which are shown in Fig. 3, spanning more than 233 kb of genomic DNA. The genomic structure

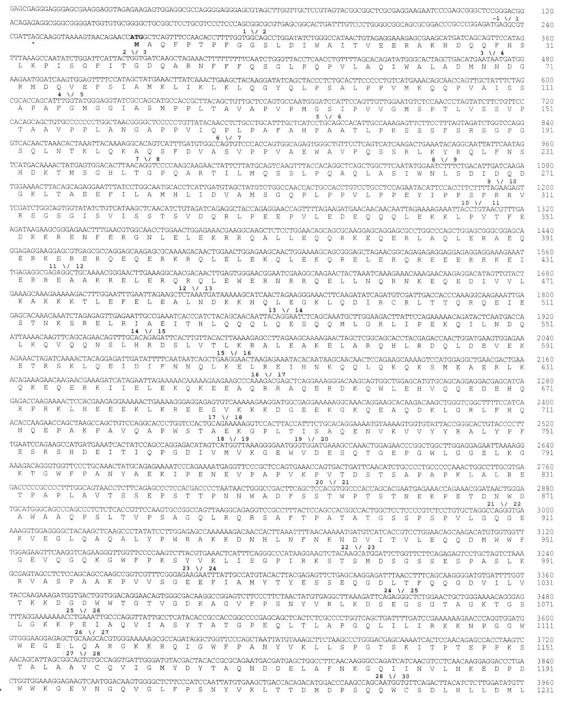

Fig. 3. Nucleotide and amino acid sequence of the human *Intersectin* isoforms. **A** Nucleotide and deduced amino acid sequence of *ITSN-L*. Nucleotides and amino acid residues are numbered in the right margin. Potential initiation and stop codon are in bold. The stop codon upstream of the initiating ATG is indicated with a star. The original *Alu*-splice exon is underlined. Exon junctions are indicated by (\/) with the number of the corresponding exon above the nucleotide sequence. Note that exon 29 is skipped. **B** Nucleotide sequence of the 3′-end of *ITSN-S*. Joining exon 28 to exon 29 provides a stop codon at position 3,929 generating a shorter transcript encoding a putative protein of 1,220 amino acids. Putative polyadenylation signals are underlined

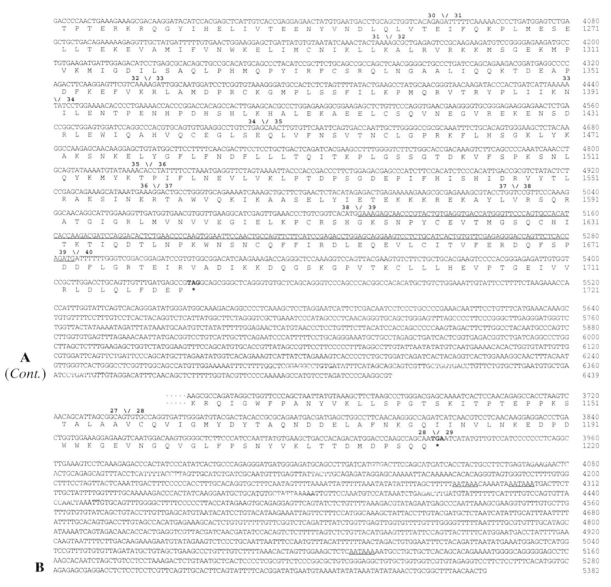

Fig. 3. *Continued*

enabled us to confirm that the generation of the long isoform is the result of an alternative splicing event in which exon 29, which provides the stop codon for the short isoform when joined to exon 28, is skipped.

Analysis of the expression pattern of human *Intersectin* detected mRNAs in all adult and foetal tissues tested, with the longer isoform present only in brain (foetal and adult). Foetal mouse Northerns indicated that this gene is expressed early in development. In situ hybridisation studies in the developing mouse brain show *ITSN* expression in both proliferating and differentiating neurones. High levels of expression were observed in many adult brain regions, including the cerebral cortex. Moreover, analysis of *ITSN* expression by semiquantitative RT-PCR indicated an approximate 2–3 fold overexpression of *ITSN-L* in the brains of DS foetuses compared to controls.

Human *Intersectin* is highly similar to *Xenopus intersectin* and encodes a putative multi-modular protein with several conserved domains, including EH, SH3, DH, PH and C2 domains (Fig. 4A). The two protein isoforms differ in the lengths of their carboxy termini and thus, the larger isoform contains extra functional domains. Both isoforms have two EH domains, followed by five SH3 domains (Fig. 4A). EH-domains (Fig. 4B) are involved in protein-protein interactions recognising Asn-Pro-Phe containing peptides (Salcini et al., 1997) and most of the EH-containing proteins have been linked to endocytic pathways. SH3 domains (Fig. 4B) are commonly found in signal transduction and cytoskeletal proteins. They mediate protein-protein interactions by binding to proline rich peptides that adopt an extended conformation (PPII helix) (Musacchio et al., 1994). *Xenopus* intersectin protein, which is homologous to human Intersectin, brings together several proteins into a macromolecular complex through its two EH and five SH3 domains (Yamabhai et al., 1998). Two of the proteins which interact with the *Xenopus* intersectin via its SH3 modules are dynamin (a GTPase) and synaptojanin 1 which were identified as components of the recycling machinery of synaptic vesicles in nerve terminals (Cremona et al., 1997). *Xenopus* intersectin also binds two mouse proteins named Ibp1 and Ibp2 through its EH domain (Yamabhai et al., 1998). Ibp1 is the murine homologue of human epsin (Chen et al., 1998), a protein concentrated in presynaptic nerve terminals which

---

Fig. 4. Analysis of human *Intersectin* amino acid sequence. **A** Schematic representation of ITSN-L showing its multi-modular structure and the relationship between *Xenopus* intersectin (GenBank accession no. AF032118, XInt) and *Drosophila* Dap160–1 (GenBank accession no. AF053957) and Dap160–2 (GenBank accession no. AF054612) proteins. The SH3P17 protein (Swiss-Prot accession no. Q15811) is included because it is likely that it represents a partial sequence of ITSN-S. Domains are represented as boxes and numbers below them show the domain position in the amino acid sequence of ITSN-L. **B** Amino acid multi-alignment of the ITSN-L domains. Domains were initially aligned using the CLUSTALW program and optimised by visual inspection. The numbers before and after the sequence indicate the positions of the regions within their respective sequences. Conserved residues are shaded in black. Residues where the chemical character is conserved are shaded in grey. (i) Multi-alignment of the two EH domains of ITSN-L (ITSN-EH1 and ITSN-EH2), *Xenopus* XInt (XInt-EH1 and XInt-EH2) and *Drosophila* Dap160–1 (Dap160.1-EH1 and Dap160.1-EH2) and two classical EH domains from human Eps15 (Swiss-Prot accession no. P42566, hEps15-EH2) and mouse eps15-EH2 (Swiss-Prot accession no. P42567, mEps15-EH2) as models. The conserved EF-hand domain present in all of them is boxed. (ii) Amino acid sequence multi-alignment of the five SH3 domains of ITSN (named A to E). The SH3 domain of human Src (Swiss-Prot accession no. P12931, hSrc-SH3) is included in the comparison. The underlined sequence corresponds to exon 19. (iii) Multi-alignment of the DH domain of ITSN-L with several human proteins containing DH domains: p115-RhoGEF (GenBank accession no. U64105), FGD1 (Swiss-Prot accession no. P98174), NET1 (GenBank accession no. U02081) and Bcr (Swiss-Prot accession no. P11274). Conserved amino acids according to Whitehead et al. (1997) are marked with a star

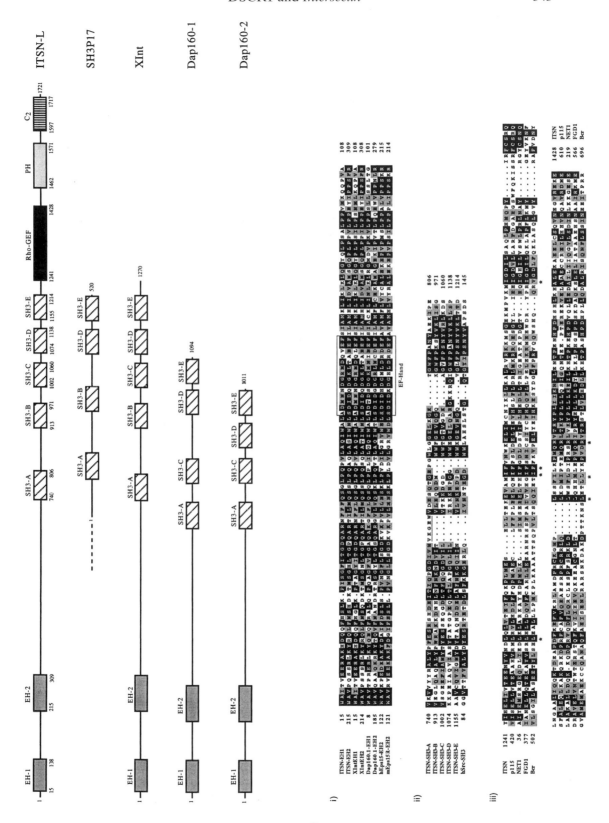

binds the α-adaptin subunit of AP-2 and the clathrin-coat-associated protein Eps15. Moreover, human Intersectin showed a great similarity to two *Drosophila* proteins, Dap 160-1 and Dap 160-2 (Roos et al., 1998). The sequence similarity between the *Drosophila* Daps and the vertebrate homologs is more pronounced in the EH and SH3 domains, with a striking conservation in the distances between the SH3 domains. Dap 160 was isolated as a dynamin associated protein and identified as a membrane protein present in resting endocytic hot spots in *Drosophila* nerve terminals (Roos et al., 1998). Taken together, it is likely that Dap 160, *Xenopus* intersectin and human Intersectin are members of a family of proteins which may serve as scaffolding proteins required for clathrin-mediated endocytosis.

In addition to the EH and SH3 domains, the long isoform of human intersectin has the capacity to encode DH, PH and C2 motifs. The DH (or Dbl) domain (Fig. 4B) is the region in guanine-nucleotide exchange factors (GEFs) that catalyses an exchange of GTP for GDP on Rho-like G proteins (Whitehead et al., 1997). Interestingly, Rho and Rac have been shown to function in regulating endocytosis (Lamaze et al., 1996). As in all the other DH domain-containing proteins, the ITSN DH domain is followed by a pleckstrin homology (PH) domain. The intersectin PH domain belongs to the low affinity binding class (Isakoff et al., 1998) and thus, its role might be to modulate the guanine nucleotide exchange activity of the associated DH domain, as has been described for the GEF Sos (Nimnual et al., 1998). The most carboxy region of the long isoform has sequence homology with C2 domains which are regulatory sequence motifs that occur widely in nature and although they were initially described as calcium-binding motifs, they have also been shown to bind phosphoinositides (Nalefski et al., 1996). Synaptotagmin I, a synaptic vesicle protein involved in calcium regulation of endocytosis, contains two C2 domains, one of which acts as a calcium sensor (Sutton et al., 1995). We predict that the binding of Intersectin to phospoinositides might contribute to its function in membrane anchoring. Although it is difficult to speculate about how these domains may come together in a functional Intersectin protein, their unifying feature is that they are all found in proteins involved in membrane trafficking.

The primary function of the synaptic vesicle is neurotransmitter release, or exocytosis. After exocytosis, the components of the synaptic vesicle membrane are selectively recovered by endocytosis and recycled within the nerve terminal to generate new synaptic vesicles. This local recycling pathway allows neurones to maintain a constant supply of synaptic vesicles and this recycling capacity is essential for the consistent release of neurotransmitter. The processes of docking, fusion and endocytosis are not restricted to the nerve terminal, they are elements of membrane-trafficking pathways in all eukaryotic cells. However, in non-neural cells, most of these events are constitutive (Zigmond et al., 1999), therefore, additional mechanisms must account for the specific temporal, spatial and calcium-sensitive properties of transmitter release and recycling, and the speed at which they occur. Consistent with the ubiquitous expression of the short *Intersectin* isoform, we propose that *ITSN-S* is involved in the generalised endocytic pathways of all cells and that the long isoform, *ITSN-L*, has been endowed with the more specialised properties required for brain-specific synaptic vesicle recycling.

Genes which show expression during development or more particularly, in the developing brain, may be of special importance in DS. Moreover, processes involving proteins with the capacity to interact with a myriad of other gene products, may be more susceptible to the effects of dosage imbalance. The long isoform was overexpressed in the developing brains of foetuses with DS. At present, we can only speculate on the significance of this. Abnormal synaptic parameters have been reported in DS brains, including changes that could lead to a reduced efficiency of synaptic transmission (Becker et al., 1993; Wisniewski, 1990). In the brain of the Ts65Dn mouse (a mouse model for DS) impaired cAMP production, alterations in central noradrenergic transmission and reduced LTP (long term potentiation) have been found (Dierssen et al., 1996, 1997; Siare et al., 1997). Both LTP and cAMP responses have been implicated in synaptic plasticity. LTP is associated with synaptic enhancement, and enhanced synaptic transmission involves either increased transmitter release or increased receptivity to released transmitter (Zigmond et al., 1999). If *Intersectin* is involved in synaptic transmission, then it is feasible that the abnormal stoichiometry of such a multi-modular protein may have detrimental consequences in DS. Intriguingly, the isolation of the mouse *Intersectin* gene (*Ese1*) has recently been reported and functional studies have demonstrated that overexpression of the Ese1 short isoform in vitro is sufficient to block clathrin-mediated endocytosis (Senger et al., 1999).

DSCR1

DSCR1 was isolated by *Alu*-splice PCR (Fuentes et al., 1995). The function of *DSCR1* is unknown but it encodes a novel protein which has an acidic domain, a serine-proline motif, a putative DNA binding domain and a proline-rich region with the characteristics of a SH3 domain ligand. These features suggest that DSCR1 could be involved in transcriptional regulation and/or signal transduction. DSCR1 shares sequence similarity (62%) with *ZAKI-4*, a thyroid hormone responsive gene (Miyazaki et al., 1996) and more recently, the hamster homologue of *DSCR1* (*Adapt78*) was isolated by differential display after the induction of its expression by oxidative stress in a hamster cell line (Crawford et al., 1997).

On Northern blots *DSCR1* was highly expressed in human brain and heart and expression was detected in situ (using antisense oligonucleotides) in several regions of the rat brain, including the olfactory bulb, the cerebellum and the hippocampus. For a more detailed anatomical study of expression, we used digoxigenin labelled riboprobes to detect *Dscr1*. Early expression is detected in the mouse embryonic heart at day 9 (E9), and thereafter in the central nervous system at E10 where expression persists through adult life. In the facial motor nucleus at E11 and E12 there is very intense expression. In general, expression is high during foetal development, with the *Dscr1* mRNA widely distributed in the prenatal CNS. In postnatal sections the expression becomes progressively more condensed, and is restricted to specific cells in adult brain sections: neocortex cell layers II, III and V, pyriform cortex, dentate granule cell layer, pyramidal cell layer of the hippocampus, striatum and Purkinje cell layer of the cerebellar cortex. The pattern is predominantly neuronal, although expression in some types of glia cannot be ruled out. Its structural characteristics, together with its particular expression in brain and heart (Casas, unpublished results), encourage us to suggest that the overexpression of *DSCR1* may be involved in the pathogenesis of DS, in particular in the mental retardation and/or cardiac defects (Fuentes et al., 1995).

DSCR1 is organised in 7 exons of which 4 are alternative first exons with characteristic expression patterns (Fuentes et al., 1997b). Exons 1 and 4 are most abundantly expressed and were detected in foetal tissues but the levels of expression are different across tissues. Exon 4 is expressed at the lowest level in brain while exon 1 is highly expressed in brain as compared to the other tissues. In adults, both the pattern and level of expression are different for exons 1 and 4. Notably, by Northern, in adult kidney transcripts containing exon 1 are undetectable and in adult brain transcripts containing exon 4 are not detected. (Fuentes et al., 1997b). We concluded that the expression of exons 1 and 4 is temporally regulated, with both alternative exons expressed in foetus and in most adult tissues, albeit differentially.

As yet, we do not understand the functional significance of the putative alternative proteins. The proteins encoded by exons 1 and 4 differ only in their first 28 amino acids and both forms are presumably present in most tissues. Since both isoforms have been conserved throughout evolution (both are present in mouse), presumably both products are essential.

The murine *Dscr1* gene encodes a protein of 198 amino acids which is highly similar to its human counterpart (97% amino acid identity). Similar to the human gene, the mouse gene has alternative first exons which are differentially expressed. As occurs in humans, exon 1 is not expressed in adult kidney and exon 3 is not expressed in adult brain (Fuentes, unpublished results).

To evaluate the functional consequences of in vivo overexpression of *DSCR1*, transgenic mice were generated. Chimaeric genes containing the *DSCR1* cDNA with the alternative exons 1 or 3 under the control of the brain-specific PDGF-β promoter were constructed and injected into embryos at the one cell-stage collected from the oviducts of B6SJL F1 females. Four independent transgenic lines were obtained containing 2, 3, 19 and 22 copies of the integrated *DSCR1* transgene containing exon 3. Analysis of mice containing *DSCR1* exon 1 is underway. By Northern blot, it was determined that only the transgenic line containing 19 copies expressed *DSCR1*. The behavioural characterisation of the *DSCR1* transgenic animals has been initiated and some preliminary experiments have been performed on the transgenic line with 19 copies. Preliminary behavioral characterisation included the measurement of sensorimotor reflexes, locomotor and exploratory activity, learning and memory. No differences were found in the sensorimotor reflexes measured (visual placing and vibrissa placing reflexes, equilibrium, prehensile reflex, traction capacity, motor coordination or negative geotaxis) between *TgDSCR1* and control mice of either sex. However, overexpression of *DSCR1* leads to a lower spontaneous locomotor activity as measured by actimetry, during the first hours of the dark period of the light/dark cycle. Also, male Tg*DSCR1* showed worse motor coordination in the acceleration cycle of the rotarod test, with respect to non-transgenic controls (Martínez-Cué et al., 1997), but presented no differences in activity and exploration as measured in the open field and hole board tests.

One of the most important concerns in DS is the alteration in learning and memory that gives rise to mental retardation. No single neuropathological process can be related to mental retardation, and therefore, it is probably the synergistic effects of various genes that induces the phenotype. Therefore, it is important to compare the results obtained in the learning and memory trials with those observed in trisomic murine models, in which *Dscr1* is triplicated in a trisomic environment, thus providing two different models of overexpression. The comparison with Ts65Dn indicates a quite distinct behavioral profile in Tg*DSCR1*, but some aspects regarding learning and memory may overlap in both models (unpublished data). It is however, important to stress that these results are preliminary. Until we have analysed at least another independent line, we are unable to discount the possibility that the effect we have observed is due to the position of integration of the transgene.

The studies described on *DSCR1* and *Intersectin* are unable to directly address the contribution of either of these genes to the DS phenotype, but these types of analyses i.e. the gradual accumulation of information about the normal biological functions of these genes, will allow an assessment of their possible involvement in DS and reveal avenues for investigation to more directly determine their roles in pathology.

References

Antonarakis SE, Avramopoulos D, Blouin J-L, Talbot CC, Schinzel AA (1993) Mitotic errors in somatic cells cause trisomy 21 in about 4.5% of cases and are not associated with advanced maternal age. Nature Genet 3: 146–150

Becker LE, Mito T, Takashima S, Onodera K, Friend WC (1993) Association of phenotypic abnormalities of Down syndrome with an imbalance of genes on chromosome 21. APMIS [Suppl] 40: 57–70

Buckler AJ, Chang DD, Graw SL, Brook JD, Haber DA, Sharp PA, Housman DE (1991) Exon amplification: a strategy to isolate mammalian genes based on RNA splicing. Proc Natl Acad Sci USA 88: 4005–4009

Chen H, Fre S, Slepnev VI, Capua MR, Takei K, Butler MH, Di Fiore PP, De Camilli P (1998) Epsin is an EH-domain-binding protein implicated in clathrin-mediated endocytosis. Nature 394: 793–797

Church DM, Stotler CJ, Rutter JL, Murrell JA, Trofatter JA, Buckler AJ (1994) Isolation of genes from complex sources of mammalian genomic DNA using exon amplification. Nature Genet 6: 98–105

Chumakov IM, Le Gall I, Billault A, Ougen P, Soularue P, Giullou S, Rigault P, Bui H, De Tand M-F, Barillot E, Abderrahim H, Cherif D, Berger R, Le Paslier D,

Choen D (1992a) Isolation of chromosome 21-specific yeast artificial chromosomes from a total human genome library. Nature Genet 1: 222–225

Chumakov I, Rigault P, Guillou S, Ougen P, Billaut A, Guasconi G, Gervy P, LeGall I, Soularue P, Grinas L, Bougueleret L, Bellanné-Chantelot C, Lacroix B, Barillot E, Gesnouin P, Pook S, Vaysseix G, Frelat G, Schmitz A, Sambucy JL, Bosch A, Estivill X, Weissenbach J, Vignal A, Riethman H, Cox D, Patterson D, Gardiner K, Hattori M, Sakaki Y, Ichikawa H, Ohki M, Le Paslier D, Heilig R, Antonarakis S, Cohen D (1992b) Continuum of overlapping clones spanning the entire human chromosome 21q. Nature 359: 380–387

Crawford DR, Leahy KL, Abramova N, Lan L, Wang Y, Davies KJA (1997) Hamster adapt78 mRNA is a Down syndrome critical region homologue that is inducible by oxidative stress. Arch Biochem Biophys 342: 6–12

Cremona O, De Camilli P (1997) Synaptic vesicle endocytosis. Curr Opin Neurobiol 7: 323–330

Davisson MT, Schmidt C, Reeves RH, Irving NG, Akeson EC, Harris BS, Bronson RT (1993) Segmental trisomy as a mouse model for Down syndrome. Prog Clin Biol Res 384: 117–133

Dierssen M, Vallina IF, Baamonde C, Lumbreras MA, Martinez-Cue C, Calatayud SG, Florez J (1996) Impaired cyclic AMP production in the hippocampus of a Down syndrome murine model. Brain Res Dev Brain Res 95: 122–124

Dierssen M, Vallina IF, Baamonde C, Garcia-Calatayud S, Lumbreras MA, Florez J (1997) Alterations of central noradrenergic transmission in Ts65Dn mouse, a model for Down syndrome. Brain Res 749: 238–244

Duyk GM, Kim SW, Myers RM, Cox DR (1990) Exon trapping: a genetic screen to identify candidate transcribed sequences in cloned mammalian genomic DNA. Proc Natl Acad Sci USA 87: 8995–8999

Epstein CJ (1986) The consequences of chromosome imbalance: principles, mechanisms, and models. Cambridge University Press, New York

Epstein CJ (1995) The metabolic and molecular basis of inherited disease. In: Scriver CR, Beaudet AL, Sly WS, Valle D (eds) Down syndrome, 7th edn. Mac Graw-Hill, New York

Estivill X, Williamson R (1987) A rapid method to identify cosmids containing rare restriction sites. Nucl Acids Res 15: 1415–1425

Fuentes JJ, Pritchard MA, Planas AM, Bosch A, Ferrer I, Estivill X (1995) A new human gene from the Down syndrome critical region encodes a proline-rich protein highly expressed in fetal brain and heart. Hum Mol Genet 10: 1935–1944

Fuentes JJ, Pucharcos C, Pritchard M, Estivill X (1997a) Alu-splice PCR: a simple method to isolate exon-containing fragments from cloned human genomic DNA. Hum Genet 101: 346–350

Fuentes JJ, Pritchard M, Estivill X (1997b) Genomic organisation, alternative splicing and expression patterns of the DSCR1 (Down Syndrome Candidate Region 1) gene. Genomics 44: 358–361

Green MR (1986) Pre-mRNA splicing. Annu Rev Genet 20: 671–708

Guipponi M, Scott HS, Chen H, Schebesta A, Rossier C, Antonarakis SE (1998) Two isoforms of a human intersectin (ITSN) protein are produced by brain-specific alternative splicing in a stop codon. Genomics 53: 369–376

Hassold T, Jacobs P (1984) Trisomy in man. Annu Rev Genet 18: 69–97

Isakoff SJ, Cardozo T, Andreev J, Li Z, Ferguson KM, Abagyan R, Lemmon MA, Aronheim A, Skolnik EY (1998) Identification and analysis of PH domain-containing targets of phosphatidylinositol 3-kinase using a novel in vivo assay in. Embo J 17: 5374–5387

Jacobs PA, Baikie AG, Court-Bourn WM, Strong JA (1959) The somatic chromosomes in mongolism. Lancet i: 710–711

Korn B, Sedlacek Z, Manca A, Kioschis P, Konecki D, Lehrach H, Poustka A (1992) A strategy for the selection of transcribed sequences in the Xq28 region. Hum Mol Genet 1: 235–242

Krawczak M, Reiss J, Cooper DN (1992) The mutational spectrum of single base-pair substitutions in mRNA splice junctions of human genes: causes and consequences. Hum Genet 90: 41–54

Lamaze C, Chuang TH, Terlecky LJ, Bokoch GM, Schmid SL (1996) Regulation of receptor-mediated endocytosis by Rho and Rac. Nature 382: 177–179

Lejeune J, Gautier M, Turpin R (1959) Etudes chromosomiques sometiques de neuf enfants mongoliens. CR Hebd Séances Acad Sci 248: 409–411

Lovett M, Kere J, Hinton L (1991) Direct selection: a method for the isolation of cDNAs encoded by large genomic regions. Proc Natl Acad Sci USA 88: 9628–9422

Martínez-Cué C, Vallina IF, Baamonde C, Dierssen M, Fillat C, Fuentes JJ, Pritchard M, Estivill X, Flórez J (1997) Transgenic mice overexpressing the human DSCR1 gene: behavioral characterization of two lines. Proc Int Chromosome 21 Workshop (Berlin), pp 45

Melmer G, Sood R, Rommens J, Rego D, Lap-Chee T, Buchwald M (1990) Isolation of clones on chromosome 7 that contain recognition sites for rare-cutting enzymes by oligonucleotide hybridization Genomics 7: 173–181

Miyazaki T, Kanou Y, Murata Y, Ohmori S, Niwa T, Maeda K, Yamamura H, Seo H (1996) Molecular cloning of a Novel Thyroid Hormone-responsive Gene, ZAKI-4, in human skin fibroblasts. J Biol Chem 24: 14567–14571

Musacchio A, Saraste M, Wilmanns M (1994) High-resolution crystal structures of tyrosine kinase SH3 domains complexed with proline-rich peptides. Nat Struct Biol 1: 546–551

Nalefski EA, Falke JJ (1996) The C2 domain calcium-binding motif: structural and functional diversity. Protein Sci 5: 2375–2390

Nimnual AS, Yatsula BA, Bar-Sagi D (1998) Coupling of Ras and Rac guanosine triphosphatases through the Ras exchanger Sos. Science 279: 560–563

Overhauser J, Radic MZ (1987) Encapsulation of cells in agarose beads for use with pulsed-field gel electrophoresis. Focus 9: 8–9

Padgett RA, Grabowski PJ, Konarska MM, Seiler S, Sharp A (1986) Splicing of messenger RNA precursors. Annu Rev Biochem 55: 1119–1150

Pucharcós C, Fuentes JJ, Casas C, de la Luna S, Alcántara S, Arbonés ML, Soriano E, Estivill X, Pritchard M (1999) *Alu*-splice cloning of human Intersectin (ITSN), a putative multivalent binding protein expressed in proliferating and differentiating neurons and overexpressed in Down syndrome. Eur Hum Gene (in press)

Roos J, Kelly RB (1998) Dap160, a neural-specific Eps15 homology and multiple SH3 domain-containing protein that interacts with Drosophila dynamin. J Biol Chem 273: 19108–19119

Salcini AE, Confalonieri S, Doria M, Santolini E, Tassi E, Minenkova O, Cesareni G, Pelicci PG, Di Fiore PP (1997) Binding specificity and in vivo targets of the EH domain, a novel protein-protein interaction module. Genes Dev 11: 2239–2249

Sengar AS, Wang W, Bishay J, Cohen S, Egan SE (1999) The EH and SH3 domain Ese proteins regulate endocytosis by linking to dynamin and Eps 15. EMBO J 18: 1159–1171

Siarey RJ, Stoll J, Rapoport SI, Galdzicki Z (1997) Altered long-term potentiation in the young and old Ts65Dn mouse, a model for Down Syndrome. Neuropharmacology 36: 1549–1554

Sutton RB, Davletov BA, Berghuis AM, Sudhof TC, Sprang SR (1995) Structure of the first C2 domain of synaptotagmin I: a novel Ca^{2+}/phospholipid-binding fold. Cell 80: 929–938

Tagle DA, Collins FS (1992) An optimized *Alu*-PCR primer pair for human-specific amplification of YACs and somatic cell hybrids. Hum Mol Genet 1: 121–122

Ullu E, Murphy S, Melli M (1982) Human 7SL RNA consists of a 140 nucleotide middle-repetitive sequence inserted in an Alu sequence. Cell 29: 195–202

Warren AC, Chakravarti A, Wong C, Slaugenhaupt SA, Halloran SL, Satkins PC, Metazotou C (1987) Evidence for reduced recombination on the nondisjoined chromosome 21 in Down syndrome. Science 237: 652–654

Whitehead IP, Campbell S, Rossman KL, Der CJ (1997) Dbl family proteins. Biochim Biophys Acta 1332: 11–23

Wilson R, Ainscough R, Anderson K, Baynes C, Berks M, Bonfield J, Burton J, Connell M, Copsey T, Cooper J, Coulson A, Craxton M, Dear S, Du Z, Durbin R, Favello A, Fraser A, Fulton L, Gardner A, Green P, Hawkins T, Hillier L, Jier M, Johnston L, Jones M, Kershaw J, Kirsten J, Laisster N, Latreille P, Lightning J, Lloyd C, Mortimore B, O'Callaghan M, Parsons J, Percy C, Rifken L, Roopra A, Saunders D, Shownkeen R, Sims M, Smaldon N, Smith A, Smith M, Sonnhammer E, Staden R, Sulston J, Thierry-Mieg J, Thomas K, Vaudin M, Vaughan K, Waterston R, Watson A, Weinstock L, Wilkinson-Sproat J, Wohldaman P (1994) 2.2 Mb of contigous nucleotide sequence from chromosome III of C. elegans. Nature 368: 32–38

Wisniewski KE (1990) Down syndrome children often have brain with maturation delay, retardation of growth, and cortical dysgenesis. Am J Med Genet [Suppl] 7: 274–281

Yamabhai M, Hoffman NG, Hardison NL, McPherson PS, Castagnoli L, Cesareni G, Kay BK (1998) Intersectin, a novel adaptor protein with two eps15 homology and five src homology 3. J Biol Chem 273: 31401–31407

Zigmond MJ, Bloom FE, Landis SC, Roberts JL, Squire LR (1999) Fundamental neuroscience. Academic Press, San Diego

Authors' address: Dr. M. Pritchard, IRD, Monash University, 27-31 Wright Street, Clayton, Victoria, 3168 Australia, e-mail: melanie.pritchard@med.monash.edu.au

J Neural Transm (1999) [Suppl] 57: 353–362

Overexpression of DNAse I in brain of patients with Down Syndrome

D. Schatzmann-Turhani[1], O. Labudova[1], K. Yeghiazaryan[2], H. Rink[2], E. Hauser[1], N. Cairns[3], and G. Lubec[1]

[1] Department of Pediatrics, University of Vienna, Austria
[2] Department of Radiobiology, University of Bown, Federal Republic of Germany
[3] Brain Bank, Institute of Psychiatry, London, United Kingdom

Summary. Human DNAse I (EC 3.1.21.1) is an enzyme most probably involved in apoptotic processes. Splicing of the DNAse I primary transcript in normal and apoptotic cells into up to 20 splicing forms and the recent description of a different family of caspase-activated DNAses, hampered studies on the role of DNAse I in apoptosis research.

Performing gene hunting in fetal brain of patients with DS we found a sequence with 100% homology to DNAse I and this formed the Rationale for studies in adult DS brain. It was therefore the aim of the study to evaluate DNAse I — mRNA steady state levels in DS brain using adult brain without brain pathologies and Alzheimer's Disease (AD) brain as control, in order to rule out that DNAse I — overexpression may not be specific for DS but rather reflecting apoptosis per se, a hallmark of both disorders.

Determination of DNAse I — mRNA steady state levels was carried out by a blotting method in frontal, parietal, temporal occipital lobe and cerebellum.

We found significantly increased DNAse I transcripts in brain of DS and AD both, when normalized versus the house-keeping gene beta actin or total RNA.

We demonstrate the significant increase of DNAse I — transcript in the pathogenesis of DS and AD suggesting a role for this enzyme in the apoptotic process known to occur in both disorders. We are now going to carry out protein and enzyme activity levels in our laboratory to confirm our findings at the transcriptional level.

Introduction

Human deoxyribonuclease I (DNAse I; EC 3.1.21.1), an enzyme with genetic polymorphism representing 15 phenotypes and controlled by 5 autosomal codominant alleles, is one of the candidate nucleases involved in the apoptotic process Tsutsumi (1998).

The DNAse I primary transcript is alternatively spliced in both normal and apoptotic cells with up to 20 alternatively spliced transcripts in some cells, a fact that may hamper studies on the role of this enzyme in apoptosis Liu (1997).

A series of DNAses has been reported to be involved in the apoptotic process in several tissues (Nitahara, 1998; Waring, 1997; Shiokawa, 1997; Schroter, 1996; Didenko, 1996; Polzar, 1994; Tanuma, 1994; Barry, 1993), but information on this subject is still limited.

A major contribution to the subject was published recently Enari (1998) identifying a caspase — activated deoxyribonuclease (CAD) and its inhibitor (ICAD): CAD is a protein of 343 amino acids carrying a nuclear-localization signal. ICAD containing two caspase-3-binding sites, specifically inhibits CAD-induced degradation of nuclear DNA and its DNAse activity. Caspase 3 cleaves ICAD and inactivates its CAD-inhibitory effect leading to inter-nucleosomal DNA degradation during apoptosis (Sakahira, 1998).

Information on DNAses in human brain consists of two reports describing identification, localization and expression of three novel human genes similar to deoxyribonuclease I (Rodriguez, 1997; Pergolizzi, 1996) and the finding of a DNAse I transcript in human brain (Chen X et al., unpublished, available at gene bank EMBL, Heidelberg).

In this report we are adding data on expression of a sequence 100% homologous to the human DNAse I and DNAse I precursor in brain of controls and significantly upregulated in patients with Down Syndrome (DS) and Alzheimer's disease (AD) as detected by a gene hunting method and quantified by a blotting method.

Methods

Subtractive hybridization protocol (Labudova, 1998)

Fetal brain samples of two fetuses with Down Syndrome and two age and sex matched control, 23rd week of gestation, were obtained from the Brain Bank of the Institute of Psychiatry (Denmark Hill, London, UK). Hippocampus (gyrus parahippocampalis) was taken into liquid nitrogen and ground for the isolation of mRNA.

Isolation of mRNA was performed using the Quick Prep Micro mRNA purification kit (Pharmacia Biotech Inc., Uppsala, Sweden, cat.27 92 55 01).

1 microgram of mRNA from each (of the two) preparation was quality — checked by cDNA cloning kit (Gibco, Life Technologies, Eggenstein, Germany, cat. 18248-013) using the incorporation of [alpha — 32P] dATP (Amersham, Buckinghamshire, UK, cat AA0004) with subsequent electrophoresis on 1% agarose followed by autoradiography. The Reflection film (Dupont NEF 496) was exposed to the gel for a period of two hours at room temperature.

Construction of the subtractive library: 10 micrograms each of mRNA from brain of DS and control were biotinilated by UV irradiation at 360nm according to the instructions supplied in the subtractor kit (Invitrogen, Leek, Netherlands, cat K4320-01). 1 microgram of mRNA — pools each from the DS brain sample was subject to reverse transcriptase reaction (subtractor kit, Invitrogen) and the cDNA — pools were hybridized with the corresponding biotinilated mRNAs from controls. The subtractive hybridization mixture was incubated with streptavidin according to the subtractor kit given above and thus the biotinilated molecules (non-induced biotinilated mRNAs and the hybrid [biotinilated mRNAs/cDNAs]) complexed. The streptavidine complexes were removed by repeated phenol-chloroform extraction and subtracted cDNAs were separated from the aquous phase by alcohol precipitation (subtractor kit).

In order to amplify and clone subtracted cDNAs, they were ligated with Not I — linkers followed by Not I — digestion. These Not I linked cDNAs were ligated to Not I site of sPORT 1 cloning vector (cDNA cloning kit, Gibco).

To enable visualization of subtracted cDNAs the cloned cDNAs were amplified using universal primers

I   5′-GTAAAACGACGGCCAGT-3′
II 5′-ACAGCTATGACCATG-3′

from multiple cloning site of the sPORT-1 vector (cDNA cloning kit, Gibco). Amplified cDNAS were analysed on 1% agarose electrophoresis.

Cloning of subtracted Not I — linked cDNAs: Not I linked cDNAs ligated with sPORT 1 vector were used for the transformation of highly competent INFalpha F′ E. coli cells (Invitrogen, Leek, Netherlands, cat C2020-03) and plated clones were analysed by plasmid isolation kit (Quiagen, Hilden, Germany, cat 12245) and digestion with Eco RI/ Hind III.

Recombinant clones were sequenced by K. Granderath, MWG — Biotech (Ebersberg, Germany).

Homologies were determined by computer assisted comparison of data from the genebank sequence library: fastA@ebi.ac.uk (GBALL, EMBL Gen Bank Heidelberg, Germany).

Subtractive hybridization was performed cross-wise i.e. DS sample mRNA subtraction from control and vice versa at the 1:3 level (DSmRNA:control mRNA).

Northern and dot blot determination and quantification of DNAse I transcript

Northern and dot blot analyses were carried out on postmortem brain samples from adult patients with DS, AD and controls.

The brain regions temporal, frontal, occipital, parietal cortex and cerebellum of patients with DS (n = 9; 3 females, 6 males; 56.1 ± 7.1 years old), AD (n = 9; 6 females, 3 males; 72.3 ± 7.6 years old) and controls (n = 9; 5 females, 4 males; 72.6 ± 9.6 years old) were used and characterized in a previous publication Seidl (1997) were used for the studies at the transcriptional level. Briefly, post mortem brain samples were obtained from the MRC London Brain Bank for Neurodegenerative Diseases, Institute of Psychiatry). In all DS brains there was evidence of abundant beta A plaques and neurofibrillary tangles. The AD patients fulfilled the National Institute of neurological and Communicative Disorders and Stroke and Alzheimer Disease and Related Disorders Association (NINCDS/ADRDA) criteria for probable AD (Mirra, 1991). The histological diagnosis of AD was established and was consistent with the CERAD criteria (Tierney, 1988) for a "definite" diagnosis of AD. The controls were brains from individuals with no history of neurological or psychiatric illness. The major cause of death was bronchopneumonia in DS and AD patients and heart disease in controls. Post mortem interval of brain dissection in AD, DS and controls was 34.1 ± 13.7, 30.6 ± 17.5 and 34.8 ± 15.0 hrs. Tissue samples were stored at −70°C and the freezing chain was never interrupted.

Biopsies from frontal, temporal, parietal, occipital cortex and cerebellum were obtained at autopsy and taken immediately into liquid nitrogen.

Northern and slot blots were performed according to the principle given previously (Hardmeier, 1997): Frozen samples were ground and mRNA extraction was performed using the TotalRNA Purification kit (Pharmacia). Subsequently, RNA was applied onto 1.4% agarose gel after denaturation with glyoxal and dimethylsulfoxide according to the method by McMaster (1997) and electrophoresed at 3–4 V/cm for 2.5 hrs in circulating 0.01 M phosphate buffer pH 7.0. RNA was then transfered to a positively charged nylon membrane (Hybond N+, Dupont, NEF 986) by capillary blotting (Southern, 1975) and fixed with 0.05 N sodium hydroxide for 5 min at room temperature and finally equilibrated at pH 7.0 with 3 washes in 2XSSC.

The probe for human beta-actin (ATCC 9800) was bought and the cDNA human DNAse I precursor from the subtractive library was purified and labeled (Sambrook, 1989).

The probe was denatured prior to labelling by boiling for 5 min and subsequent colling on ice, and labeled with fluorescein-12-dUTP using the Renaissance Random Primer Fluorescein-12 d-UTP labeling kit (Dupont, NEL 203).

After fixation of bound RNA, the nylon membrane was incubated in pre-hybridization solution (0.25 M phosphate buffer pH 7.2, containing 5% SDS w/v, 1 mM EDTA and 0.5% blocking reagent [from Dupont NEL 203]) for 12 hrs at 65°C in a hybridization oven. The blots were hybridized overnight at 65°C with the labelled probes each (50 ng/ml of pre-hybridization buffer).

After hybridization, nonspecifically bound material was removed by post-hybridization wash with 0.5× and 0.1× pre-hybridization buffer 2 × 10 min each at 65°C. The 0.5X and 0.1X pre-hybridization buffer was brought up to 65°C prior to use and the second wash was performed at room temperature.

Hybridized blots were blocked with 0.5% blocking reagent in 0.1 M Tris HCl pH 7.2 and 0.15 M NaCl for 1 hr at room temperature. Membranes were then incubated with antifluorescein HRP antibody (Dupont NEL 203) at a 1:1000 dilution in the solution given above for 1 hr under constant shaking.

Membranes were washed 4× for 5 min each in the solution given above.

The Nucleic Acid Chemiluminescence Reagent (Dupont NEL 201) was added to the membranes and incubated for 1 min. Excess detection reagent was removed by the use of filter paper, the membrane was placed in Sarawrap paper and exposed to autoradiography Reflectio films (Dupont NEF 496) for 15 min at room temperature.

Slot blots were performed according to the method of White (1982). This procedure consisted of placing 2 ug of total RNA dissolved in 10 ul double distilled water mixed with 500 ul of 100% formamide, 162 ul of 37% formaldehyde, 100 ul of 10× MOPS(4-morpholinepropanesulfonic acid) buffer. This mixture was incubated for 10 min at 65°C and cooled down subsequently on ice. Samples were placed onto the membrane by the Manifolds filtration equipment (slot blot apparatus Bio slot TM, BIORAD) and the hybridization was performed as described above.

Denistometry of films was performed using the Hirschmann elscript 400 densitometer and calibration performed as given in the instrument's software program (Germany).

Statistical calculations were performed using the ANOVA followed by Kruskal-Wallis or Wilcoxon test, when appropriate. $P < 0.05$ was considered as statistical significant and is indicated as an asterisks in the figures.

Results

Subtractive hybridization

The corresponding sequence found by subtractive hybridization is given in Fig. 1 and reveals 100% homology with human DNAse I DNAse I and the precursor (Chen X et al. Construction of a transcription map for 660 kb of genomic DNAs on chromosome 16p13.13 by cDNA selection. Unpublished, data from EMBL NETWORK File Server <netserv@ebi.ac.uk>).

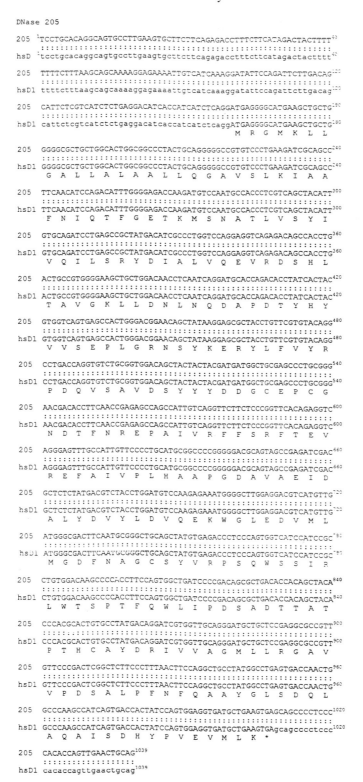

Fig. 1. Alignment of the nucleic acid sequence found by subtractive hybridization with the DNAse I sequence obtained from the Gen Bank Heidelberg. Asterisks show identical and open circles show conserved amino acids

The alignment of DNAse I with the caspase — activated deoxyribonuclease (Halenbeck, 1998) with the sequence of DNAse I found in our study revealed 48% homology (not shown).

Northern blots and slot blots

Northern blots revealed a single band indicating the specificity of the hybridization process.

Densitometry of slots revealed the following:

As shown in Fig. 2a DNAse I was manifold and significantly overexpressed in frontal, temporal, parietal, occipital lobe and cerebellum of patients with DS and in temporal, parietal, occipital lobe and cerebellum of in patients with AD, when expressed as mRNA DNAse I normalized versus the housekeeping gene beta actin.

Figure 2b shows the manifold and significant overexpression of DNAse I normalized versus 50 micrograms of total RNA in frontal, temporal, parietal, occipital lobe and cerebellum of patients with DS. In patients with AD DNAse I was overexpressed in temporal, parietal lobe and cerebellum, downregulated in occipital lobe and comparable to controls in frontal lobe.

Discussion

As shown in the Results we detected a human DNAse I transcript by subtractive hybridization in human brain. This transcript was upregulated at least threefold as the mRNA level set at mRNA subtraction in controls was three times higher. This information from gene hunting performed in fetal brain when secondary pathological changes could be ruled out, was tested in brain from normal adults, adults with DS and AD — pathology and in adults with AD. Using this experimental design we could show that upregulation of the DNAse I was found in both neurodegenerative disorders and was not specific for DS pathology.

It is well-documented and widely accepted that all patients with DS develop AD pathology in brain from the fourth decade (Burger, 1973; Wisniewski, 1985). Information drawn from the two fetal brain samples with DS would suggest that upregulation of DNAse I possibly reflecting apoptosis is an event taking place early in (prenatal) life. And indeed there is sufficient evidence to support this hypothesis as a list of prenatal brain dysmorphologies and pathologies compatible with increased apoptosis were reported in Down Syndrome (Epstein, 1992). Overexpression of DNAse I was found when mRNA steady state levels obtained from slot blots were expressed per RNA or when normalized versus the housekeeping gene beta-actin. This housekeeping gene was also determined to show the integrity of RNA.

Increased apoptosis was suggested and shown to take place in both disorders, DS and AD:

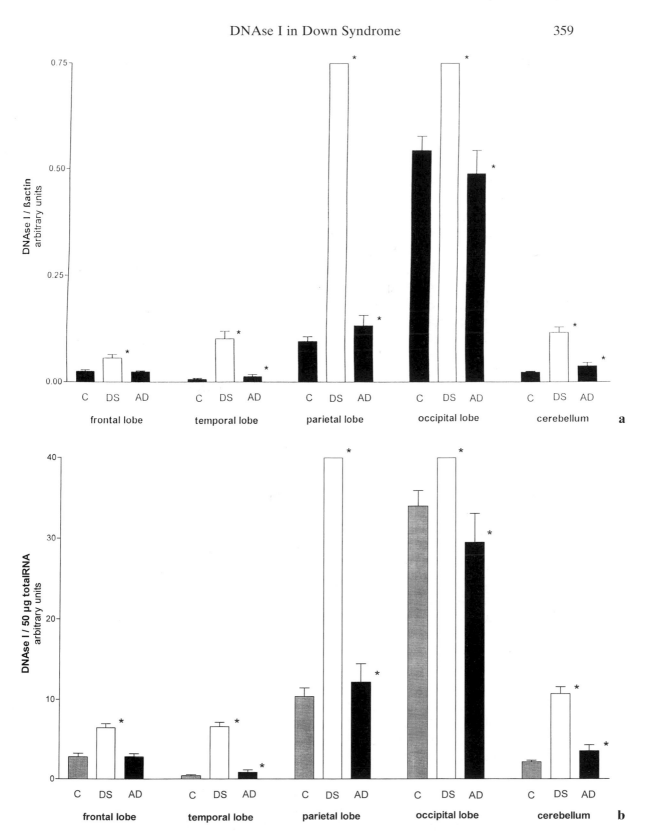

Fig. 2. DNAse I-mRNA steady state levels in the individual brain regions

Sawa and coworkers (1997) found increased expression of bcl-2 with most prominent changes in temporal lobe in brain of patients with DS. Busciglio and Yankner (1995) examined DS — cortical neurons in vitro revealing apoptotic mechanisms.

Apoptosis related protein expression in hippocampus of patients with AD was reported by Nagy (1997) they studied the expression of bcl-2, bcl-x and bax and found that bax, a protein believed to promote apoptosis, was widely expressed in neurons. Kitamura and coworkers published the alterations of proteins regulating apoptosis as bcl-2, bcl-x, bax, bak, bad and the members of the caspase family ICH-1 and CPP32 (Kitamura, 1998). Bak, bad, bcl-2 and bcl-x were upregulated in temporal cortex of AD patients whereas the caspase family members were unaffected. There was significant overexpression of DNAse I in several brain regions with AD in our study, which may suggest that apoptosis does occur in AD although the involvement of the involvement of a different DNAse, probably not involving the caspase pathway, may also participate in the apoptotic mechanisms (Kim, 1997; Grunberg, 1998; Okuchi, 1997). The DNAse I transcript found by our gene hunting approach in DS brain, however, showed homology of 48% to the caspase — activated DNAse, which at least suggests the involvement of DNAse I in AD.

We are currently further characterizing the DNAse I transcript found by subtractive hybridization in our laboratory and extending research at the protein and activity level.

Acknowledgement

This study was supported generously by the Red Bull Company, Salzburg, Austria.

References

Barry MA, Eastman A (1993) Identification of deoxyribonuclease II as an endonuclease involved in apoptosis. Arch Biochem Biophys 300: 440–450

Busciglio J, Yankner BA (1995) Apoptosis and increased generation of reactive oxygen species in Down's syndrome neurons in vitro. Nature 378: 776–779

Burger PC, Vogel FS (1973) The development of pathologic changes of Alzheimer's disease and senile dementia in patients with Down's syndrome. Am J Pathol 73: 457–476

Didenko VV, Hornsby PJ (1996) Presence of double-strand breaks with single-base 3' overhangs in cells undergoing apoptosis but not necrosis. J Cell Biol 135: 1369–1376

Enari M, Sakahira H, Yokoyama H, Okawa K, Iwamatsu A, Nagata S (1998) A caspase-activated DNAse that degrades DNA during apoptosis, and its inhibitor ICAD. Nature 391: 43–50

Epstein CJ (1992) Down Syndrome (Trisomy 21). In: Scriver CR, Beaudet AL, Sly WS, Valle D (eds) The metabolic and molecular basis of inherited disease. McGraw Hill, New York, pp 749–794

Grunberg J, Walter J, Loetscher H, Deuschle U, Jacobsen H, Haass C (1998) Alzheimer's disease associated presenilin-1 holoprotein and its 18–20 kDa C-terminal fragment are death substrates for proteases of the caspase family. Biochemistry 37: 2263–2270

Halenbeck R, MacDonald H, Roulston A, Chen T, Conroy L, Williams LT (1998) CPAN, a human nuclease regulated by the caspase-sensitive inhibitor DFF45. Curr Biol 8: 537–540

Hardmeier R, Hoeger H, Fang-Kircher S, Khoshsorur A, Lubec G (1997) Transcription and activity of superoxide dismutase, catalase and glutathione peroxidase following irradiation in radiation resistant and radiation sensitive mice. Proc Natl Acad Sci (USA) 94: 7572–7576

Huang P, Ballal K, Plunkett W (1997) Biochemical characterization of the protein activity responsible for high molecular weight DNA fragmentation during drug-induced apoptosis. Cancer Res 57: 3407–3414

Kim TW, Pettingell WH, Jung YK, Kovacs DM, Tanzi RE (1997) Alternative cleavage of Alzheimer-associated presenilins during apoptosis by a caspase — 3 family protease. Science 277: 373–376

Kitamura Y, Shimohama S, Kamoshima W, Ota T, Matsuoka Y, Nomura Y, Smith MA, Perry G, Whithouse PJ, Taniguchi T (1998) Alteration of proteins regulating apoptosis, bcl-2, bcl-x, bax, bak, bad, ICH-1 and CPP32 in Alzheimer's disease. Brain Res 780: 260–269

Labudova O, Lubec G (1998) cAMP upregulates the transposable element mys-1: a possible link between signaling and mobile DNA. Life Sci 62: 431–437

Labudova O, Fang-Kircher S, Cairns N, Lubec G (1998) Upregulation of vasopressin in brain of patients with Down Syndrome. Brain Res 806: 55–59

Liu QY, Ribecco M, Hou Y, Walker PR, Sikorska M (1997) DNAse I primary transcript is alternatively spliced in both normal and apoptotic cells: no evidence of upregulation in apoptosis. DNA-Cell Biol 16: 911–918

McMaster GK, Carmichael GG (1977) Analysis by single and double stranded nucleic acids on polytacrylamide and agarose gels by using glyoxal and acridine orange. Proc Natl Acad Sci (USA) 74: 4835–4841

Mirra SS, Heyman A, McKeel D, Sumi S, Crain BJ (1991) The consortium to establish a registry for Alzheimer disease (CERAD). II. Standardisation of the neuropathological assessment of Alzheimer's disease. Neurology 41: 479–486

Nagy ZS, Esiri MM (1997) Apoptosis — related protein expression in the hippocampus in Alzheimer's disease. Neurobiol Aging 18: 565–571

Nitahara JA, Cheng W, Liu Y, Li B, Li P, Mogul D, Gambert SR, Kajstura J, Anversa P (1998) Intracellular calcium, DNAse activity and myocyte apoptosis in aging Fisher 344 rats. J Mol Cell Cardiol 30: 519–535

Okuchi M, Ishii K, Usami M, Sahara N, Kametani F, Tanaka K, Fraser PE, Ikeda M, Saunders AM, Hendriks L, Shoji SI, Nee LE, Martin JJ, Van Broeckhoven C, St. George-Hyslop PH, Roses AD, Mori H (1997) Proteolytic processing of presenilin-1 is not associated with Alzheimer's disease with or without PS-1 mutations. FEBS Lett 418: 162–166

Pergolizzi R, Appierto V, Bosetti A, DeBellis GL, Rovida E, Biunno I (1996) Cloning of a gene encoding a DNAse like endonuclease in the human Xq28 region. Gene 168: 267–270

Polzar B, Zanotti S, Stephan H, Rauch F, Peitsch MC, Irmler M, Tschopp J, Mannherz HG (1994) Distribution of deoxyribonuclease I in rat tissues and its correlation to cellular turnover and apoptosis. Eur J Cell Biol 64: 200–210

Rodriguez AM, Rodin D, Nomura H, Morton CC, Weremowicz S, Schneider MC (1997) Identification, localization, and expression of two novel human genes similar to deoxyribonuclease I. Genomics 42: 507–513

Sakahira H, Enari M, Nagata S (1998) Cleavage of CAD inhibitor in CAD activation and DNA degradation during apoptosis. Nature 391: 20–21

Sambrook J, Fritsch EF, Maniatis T (1989) A laboratory manual. Cold Spring Harbor Laboratory, New York, pp 687–701

Sawa A, Oyama F, Cairns NJ, Amano N, Matsushita M (1997) Aberrant expression of bcl-2 gene family in Down's syndrome brains. Brain Res-Mol Brain Res 48: 53–59

Schroter M, Peitsch MC, Tschopp J (1996) Increased p34cdc2-dependent kinase activity during apoptosis: a possible activation mechanism of DNAse I leading to DNA breakdown. Eur J Cell Biol 69: 143–150

Seidl R, Greber S, Schuller E, Bernert G, Cairns N, Lubec G (1997) Evidence against increased oxidative DNA damage in Down Syndrome. Neurosci Lett 235: 137–140

Shiokawa D, Ohyama H, Yamada T, Tanuma S (1997) Purification and properties of DNAse gamma from apoptotic rat thymocytes. Biochem J 326: 675–681

Southern EM (1975) Detection of specific sequences among DNA fragments separated by gel electrophoresis. J Mol Biol 98: 503–511

Tanuma S, Shiokawa D (1994) Multiple forms of nuclear deoxyribonucleases in rat thymocytes. Biochem Biophys Res Commun 203: 789–797

Tierney MC, Fisher RH, Lewis AJ, Torzitto ML, Snow WG, Reid DW, Nieuwstraaten, P, Van Rooijen LAA, Derks HJGM, Van Wijk R, Bischop A (1988) The NINCDA-ADRDA work group criteria for the clinical diagnosis of probable Alzheimer's disease. Neurology 38: 359–364

Tsutsumi S, Asao T, Nagamachi Y, Nakajima T, Yasuda T, Kishi K (1998) Phenotype 2 of deoxyribonuclease I may be used as a risk factor for gastric carcinoma. Cancer 82: 1621–1625

Waring P, Khan T, Sjaarda A (1997) Apoptosis induced by gliotoxin is preceded by phosphorylation of histone H3 and enhanced sensitivity of chromatin to nuclease digestion. J Biol Chem 272: 17929–17936

White BA, Bancroft FC (1982) Cytoplasmic dot hybridization. Simple analysis of mRNA levels in multiple small cell or tissue samples. J Biol Chem 257: 8569–8574

Wisniewski KE, Wisniewski HM, Wen GY (1985) Occurrence of neuropathological changes and dementia of Alzheimer's disease in Down's syndrome. Ann Neurol 17: 278–282

Zanotti S, Polzar B, Stephan H, Doll U, Niessing J, Mannherz HG (1995) Localization of deoxyribonuclease I gene transcripts and protein in rat tissues and its correlation with apoptotic cell elimination. J Histochem Cell Biol 103: 369–377

Zhang C, Robertson MJ, Schlossmann SF (1995) A triplet of nuclease proteins (NP42-50) is activated in human Jurkat cells undergoing apoptosis. Cell Immunol 165: 161–167

Authors' address: Prof. Dr. G. Lubec, Department of Pediatrics, University of Vienna, Währinger Gürtel 18, A-1090 Vienna, Austria, e-mail: gert.lubec@akh-wien.ac.at

SpringerNeurology

Horst Przuntek, Thomas Müller (eds.)

Diagnosis and Treatment of Parkinson's Disease – State of the Art

1999. VII, 220 pages. 32 partly coloured figures.
Hardcover DM 185,–, öS 1295,–
(recommended retail price)
ISBN 3-211-83276-9
Special edition of "Journal of Neural Transmission, Suppl. 56, 1999"

Expert clinicians and basic scientists with a special interest in Parkinson's disease review the current state of science and clinical therapeutics of the disease. Therefore these articles represent an authorative review of the current state of knowledge regarding preclinical course and symptomatology, subtypes with their impact on the pathology, genetic alterations, novel mechanisms of neuronal cell death, diagnostic tools and old and novel therapeutic approaches with respect to neuroprotection and neuroregeneration in Parkinson's disease.

Particular emphasis has been placed on a novel antiparkinsonian drug called budipine with various modes of action also influencing altered non dopaminergic systems in Parkinson's disease. It is evident, that many questions on the cause, course and treatment of Parkinson's disease are still unanswered and therefore the ideal way to treat a parkinsonian patient remains to be defined.

Contents

SpringerWienNewYork

Sachsenplatz 4–6, P.O.Box 89, A-1201 Wien, Fax +43-1-330 24 26, e-mail: books@springer.at, Internet: **www.springer.at**
New York, NY 10010, 175 Fifth Avenue • D-14197 Berlin, Heidelberger Platz 3 • Tokyo 113, 3–13, Hongo 3-chome, Bunkyo-ku

SpringerNeurology

W. Poewe, G. Ransmayr (eds.)

Advances in Research on Neurodegeneration

Volume 6

1999. VIII, 147 pages. 21 figures.
Hardcover DM 168,–, öS 1176,–
Special edition of "Journal of Neural Transmission, Suppl. 55, 1999"
ISBN 3-211-83262-9

The book summarizes the presentations of outstanding scientists in the field of neuro-immunology and neurodegeneration, in particular pathogeneity, neuroprotection and restoration in multiple sclerosis and neurodegenerative disorders, such as Parkinson's disease, amyotrophic lateral sclerosis, Huntington's disease and multisystem atrophy. It is demonstrated that neuroimmunology and neurodegeneration have several aspects in common. Recent trends in preclinical and clinical research are presented.

H. J. Gertz, T. Arendt (eds.)

Alzheimer's Disease –
From Basic Research to Clinical Applications

1998. VIII, 315 pages. 52 figures.
Hardcover DM 198,–, öS 1386,–
Special edition of "Journal of Neural Transmission, Suppl. 54, 1998"
ISBN 3-211-83113-4

This volume brings together the reports of basic scientists and clinical investigators on Alzheimer's disease. The issue bridges the gap between laboratory work in basic science and the development of urgently needed therapeutic strategies. Areas presented are the molecular and cellular biology of the disease, pathogenetic mechanisms and potential therapeutic targets, genetics, risk factors, strategies of prevention and treatment as well as practical aspects of medical and social care for patients with Alzheimer's disease.

All prices are recommended retail prices

 SpringerWienNewYork

Sachsenplatz 4–6, P.O.Box 89, A-1201 Wien, Fax +43-1-330 24 26, e-mail: books@springer.at, Internet: **www.springer.at**
New York, NY 10010, 175 Fifth Avenue • D-14197 Berlin, Heidelberger Platz 3 • Tokyo 113, 3–13, Hongo 3-chome, Bunkyo-ku

SpringerNeurology

K. Jellinger, F. Fazekas, M. Windisch (eds.)

Ageing and Dementia

1998. VIII, 406 pages. 65 partly coloured figures.
Hardcover DM 248,–, öS 1736,–. ISBN 3-211-83115-0
Special edition of "Journal of Neural Transmission, Suppl. 53, 1998"

Ageing and dementia are closely related conditions. Increasing age of the general population causes increasing incidence of dementing disorders in later life, although cognitive impairment is not necessarily a consequence of advancing age.

"... this book will be of interest to clinicians with previous experience of clinical dementia assessements, and to researchers who want a comprehensive update on research areas of dementia with which they are less familiar. It will also be of interest to those following the development of neurotrophe factors for treatment of dementia who need an extensive introduction to the preclinical studies of Cerebrolysin®. The book will be fairly useful as a textbook for clinicians who are learning about clinical dementia assessments for the first time."

Acta Psychiatrica Scandinavica

J. P. M. Finberg, M. B. H. Youdim,
P. Riederer, K. F. Tipton (eds.)

MAO – The Mother of all Amine Oxidases

1998. XIX, 355 pages. 76 partly coloured figures.
Hardcover DM 248,–, öS 1736,–. ISBN 3-211-83038-3
Special edition of "Journal of Neural Transmission, Suppl. 52, 1998"

Monoamine oxidase (MAO) is linked to psychiatric and neurological disorders, because inhibitors of the enzyme are used clinically for treatment of affective disorders and Parkinson's disease.

"... the book is of particular value and a rich source of information for basic neuroscientists in the field. But it also can be of substantial interest to the rest of the neuroscientific community to get a glimpse of the current status of research in the field of MAO."

Acta Neurologica Belgica

All prices are recommended retail prices

SpringerWienNewYork

Sachsenplatz 4–6, P.O.Box 89, A-1201 Wien, Fax +43-1-330 24 26, e-mail: books@springer.at, Internet: **www.springer.at**
New York, NY 10010, 175 Fifth Avenue • D-14197 Berlin, Heidelberger Platz 3 • Tokyo 113, 3–13, Hongo 3-chome, Bunkyo-ku

Springer-Verlag
and the Environment

WE AT SPRINGER-VERLAG FIRMLY BELIEVE THAT AN international science publisher has a special obligation to the environment, and our corporate policies consistently reflect this conviction.

WE ALSO EXPECT OUR BUSINESS PARTNERS – PRINTERS, paper mills, packaging manufacturers, etc. – to commit themselves to using environmentally friendly materials and production processes.

THE PAPER IN THIS BOOK IS MADE FROM NO-CHLORINE pulp and is acid free, in conformance with international standards for paper permanency.